Impact of AI and Robotics on the Medical Tourism Industry

Viana Hassan
University of Malta, Malta

Ahmad Rasmi Albattat
Management and Science University, Malaysia

Shakeel Basheer
Lovely Professional University, India

A volume in the Advances in Hospitality, Tourism, and the Services Industry (AHTSI) Book Series

IGI Global

Published in the United States of America by
IGI Global
Business Science Reference (an imprint of IGI Global)
701 E. Chocolate Avenue
Hershey PA, USA 17033
Tel: 717-533-8845
Fax: 717-533-8661
E-mail: cust@igi-global.com
Web site: http://www.igi-global.com

Library of Congress Cataloging-in-Publication Data

CIP Pending
ISBN: 979-8-3693-2248-2
EISBN: 979-8-3693-2249-9

This book is published in the IGI Global book series Advances in Hospitality, Tourism, and the Services Industry (AHTSI) (ISSN: 2475-6547; eISSN: 2475-6555)

British Cataloguing in Publication Data
A Cataloguing in Publication record for this book is available from the British Library.

Advances in Hospitality, Tourism, and the Services Industry (AHTSI) Book Series

Maximiliano Korstanje
University of Palermo, Argentina

ISSN:2475-6547
EISSN:2475-6555

MISSION

Globally, the hospitality, travel, tourism, and services industries generate a significant percentage of revenue and represent a large portion of the business world. Even in tough economic times, these industries thrive as individuals continue to spend on leisure and recreation activities as well as services.

The Advances in Hospitality, Tourism, and the Services Industry (AHTSI) book series offers diverse publications relating to the management, promotion, and profitability of the leisure, recreation, and services industries. Highlighting current research pertaining to various topics within the realm of hospitality, travel, tourism, and services management, the titles found within the AHTSI book series are pertinent to the research and professional needs of managers, business practitioners, researchers, and upper-level students studying in the field.

COVERAGE

- Service Design
- Hotel Management
- International Tourism
- Travel Agency Management
- Cruise Marketing and Sales
- Service Management
- Leisure & Business Travel
- Sustainable Tourism
- Customer Service Issues
- Health and Wellness Tourism

IGI Global is currently accepting manuscripts for publication within this series. To submit a proposal for a volume in this series, please contact our Acquisition Editors at Acquisitions@igi-global.com or visit: http://www.igi-global.com/publish/.

The Advances in Hospitality, Tourism, and the Services Industry (AHTSI) Book Series (ISSN 2475-6547) is published by IGI Global, 701 E. Chocolate Avenue, Hershey, PA 17033-1240, USA, www.igi-global.com. This series is composed of titles available for purchase individually; each title is edited to be contextually exclusive from any other title within the series. For pricing and ordering information please visit http://www.igi-global.com/book-series/advances-hospitality-tourism-services-industry/121014. Postmaster: Send all address changes to above address.

Titles in this Series

For a list of additional titles in this series, please visit:
http://www.igi-global.com/book-series/book-series/advances-hospitality-tourism-services-industry/121014

Service Innovations in Tourism Metaverse, Immersive Technologies, and Digital Twin
Surabhi Singh (G.L. Bajaj Institute of Management and Research, India)
Business Science Reference • © 2024 • 295pp • H/C (ISBN: 9798369311035) • US $275.00

Utilizing Smart Technology and AI in Hybrid Tourism and Hospitality
Sanjeev Kumar (Lovely Professional University, India) Mohammad Badruddoza Talukder (Daffodil Institute of IT, Bangladesh) and Ana Pego (Nova University of Lisbon, Portugal)
Business Science Reference • © 2024 • 367pp • H/C (ISBN: 9798369319789) • US $290.00

Innovative Technologies for Increasing Service Productivity
Sridhar Manohar (Chitkara Business School, Chitkara University, India) Amit Mittal (Chitkara Business School, Chitkara University, India) Sekar Raju (Ivy College of Business, Iowa State University, USA) and Arjun J. Nair (Chitkara University, India)
Business Science Reference • © 2024 • 404pp • H/C (ISBN: 9798369320198) • US $290.00

Supporting Environmental Stability Through Ecotourism
Muhammad Abrar (Lyallpur Business School, Government College University, Faisalabad, Pakistan) Muhammad Asim Nawaz (Lyallpur Business School, Government College University, Faisalabad, Pakistan) Faiqa Kiran (Lyallpur Business School, Government College University, Faisalabad, Pakistan) and Fahad Asmi (School of Business, Charles Sturt University, Bathurst, Australia)
Business Science Reference • © 2024 • 296pp • H/C (ISBN: 9798369310304) • US $265.00

Career Pathways and Professional Identities for Front-Line Workers in the Service Industries
Mark Peter Vitale (University of Phoenix, USA)
Business Science Reference • © 2024 • 370pp • H/C (ISBN: 9781668425282) • US $255.00

IGI Global
PUBLISHER of TIMELY KNOWLEDGE

701 East Chocolate Avenue, Hershey, PA 17033, USA
Tel: 717-533-8845 x100 • Fax: 717-533-8661
E-Mail: cust@igi-global.com • www.igi-global.com

Table of Contents

Detailed Table of Contents

The current concept of medical tourism will be significantly impacted by the integration of robotics, AI, the growing demand for cosmetic operations in the United Arab Emirates. This abstract examines how cutting-edge technology are revolutionizing healthcare in the United Arab Emirates, with a focus on how robotics and artificial intelligence might improve accuracy and efficiency in medical procedures, especially when it comes to cosmetic procedures. The expansion of the cosmetic tourism industry is indicative of the economic consequences, which include increased employment opportunities, economic diversification, and improved public perception worldwide. Simultaneously, the combined impact of these developments establishes the United Arab Emirates as a top choice for foreign patients looking for cutting edge healthcare. The ramifications discussed here highlight the complex effects of artificial intelligence, robotics, and cosmetic operations on the changing nature of medical tourism in the United Arab Emirates.

The Medical Tourism Industry is undergoing a transformative shift with the introduction of an innovative AI-based mental health care framework. This framework integrates wearable technology and advanced algorithms, enabling real-time analysis

The internet and its proliferation have had a profound impact on various fields such as science, industry, and medical computer networks. This has resulted in improved medical treatments. Telemedicine serves as a connection between the fields of medicine and engineering, allowing the medical community to utilize engineering resources in order to enhance the well-being of the population. Smart surgery is a form of remote surgery where a surgeon operates on a patient from a distance. The surgeon operates remotely via communication tools, highlighting the intriguing potential of this technology. Telemedicine can be employed during surgery to establish communication with another specialist regarding the feasibility of the procedure. Televisions can serve as a means for postoperative monitoring, while robotic surgeons enable proficient medical professionals to do minimally invasive surgery (MIS) with remote-controlled surgical robots, delivering expert treatment from distant sites.

This study examines how AI and robots shift competitive advantage across industries. Through a literature analysis, case studies, and a robust methodological framework, this study shows how new technologies change operational paradigms, redefine innovation, and put organizations at the forefront of industry dynamics. Strategic AI and robot integration is essential for competitive advantage in manufacturing, healthcare, banking, logistics, and tourism. This theoretical framework helps scholars, practitioners, and policymakers navigate the intricate relationships between AI, robots, and industry dynamics. Finally, the chapter suggests further research on ethical issues, interdisciplinary studies, long-term effects, global perspectives, and human-AI collaboration. This research provides a solid foundation for understanding how AI and robotics are changing the commercial landscape at the technological frontier.

Health and tourism are benefiting from AI technology. This study examines medical tourism's positive outlook and emphasizes AI's role in improving the complete experience for foreign patients. Qualitative desk research and thorough review of contemporary research literature, including content analysis, identified AI and smart healthcare system (SHS) dimensions in medical tourism. The study found that AI is helping medical agencies connect physicians with prospective medical tourists via telehealth and virtual assistance. Results demonstrate that medical tourists utilize AI to research hospitals, treatments, check-ups, and consultations. AI secures medical data exchange and blockchain-based cashless transactions. Medical travelers find AI-enabled smart health devices useful for tracking health data. AI problems in medical tourism are highlighted in the study. The study concludes with insights on AI's transformative impact on medical tourism. The chapter helps comprehend the changing dynamics of AI, healthcare, AI's role in improving medical tourism and addresses its issues.

The popularity of medical tourism has matured substantially over the last few decades due to escalating medical costs, advancements in medical technology, and reduced travel costs. The significance of medical marketing has thus become vital for hospitals to lure domestic and global patients, designate their brand, and boost awareness about life-saving treatments delivered by distinctive hospitals across the globe. Even though medical tourism has acquired growing interest, prior research has been limited in scope, particularly in studying the industry's marketing aspects. Because of this, comprehending the target market is vital. To address this need, this chapter aims to answer two research questions: RQ1, What key themes, study characteristics, study contexts, and methodologies are investigated in medical tourism marketing research studies?; RQ2, What future research avenues in medical tourism marketing exist for scholars across diverse fields? This chapter thoroughly synthesizes existing medical tourism marketing literature and pinpoints future research opportunities.

The proposed chapter delves into the transformative repercussions of Robotics and Artificial Intelligence (AI) on the burgeoning Indian Medical Travel and Tourism Sector. As India emerges as a global hub for medical tourism, the integration of innovative technologies promises to reshape patient experiences and healthcare delivery. The chapter meticulously investigates the expansion of the medical travel sector in India, the role of AI in healthcare, the impact of robotics, and the intersection of these technologies in the context of medical tourism. It not only highlights the positive outcomes and success stories but also addresses the challenges and ethical concerns related to healthcare robots and artificial intelligence. Furthermore, the chapter provides a forecast for the future of these technologies in India's wellness sector, offering recommendations for stakeholders to navigate the evolving landscape successfully.

The revolutionary impact of virtual reality (VR) and augmented reality (AR) on medical tourism is examined in this article. AR and VR are transforming the patient experience and changing the face of healthcare tourism destinations as technology advances. The abstract explores the use of AR and VR in medical tourism, emphasizing how they can be used for patient education, pre-treatment consultations, and post-treatment care. Patients can virtually visit medical facilities, comprehend intricate medical processes, and participate in therapeutic virtual reality interventions through immersive experiences, which improves their decision-making process and general well-being. The study also looks at case studies from top medical tourism locations that highlight effective AR and VR technology applications. The study addresses how these technologies could draw in a worldwide patient population by providing a special fusion of state-of-the-art medical treatment with innovative experiential elements. The incorporation of AR and VR into medical tourism offers new prospects for improved patient involvement and comprehensive healthcare services as these technologies continue to develop.

The proposed chapter delves into the transformative repercussions of Robotics and Artificial Intelligence (AI) on the burgeoning Indian Medical Travel and Tourism Sector. As India emerges as a global hub for medical tourism, the integration of innovative technologies promises to reshape patient experiences and healthcare delivery. The chapter meticulously investigates the expansion of the medical travel sector in India, the role of AI in healthcare, the impact of robotics, and the intersection of these technologies in the context of medical tourism. It not only highlights the positive outcomes and success stories but also addresses the challenges and ethical concerns related to healthcare robots and artificial intelligence. Furthermore, the chapter provides a forecast for the future of these technologies in India's wellness sector, offering recommendations for stakeholders to navigate the evolving landscape successfully.

The revolutionary impact of virtual reality (VR) and augmented reality (AR) on medical tourism is examined in this article. AR and VR are transforming the patient experience and changing the face of healthcare tourism destinations as technology advances. The abstract explores the use of AR and VR in medical tourism, emphasizing how they can be used for patient education, pre-treatment consultations, and post-treatment care. Patients can virtually visit medical facilities, comprehend intricate medical processes, and participate in therapeutic virtual reality interventions through immersive experiences, which improves their decision-making process and general well-being. The study also looks at case studies from top medical tourism locations that highlight effective AR and VR technology applications. The study addresses how these technologies could draw in a worldwide patient population by providing a special fusion of state-of-the-art medical treatment with innovative experiential elements. The incorporation of AR and VR into medical tourism offers new prospects for improved patient involvement and comprehensive healthcare services as these technologies continue to develop.

The aim of the study is to explore the growth and opportunity of medical tourism in context to Indian healthcare. Advancements in science and technology have made it possible for transcontinental healthcare services to exist. In addition to offering standard tourist amenities, medical tourism aims to attract visitors by purposefully promoting its healthcare offerings. Based on extensive literature review, factors impacting inclusive growth and medical tourism in India. Medical tourism—also known as wellness tourism, medical travel, or health tourism—was first used by travel firms and the media to refer to the quickly expanding trend of people traveling abroad for medical treatment. The health care industry is becoming more globalized, and many developing nations like India which have the means and potential to meet this demand. Furthermore, have become interested in the recent surge in demand for affordable, high-quality medical care. Nowadays, because medical care in India is so affordable, people from other nations travel there for procedures including knee transplants, hip replacements, heart surgery, cosmetic surgery, and dental work. While medical tourism is a relatively new phenomenon, it is growing at an exponential rate and is becoming a significant contributor to the global expansion of services exports. Among the leading nations in this sector is India. Medical tourism is a crucial source of income, particularly for underdeveloped countries. Hence, it can be concluded that both the destination countries and the patients benefit from it.

Preface

Welcome to the comprehensive exploration of the dynamic intersection between artificial intelligence (AI), robotics, and the flourishing domain of medical tourism. As the editors of this edited reference book, we are thrilled to present a collection that delves into the transformative influence of AI and Robotics on the ever-evolving landscape of the Medical Tourism industry.

The concept of medical tourism, with its historical roots tracing back to antiquity, has undergone a remarkable metamorphosis in the contemporary era. The willingness of individuals to traverse vast distances for health-related purposes has evolved into a global trend, generating substantial economic impact within the broader tourism industry. This book encapsulates the amalgamation of traditional healthcare tourism with the revolutionary integration of AI and Robotics, offering a profound understanding of the synergies and challenges that define this fascinating convergence.

In the current healthcare landscape, the role of AI is paramount in fostering individual empowerment and enhancing healthcare practitioners' capabilities. The insights derived from AI enable practitioners to comprehend the nuanced patterns and requirements of patients, thereby facilitating improved feedback, guidance, and support for optimal health maintenance. Concurrently, Robotics has found its place as a transformative force, with notable examples such as the integration of robotic surgery at institutions like Cleveland Clinic Abu Dhabi.

This book is meticulously crafted to serve as a valuable resource for professionals seeking an in-depth comprehension of the intricate relationship between AI, Robotics, and the medical tourism industry. We aim to provide theoretical frameworks and the latest empirical research findings that shed light on the impact of AI and Robotics on the reshaping of Medical Tourism, along with the associated challenges and opportunities.

Designed for both tourism industry professionals and academics, this book covers a broad spectrum of topics, ranging from the impact of AI on the transformation of medical tourism to the effectiveness of AI in the field. The chapters encompass areas such as telemedicine, medical tourism marketing, regulatory and ethical challenges, the economic expansion driven by AI and Robotics, and much more.

We invite readers to embark on this intellectual journey, exploring each chapter that contributes to a deeper understanding of the symbiotic relationship between AI, Robotics, and the ever-evolving Medical Tourism industry. The insights provided will be invaluable for medical travel agents, tour operators, healthcare professionals, and researchers keen on expanding their knowledge base and navigating the challenges and opportunities in this dynamic sector.

The chapters are meticulously curated to unravel the complexities, offering a comprehensive insight into the multifaceted impact of AI and robotics on the medical tourism industry. The contributors, experts in their respective fields, present a compelling synthesis of theoretical perspectives and empirical evidence, making this edited reference book an indispensable resource for those at the forefront of shaping the future of healthcare tourism.

Chapter 1: The Impact of Advanced Technology on Medical Tourism Transformation: Case Study of United Arabs of Emirates

In this chapter, Viana Hassan from the University of Malta and Shakeel Basheer from Lovely Professional University, India explore the transformative impact of cutting-edge technology, including robotics and AI, on the concept of medical tourism in the United Arab Emirates (UAE). Focusing on the growing demand for cosmetic procedures, the authors delve into how these advancements enhance accuracy and efficiency, especially in cosmetic operations. The economic consequences, such as increased employment opportunities and economic diversification, position the UAE as a leading choice for foreign patients seeking state-of-the-art healthcare. The chapter illuminates the intricate effects of artificial intelligence, robotics, and cosmetic operations on the evolving nature of medical tourism in the UAE.

Chapter 2: Harmonizing Minds and Machines: A Transformative AI-Based Mental Health Care Framework in Medical Tourism

This chapter introduces an innovative AI-based mental health care framework that revolutionizes the medical tourism industry. Rita Komalasari, from Yarsi University, Indonesia, outlines how this framework integrates wearable technology and advanced algorithms for real-time analysis, providing personalized guidance to medical tourists. Emphasizing the critical need for personalized mental health supervision, the chapter showcases the successful implementation of AI-driven data analysis to address this imperative. The result is a comprehensive and personalized mental health support system that promises to enhance the overall medical tourism experience.

Chapter 3: Food Challenges and Opportunities for Medical Tourist in Serbia

Miloš Zrnic, from the Academy of Applied Studies in Serbia, explores the intersection of culinary experiences and the medical tourism industry in Serbia. Focused on spa destinations like Prolom Banja, Vrnjačka Banja, and Soko Banja, the research aims to provide contextually relevant insights reflective of the broader medical tourism landscape. With a sample of 592 diverse medical tourists, the study emphasizes the significance of food in their spa experience. Structured interviews and surveys uncover perspectives on food challenges and opportunities, forming the basis for understanding and improving culinary aspects in Serbian spas.

Chapter 4: Perceptions of Participants Receiving Health Services about the Effects of AI in the Health Sectors: The Effects of AI in the Health Sectors

Muhammed Yetgin from the University of Karabuk, Turkey, investigates the perceptions of Turkish citizens and foreign patients regarding the use of AI and technology in the healthcare sector. The study employs the Diffusion of Innovation theory to analyze the experiences of 20 participants. By applying interview analysis and content analysis with the Nvivo 11 program, the chapter aims to provide insights into the evolving landscape of health services, disease diagnosis, and treatment methods in Turkey, a country where health tourism is developing in terms of technology.

Chapter 5: The strategic efficacy of artificial intelligence (AI) in medical tourism

In this chapter, Olabode Gbobaniyi, Daniela Tincani, and Peter Emelone from the Global Banking School explore the strategic efficacy of AI in the context of medical tourism. The authors delve into the emergence of AI in the industry, discussing its efficiency and effectiveness in achieving strategic outcomes. Covering categories of AI-system devices and applications to major diseases in medical tourism, the chapter aims to ensure that AI strategies not only look good on paper but continue to produce high success for global practice.

Chapter 6: Artificial Intelligence Driven Enhancements in Medical Tourism Opportunities, Challenges, and Future Prospects

Dr Gaurav Bathla, Ashish Raina, and Varinder Rana explore the opportunities, challenges, and future prospects associated with AI-driven enhancements in medical tourism. Integrating AI technologies, especially in the context of globalization, the authors uncover uncharted territories that enhance the overall patient experience, medical outcomes, and operational efficiency. Telemedicine and AI-powered virtual consultations are discussed as facilitators of uninterrupted communication between patients and medical practitioners, overcoming the limitations imposed by physical distance.

Chapter 7: Impact of AI and Robotics on the Medical Tourism Industry

Simon Abou Fayad, from the Lebanese University, explores the profound impact of AI and robotics on the medical tourism industry. Highlighting how the internet has improved medical treatments, the chapter delves into telemedicine, smart surgery, and robotic surgeons. The author discusses how these technologies enable remote Minimally Invasive Surgery (MIS), emphasizing their potential in enhancing medical treatments and facilitating communication between specialists. The chapter underscores the transformative potential of AI and robotics in reshaping the landscape of medical tourism.

Chapter 8: Redefining Industry Dynamics The Impact of AI And Robotics on Competitive Advantage

Shakeel Basheer, Ishfaq Tramboo, Sheezan Farooq, and Zahra Behboodi investigate how AI and robots shift competitive advantage across various industries. Through literature analysis, case studies, and a robust methodological framework, the chapter explores how these new technologies redefine innovation and operational paradigms. The authors argue that strategic integration of AI and robots is essential for competitive advantage in manufacturing, healthcare, banking, logistics, and tourism. The theoretical framework presented in this chapter aids scholars, practitioners, and policymakers in navigating the intricate relationships between AI, robots, and industry dynamics, suggesting further research avenues.

Chapter 9: Artificial Intelligence Enhanced Global Healthcare, The Future of Medical Tourism

Sheikh Najam-mu- Sahar and Hafizullah Dar from Lovely Professional University delve into the positive outlook of medical tourism, emphasizing AI's role in enhancing the complete experience for foreign patients. Through qualitative desk research and a thorough review of contemporary research literature, the chapter identifies AI and smart healthcare system (SHS) dimensions in medical tourism. The study demonstrates how AI connects physicians with prospective medical tourists, facilitates research on hospitals, treatments, check-ups, and consultations, secures medical data exchange, and enables blockchain-based cashless transactions. The chapter concludes by addressing AI's transformative impact on medical tourism, providing insights into the changing dynamics of AI, healthcare, and the associated challenges.

Chapter 10: The Ultimate Guide to Medical Tourism Marketing: A Comprehensive Systematic Review and Future Research Agenda

Prakash Singh, Viana Hassan, and Santosh Dastane present an ultimate guide to medical tourism marketing in this chapter. Examining key themes, study characteristics, contexts, and methodologies in medical tourism marketing research studies, the authors address two research questions. The chapter synthesizes existing medical tourism marketing literature, providing a comprehensive review and identifying future research opportunities across diverse fields.

Chapter 11: The Indian Medical Tourism Industry's Repercussions from AI and Robotics

Birendra Kishore Roy and Viveka Nand Sharma explore the transformative repercussions of AI and robotics on the Indian Medical Travel and Tourism Sector. As India emerges as a global hub for medical tourism, the chapter investigates the integration of innovative technologies, including AI, which promises to reshape patient experiences and healthcare delivery. The authors highlight positive outcomes, success stories, challenges, and ethical concerns related to healthcare robots and artificial intelligence. The chapter concludes by providing a forecast for the future of these technologies in India's wellness sector, offering recommendations for stakeholders to navigate the evolving landscape successfully.

Chapter 12: The Role of Augmented Reality (AR) And Virtual Reality (VR) in Medical Tourism

Sheezan Farooq, Murtaza Reshi, and Bazillah Farooq investigate the revolutionary impact of augmented reality (AR) and virtual reality (VR) on medical tourism. The chapter explores how AR and VR transform the patient experience and healthcare tourism destinations. Emphasizing applications in patient education, pre-treatment consultations, and post-treatment care, the study showcases case studies from top medical tourism locations. The authors discuss how AR and VR technologies could attract a global patient population by providing a unique fusion of state-of-the-art medical treatment with innovative experiential elements.

Chapter 13: Growth and Opportunity of Medical Tourism in Indian Healthcare Sector: A Critical Review

Amjad Ansari and Amrik Singh from Lovely Professional University aim to explore the growth and opportunity of medical tourism in the context of the Indian healthcare sector. The chapter, based on an extensive literature review, analyzes factors impacting inclusive growth and medical tourism in India. Addressing the surge in demand for affordable, high-quality medical care, the study sheds light on the globalized healthcare industry and India's potential to meet this demand. The chapter provides a critical review, laying the groundwork for understanding the dynamics of medical tourism in the Indian healthcare sector.

As editors of this groundbreaking edited reference book, we find ourselves exhilarated by the depth and breadth of insights presented in each chapter, collectively exploring the intricate nexus of Artificial Intelligence (AI), Robotics, and the thriving domain of Medical Tourism. This collection transcends conventional boundaries, shedding light on the transformative impact of cutting-edge technologies in reshaping the global landscape of healthcare tourism.

This edited reference book traverses a vast spectrum of topics, encompassing telemedicine, marketing strategies, regulatory challenges, economic expansions, and more. Each chapter provides a distinctive lens through which the multifaceted impact of AI and robotics on the medical tourism industry is unveiled. This compendium serves as a comprehensive resource, equipping readers with theoretical frameworks and the latest empirical research findings to navigate the challenges and opportunities presented by this dynamic sector.

The contributors to this book, experts in their respective fields, have meticulously crafted chapters that synthesize theoretical perspectives and empirical evidence. From the impact on mental health care frameworks to the repercussions on the Indian Medical Tourism Industry, each author offers a compelling exploration of their chosen theme. The result is an indispensable resource that caters to the needs of medical travel agents, tour operators, healthcare professionals, and researchers seeking to expand their knowledge base.

As editors, we invite readers to embark on an intellectual journey through the chapters of this book, exploring the intricate relationships between AI, Robotics, and the ever-evolving Medical Tourism industry. The insights provided herein are invaluable for understanding the challenges and opportunities that lie ahead. We are confident that this collection will not only serve as a benchmark for the current state of knowledge but will also inspire further exploration and innovation in the dynamic intersection of AI, robotics, and medical tourism.

Viana Hassan
University of Malta, Malta

Ahmad Rasmi Albattat
Management and Science University, Malaysia

Shakeel Basheer
Lovely Professional University, India

Acknowledgment

We appreciate the efforts of all the members of IGI Global Publisher's editing, production, and marketing teams for effectively completing the guide.

We are grateful to all the contributors who have contributed their authentic research to this book. Thank you for your support during the endeavor.

We lack the words to properly express our gratitude to our parents, family, and friends for believing in us and creating in us a persistent passion for pursuing academic research.

Viana Hassan, Ahmad Al battat, & Shakeel Basheer

Chapter 1
The Impact of Advanced Technology on Medical Tourism Transformation:
Case Study of United Arabs of Emirates

Viana Imad Hassan
https://orcid.org/0000-0002-7372-5059
University of Malta, Malta

Shakeel Basheer
https://orcid.org/0000-0001-7854-8285
Lovely Professional University, India

ABSTRACT

The current concept of medical tourism will be significantly impacted by the integration of robotics, AI, the growing demand for cosmetic operations in the United Arab Emirates. This abstract examines how cutting-edge technology are revolutionizing healthcare in the United Arab Emirates, with a focus on how robotics and artificial intelligence might improve accuracy and efficiency in medical procedures, especially when it comes to cosmetic procedures. The expansion of the cosmetic tourism industry is indicative of the economic consequences, which include increased employment opportunities, economic diversification, and improved public perception worldwide. Simultaneously, the combined impact of these developments establishes the United Arab Emirates as a top choice for foreign patients looking for cutting edge healthcare. The ramifications discussed here highlight the complex effects of artificial intelligence, robotics, and cosmetic operations on the changing nature of medical tourism in the United Arab Emirates.

DOI: 10.4018/979-8-3693-2248-2.ch001

1. INTRODUCTION

The way people seek and receive medical care around the world has changed dramatically in recent years due to the convergence of advanced technology and the healthcare sector (Heydari Fard et al., 2021a). This has had a major impact on the medical tourism industry. With an emphasis on the United Arab Emirates (UAE), this study explores how medical tourism is changing as a result of modern technologies. The United Arab Emirates (UAE) is a compelling case study to examine the dynamic interaction between technology and the emerging sector of medical tourism (Heydari Fard et al., 2021). The UAE is recognized for its quick technological breakthroughs and dedication to becoming a global powerhouse for healthcare. Modern healthcare facilities and state-of-the-art medical technology have made the United Arab Emirates a popular destination for medical tourists looking for specialized and high-quality care (Basheer, Farooq, et al., 2023). Technological developments in telemedicine, robots, artificial intelligence, and data analytics have bolstered this shift by improving healthcare delivery efficiency and broadening the range of medical tourism services offered in the area.

For instance, the use of telemedicine has made post-treatment follow-ups and remote consultations easier, increasing the accessibility of healthcare services for patients from abroad (Basheer, Walia, et al., 2023). Furthermore, a variety of patients from around the world have been drawn to these cutting-edge medical interventions due to the precision and success rates that have been increased by robotic-assisted surgeries and inventive medical procedures (Ayuningtyas & Ariwibowo, 2020). Furthermore, the UAE's calculated investments in electronic health records and health information technology have expedited the patient experience and guaranteed smooth care coordination for medical tourists. In order to shed light on the changing dynamics of global healthcare consumption and the role that technology will play in influencing medical tourism in the future, this study intends to investigate the multidimensional effects of modern technology on the transformation of medical tourism in the United Arab Emirates.

Even in ancient times, people were prepared to travel great distances to get medical treatment (Iorio et al., 2014). There is a long history of people marketing specific locations as health retreats, dating back to the spas of the ancient Greeks, Romans, and Sumerians as well as the alpine and seaside health resorts of early modern Europe (Dash, 2020). Though it only accounts for a small fraction of the tourism market, medical tourism is quickly becoming into a multi-billion dollar enterprise around the world.

More people may be encouraged to participate in medical tourism if advanced technologies like Robotics and Artificial Intelligence (AI) let people take charge of their own health and wellness. Hwang et al. (2018) cite research by Ehrbeck et al.

(2008) that found five reasons why people travel abroad for medical treatment: access to cutting-edge technology (40 percent), better quality care for essential treatments (32 percent), shorter wait times for procedures (15 percent), relatively inexpensive service (9 percent), and less expensive care for elective procedures (4 percent). Technological advancements such as the Internet of Things (IoT), automation, robots, virtual reality, artificial intelligence, cloud computing, and big data analytics are being used in medical tourism (Tyan et al., 2021). Medical operations in the UAE are highly regarded for their exceptional quality, thanks to the superior technology that the country possesses. In 2021, the Medical Tourism Index ranked Dubai at number six and Abu Dhabi at number nine among all medical tourism destinations in the globe (Medical Tourism, 2021). An evaluation and discussion of the role of technological intervention in the growth and change of medical tourism in the UAE has prompted the formulation of a number of research questions. How has cutting-edge tech, particularly AI and robots, impacted the growth of medical tourism? In what ways does this intervention help advance tourism while simultaneously involving locals and society in the plan to grow the industry? Interviews with villagers, employees, tourists, and a designated specialist in the subject were among the qualitative methods used to address these questions. The main results proved that medical tourism owes a great deal to technological advancements.

2. LITERATURE REVIEW

With an emphasis on the United Arab Emirates (UAE), we hope to fill the vacuum in the literature by offering a thorough background and literature review on the influence of cutting-edge technology on the evolution of medical tourism in this chapter. First, we examine the background of medical tourism in the United Arab Emirates and chart its development in tandem with medical technology breakthroughs. We look at significant turning points, legal frameworks, and economic forces that have influenced the development of medical tourism in the area. The literature review that follows covers recent research, reports, and scholarly articles that provide insight into the relationship between medical tourism and sophisticated technology in the United Arab Emirates. We cover subjects including the use of robotics and artificial intelligence (AI) in healthcare settings, the incorporation of virtual consultations and telemedicine, and the contribution of cutting-edge medical technology to drawing in patients from abroad.

Additionally, we showcase case studies and empirical research that provide valuable perspectives on the real-world applications of cutting-edge technology in improving the standard, affordability, and accessibility of healthcare services for both UAE locals and medical tourists. We carefully examine recent publications and reports from reliable sources, such as government agencies, business magazines, and scholarly journals, to make sure that the studies are up to date. We hope to provide readers a thorough grasp of the current situation and potential future developments of medical tourism in the United Arab Emirates, which is being pushed by innovative technology, by combining these many information sources. To summarize, this chapter aims to provide a solid framework of knowledge along with an extensive literature analysis that is complemented by new research to clarify how modern technology is influencing the way medical tourism is changing in the United Arab Emirates.

2.1 Medical Tourism

Medical tourism is defined as "traveling internationally to enhance one's health" (Bookman & Bookman, 2007, p. 1). Spas, which are still present in many parts of Europe and other areas, are an early form of tourism that specifically aims to enhance health and well-being (Connell, 2006). The practice of "taking the waters" gained popularity during the 18th century, and several spas managed to draw tourists to their sites (Cook, 2008). In the 19th century, hill stations became increasingly popular in tropical countries, including the French Pacific possession of New Caledonia. These locations were chosen for their secluded nature and the perceived therapeutic benefits of travel and recreation. (Smyth, 2005). A gradual change in the population occurred when tourism and recreational activities became accessible to not just the upper class but also the working class (Connell, 2006). Although these tourism activities never involved real medical procedures, health and leisure tourism might be considered as the precursor to medical tourism in the aforementioned manifestations. Medical tourism, the practice of traveling to receive medical care, has been a subject of discussion for a long time. However, it remains an issue that has not received much scientific attention (Smith et al., 2011). Chew and Darmasaputra (2015) assert that health and medical tourism is an expanding global trend in the 21st century, which has garnered significant interest among scholars and professionals. The emergence of this phenomenon can be attributed to the implementation of neoliberal policies and the globalization of healthcare (Hopkins et al., 2010; de la Hoz-Correa et al., 2018). Additionally, there are continuous endeavors to discover cost-effective and superior treatment options, as well as reduce waiting times (Ngamvichaikit & Beise-Zee, 2014). Previously, individuals from less developed nations would go to more developed nations in order to obtain medical care. Currently, it is common for patients from wealthy countries to seek medical treatment in underdeveloped countries due to the lower costs involved (Johnson et al., 2015; Misung et al., 2012). Medical tourism has emerged as a distinct market segment due to its varied range of offerings, which combine holiday experiences with medical services (Abubakar & Ilkan, 2016). Quantifying the market size of medical tourism and making future projections is challenging due to the difficulty in categorizing tourists. Consequently, data must be handled and utilized with caution (Connell, 2006). As per the report published by VISA and Oxford Economics, around 11 million individuals engage in medical tourism annually, resulting in a market worth of USD 100 billion (Medical Tourism, 2022). This industry demonstrates significant commercial potential, with a yearly growth rate of 25%. Moreover, with the enhancement of living standards and personal finances, there has been an increased demand for private healthcare, specifically for aesthetic procedures (Dang et al., 2020). Based on the 2019 data, the United States conducted 4.36 million cosmetic treatments, while Brazil ranked second with 2.3 million procedures (Dang et al., 2020). According to the Medical Tourism Index, which evaluates the appeal of countries for medical travel, the economic status and public perception of a nation, healthcare costs, and quality of treatment are the primary factors that contribute to the growth of medical tourism in a specific area (Stephano, n.d.). The Medical Tourism Index of 2020 identifies the following countries as the top 10 destinations for medical tourism: Canada, Singapore, Japan, Spain, United Kingdom, Dubai, Costa Rica, Israel, Abu Dhabi, and India. While Oman, Thailand, and South Korea may not be among the top ten destinations, they warrant consideration due to their increasing popularity in recent years.

Presently, the market is influenced by the affordability of high-quality treatments, but in numerous developed nations, procedure costs are exorbitant and sometimes not included in insurance coverage (Lunt et al., n.d.). Hence, marketing plays a pivotal part in enticing international patients, as an appealing and well branded online presence can be the determining factor in their choice of healthcare provider (Heung et al., 2011; Connell, 2013).Two limitations of this industry include the challenge of precise medical record transfers and the occurrence of post-operative problems. Furthermore, it is imperative to take into account the ethical concern, particularly the ramifications on local healthcare, where people frequently lack the financial resources to access medical treatment (Mehraj, Qureshi, et al., 2023). Consequently, healthcare experts are compelled to abandon the public sector in favor of the private sector due to the prospect of greater salaries. As a consequence, this leads to a lack of available treatment options for the underprivileged and extended waiting times for medical care (Connell, 2006). Additionally, there are other market prospects, including the potential for marketers to enhance the visibility of medical tourism and for governments to establish initiatives that promote medical tourism and safeguard the interests of travelers, service providers, and destinations. Another crucial group of participants include insurance firms, which establish contracts with foreign hospitals and promote medical tourism to customers, resulting in reduced waiting periods and expenses.

Within the realm of medical tourism, there are several operations. Among these, cosmetic procedures and dental treatments have gained significant popularity, prompting the investigation conducted by Rodrigues et al. in 2017.

2.2 Research Objectives of the Study

1. How are breakthroughs in artificial intelligence and robots that improve medical operations and treatments, especially cosmetic surgery, making the UAE more appealing to overseas medical tourists
2. Will the UAE's increased demand for cosmetic surgeries, AI and robotics integration, and medical tourism sector influence employment, income, and economic diversification
3. How much does the UAE's reputation as a center for cutting-edge healthcare, made possible by the convergence of cutting-edge technologies and cosmetic procedures, affect its global healthcare standing and ability to attract foreign investments and partnerships in technology and healthcare

2.3 Methodology

This book chapter delves deeply into the ways that cutting-edge technologies—specifically, robotics, artificial intelligence (AI), and the growing popularity of cosmetic procedures—are revolutionizing medical tourism in the United Arab Emirates (UAE). Utilizing a carefully designed research approach, we combine qualitative and quantitative methods to arrive at a thorough comprehension of the complex consequences that are being examined. Quantitative approaches include statistical data analysis of the increase in medical tourism in the United Arab Emirates, trends in cosmetic surgery, and relevant economic factors. To find new trends and correlations, this entails carefully going over government papers, health statistics, and economic data.

In-depth interviews with significant players in the tourist, healthcare, and technology sectors in the United Arab Emirates are a component of qualitative research methodologies, which serve to supplement the quantitative approach. The aim of these interviews is to collect incisive viewpoints, creative concepts, and firsthand accounts about the amalgamation of robots, artificial intelligence, and cosmetic procedures in the medical tourism field. In addition, we use literature evaluations and content analyses of academic journals and industry reports to contextualize our findings and offer a strong framework for comprehending the background, present developments, and future directions of the phenomena under discussion. Through the collaborative integration of various analytical techniques, our goal is to carry out a thorough examination of the complex interplay among technology, cosmetic surgery, and medical tourism in the United Arab Emirates. We provide a substantial contribution to the growing corpus of information about the changing dynamics of healthcare consumption worldwide with our research.

This study delves deeply into the intricate relationship that exists between cutting-edge technologies—robotics and artificial intelligence (AI) in particular—and the growing demand for cosmetic procedures in the context of medical tourism in the United Arab Emirates (UAE). Our study technique has been meticulously designed to offer a comprehensive comprehension of the diverse ramifications that arise from these crossing disciplines. Our quantitative approaches involve a thorough examination of statistical data related to the growing medical tourism trend in the United Arab Emirates and the changing trends seen in cosmetic procedures. To identify patterns and create correlations, this entails a methodical examination of government publications, health databases, and economic indicators.

We also use qualitative research approaches, such as in-depth interviews with important stakeholders in the tourism, healthcare, and technology sectors of the United Arab Emirates, to supplement our quantitative findings. These interviews are a priceless source of firsthand knowledge, viewpoints, and experiences about the fusion of robotics, artificial intelligence, and cosmetic surgery in the context of medical tourism. Moreover, we perform thorough literature reviews and content analyses of academic articles and industry reports to give a solid framework for our analysis and place our findings within the context of previous research. This review of earlier studies helps shed light on the background, present state of affairs, and potential future directions of the phenomenon being studied. We hope to perform a comprehensive analysis of the complex link between technology, cosmetic surgery, and medical tourism in the United Arab Emirates by utilizing this integrated method. Our study aims to add to the expanding corpus of information about the changing dynamics of healthcare consumption on a local and international scale

3. TECHNOLOGICAL PARADIGM SHIFT: AI, ROBOTICS, AND THE EVOLUTION OF MEDICAL TOURISM

3.1 Cosmetic Procedures

The term "cosmetic procedures" is employed to encompass both cosmetic surgeries and minimally invasive cosmetic treatments, such as Botulinum toxin type A injections and laser hair removal, among others. Cosmetic surgery refers to the use of surgical and medical treatments to maintain, restore, or enhance one's physical appearance (Swami et al., 2009, p.1). The American Society of Plastic Surgeons states that cosmetic procedures are performed to improve the aesthetic and physical characteristics of healthy body parts (2018). According to the American Society of Plastic Surgeons in 2019, the most sought-after cosmetic surgical procedures, prior to the epidemic, included breast augmentation, liposuction (fat removal), eyelid surgery, and rhinoplasty (nose reshaping).

Alternatively, neurotoxins, dermal fillers, deoxycholic acid, lasers, peels, and micro-needling are less invasive procedures that offer effective but temporary alternatives to surgery. These options are particularly advantageous for certain individuals (Devgan et al., 2019). The Plastic Surgery Statistics Report (2019) indicates that the most sought-after minimally-invasive procedures in 2019 were Botulinum Toxin Type A, soft tissue fillers, chemical peel, and laser hair removal. According to the American Society of Plastic Surgeons (ASPS) National Clearinghouse of Plastic Surgery Procedural Statistics, this technique has become a prominent approach for enhancing one's appearance, with more than 18 million cosmetic treatments conducted in 2019 (2020).

The influence of "popular culture" has had a notable effect on this industry, as seen by the rise in these processes following the emergence of social media photo-sharing applications (Walker et al., 2019). Body dissatisfaction, low self-esteem, high weight, social environment, and the media are identified as significant factors that drive individuals to seek cosmetic operations, as stated by several authors (Sharp, 2018; Furnham & Levitas, 2012; Thompson et al., 1999). Furthermore, Rodrigues et al. (2017) discovered that cosmetic surgery advertisements tend to specifically target women, particularly older women residing in the Western world who had the financial resources to pursue such procedures. Based on 2019 data, 92% of cosmetic procedures in the US are performed on women, while only 8% are performed on men. The bulk of these treatments are still carried out on individuals aged 40 and above (ASPS, 2019). While these operations may not be medically essential, they can still have a substantial effect on individuals' welfare, self-confidence, and general contentment (Delinsky, 2005).

3.2 Cosmetic Dentistry

The American Academy of Cosmetic Dentistry (AACD) defines cosmetic dentistry as the branch of dentistry that focuses on making improvements to the appearance of teeth and enhancing one's smile. The range of dental procedures may encompass veneers, teeth whitening, tooth-colored restorations, fillings, and various other treatments (AACD, 2022). The separation of cosmetic dentistry from other branches of dentistry is a subject of ongoing discussion due to the potential for dentists to align themselves with consumerist culture (Holden, 2017).

A significant number of individuals experience dissatisfaction with the aesthetics of their teeth and smiles, leading to negative impacts on their self-confidence and making them susceptible to social evaluation by others who assess social standing or wealth based on oral health (Rodrigues et al., 2017; Holden, 2017). Patients who have unsatisfactory smiles or facial aesthetics are typically seen as neglectful. These patients may be interested in facial aesthetic operations, but they could also benefit from more affordable dental aesthetic procedures (Schmidt & Tatum, 2006). Hence, cosmetic dentistry has a crucial function in the restoration and maintenance of a youthful appearance (Morley, 1999). Furthermore, research has revealed that dental problems impacting the facial aesthetics of teenagers and young adults have a substantial impact on their self-esteem. This finding provides an additional rationale for selecting cosmetic dentistry as a preferred option over alternative treatments (Kaur, 2017; Gavric et al., 2015).

Cosmetic dentistry plays a crucial role in the field of medical tourism. According to Herrick (2007), retirement communities in Arizona arrange excursions to Mexico in order to have dental treatment at a reduced cost. On the other hand, individuals from Northern and Western Europe travel to Eastern Europe for dental procedures that are both less expensive and of high quality. Mexico, Costa Rica, Hungary, and Poland are renowned for their dental services tailored to medical tourists (Herrick, 2007).

According to Grand View Research, the estimated value of the global cosmetic dentistry market in 2021 will be USD 29.6 billion, and it is expected to expand at a rate of 13% in 2022. These statistics indicate that the business is seeing significant growth, with medical tourism being a contributing factor to its global expansion (Grand View Research, 2022).

4. FACTORS AFFECTING MEDICAL TOURISM

The industry is seeing significant growth, and medical tourism is a key factor contributing to its global expansion (Grand View Research, 2022).

The medical tourism market represents 10% of the total foreign tourists' receipts, amounting to USD 1,481 billion (UNWTO, 2021). The majority of medical tourists are from affluent regions such as North America, Western Europe, and the Middle East, where healthcare expenses tend to be substantial (Connel, 2006). Notably, Asian countries including India, Singapore, Malaysia, Thailand, and South Korea are the primary destinations for medical tourism (Connel, 2006; Lunt & Carrera, 2010). Nevertheless, other countries are concurrently formulating their tourist strategy specifically targeting medical tourism (Mehraj, Ul Islam, et al., 2023). In addition to projected market expansion, reduced living and labor expenses, accessible medications, and backed insurance are identified as the key variables motivating tourism officials to cultivate this particular sector (Hopkins et al., 2010).

The main catalyst for the expansion of medical tourism is the expense of treatment (Hopkins et al., 2010; Lunt & Carrera, 2010). Despite the affluent backgrounds of visitors from developed nations, obtaining medical treatment overseas generally incurs lesser costs compared to their home country. Additional considerations that attract visitors are the comparatively affordable international airfares and favorable currency rates at the destination. There is also an increasing interest in less common procedures such as tattoo removal and cosmetic surgeries (Hopkins et al., 2010). In the field of medicine, the factors driving the increase in demand include the extended queue of patients awaiting specific treatments, advancements in medical care, and the implementation of bioethical laws (Lunt & Carrera, 2010). Meanwhile, bioethical law pertains to medical procedures that involve ethical considerations, such as abortion, fertility treatments, and euthanasia, which may vary across different countries.

The United Arab Emirates (UAE) recognizes the potential for a burgeoning market in medical tourism, as other countries embark on its development. The UAE has designated the world's inaugural minister of state for AI as part of its efforts to establish itself as a global medical tourism center. Wong and Hazley (2020) formulated the strategies and policies for the advancement of AI in the UAE. The UAE government introduced UAE Vision in 2021, aiming to enhance the nation's healthcare sector and bolster their medical tourism industry through the implementation of integrated IT systems and electronic data (Alrahbi et al., 2021; The UAE Government, 2021a). This essay will examine the role of AI and robotics in enhancing medical tourism in the UAE, as well as the necessary tactics for the growth of this industry.

4.1 The Development of AI and Robotics

The advancement of technology occurred decades ago with the invention of the steam engine in the first industrial revolution then, followed by electricity and mass production in the industrial revolution 2.0. Industry 3.0 is when digitalisation took place, and its development happened at the beginning of era 4.0. According to Wong & Hazley (2020, pp. 268), the technologies in this period include the "Internet of Things (IoT), automation, robotics, virtual reality, artificial intelligence (AI), cloud solutions and big data analytics". The application of technology, especially AI and robotics, can be found anywhere in any sector. Although tourism belongs to the service industry, the use of high technology is inevitable. The Henn-Na Hotel in Japan is the world's first robotics hotel that opened in 2015 with robots as its staff (Reis, 2020). In the same year in San Francisco, Eatsa restaurant operated with robots serving their dishes (Berezine et al., 2019). The use of robots in hotels and restaurants may enhance their service to a certain degree and improve the brand reputation and image (Yang et al., 2020). The service quality may be maintained simultaneously since the robots are programmed with specific repetitive tasks. Therefore, as the service is constant, the mistakes usually made by the human can be minimised or even avoided. Robots can be used for customer service and for collecting and analysing customer behaviour data based on their direct interactions (Heires, 2016). These data may act as the base for the company's following strategies toward better service and customer satisfaction.

4.2 The Application of AI and Robotics in Medical Tourism in the UAE

The use of technology can attract more people to go for medical tourism. Ehrbeck et al. (2008), as cited in Hwang et al. (2018), identified five factors that encourage medical tourism: most advanced technology (40%); better-quality care for medically necessary treatments (32%); faster access to medically necessary procedures (15%); relatively lower-cost service for medically necessary procedures (9%); and lower-cost care for discretionary procedures (4%). Internet of things (IoT), automation, robots, virtual reality, artificial intelligence, cloud solutions, and big data analytics are examples of modern technology used in medical tourism (Tyan et al., 2021). These aid healthcare providers through streamlining operations, synthesizing data, and providing real -time updates (Bernasconi, 2016). A comprehensive treatment is highly likely to shape the experience and perception of me dical tourists toward a destination. This experience will determine the decision tourists consider to return or revisit the place either as a medical or a 'normal' tourist. In this case, the UAE has advanced technology as their value-added factor, making patients prefer medical procedures in terms of quality. In 2021, Medical Tourism Index ranked Dubai and Abu Dhabi the sixth and ninth place among other medical tourism destinations globally (Medical Tourism, 2022). However, as the Covid-19 pandemic impact is inevitable, strategies need to be advised to mitigate the effect.

4.3 Dubai's AI-Driven Hair Transplant Robot Poised to Elevate Medical Tourism"

Dubai's latest technological marvel is a smart hair transplant robot, and it's drawing attention from health-conscious individuals worldwide. This innovative robot, equipped with artificial intelligence, is reshaping the landscape of cosmetic procedures. Its precise and efficient approach to hair transplants not only ensures natural-looking results but also positions Dubai as a sought-after destination for health tourism. With cutting-edge technology at the forefront, the city's commitment to marrying aesthetics and advanced healthcare is evident, making it an appealing choice for those seeking top-notch medical treatments in a cosmopolitan setting.

The smart hair transplant robot is not just a breakthrough in science; it's a testament to Dubai's dedication to becoming a global hub for innovative and high-quality healthcare services.

The implementation of the latest ARTAS iX robotic technology has enhanced the precision of transplants, hence bolstering the growth of the local business. In Dubai, surgeons are aiming to regain a portion of the prosperous hair transplant tourist industry from Istanbul, which continues to be the preferred choice for hair transplants and other aesthetic treatments. The primary objective is to provide remedies for both male pattern baldness and hair loss and balding in females. Dr. Khaled Al Nuaimi, the President of the Middle East International Dermatology and Aesthetic Medical Association, states that Turkey anticipates attracting over 2 million health tourists this year. The country's Health Ministry estimates that this influx of visitors would generate over $20 billion in revenue.

The implementation of cutting-edge technology in Dubai is enhancing the efficacy of long-term hair replacements, while simultaneously reducing expenses and minimizing the duration of clinic visits.Surgeons are utilizing software upgrades and intelligence algorithms provided by image-guided robotics to enhance current hair transplant techniques, resulting in a reduction of clinic time to approximately four to five hours.

"The integration of AI and robotics in modern hair transplant methods has led to notable progress, improving the accuracy of the procedure," stated Dr. Sanjay Parashar, the chairman of the Scientific Emirates Plastic Surgery Society. Dr. Parashar utilizes an AI-powered robot at Cocoona surgery in Al Wasl." The enhanced precision in identifying viable follicles leads to more durable outcomes and heightened rates of success. "We issue a directive to the robot to initiate the process of examining the head, prompting it to observe the hair from various perspectives and subsequently commence its removal."

The robot possesses the capacity to extract approximately 1,000 hair follicles during an hour, exhibiting a remarkable speed that mitigates human exhaustion and minimizes the likelihood of human mistakes.

5. REVIVING MEDICAL TOURISM IN THE UAE

5.1 Pre-Operative/Pre-Departure Stage

The transformation of the medical tourism experience can be observed via three distinct phases: pre-treatment, during treatment, and post-treatment. During the pre-departure stage, travelers engage in a process where they interact, either directly or indirectly, with various activities before participating in the main service of medical tourism. Prior to selecting a hospital for optimal care and service, people might seek out diverse facts, including reputation, specialized treatment, and notable tourist attractions. Given that the consumer's choice relies on a thorough investigation, it is essential to establish sufficient contact with the customer (Hwang et al., 2018). Patients have the option to directly engage with hospital medical specialists or service personnel through various means of communication such as email, telephone, or video conferencing platforms like Zoom or Google Meet. Therefore, a one-stop service is an optimal platform for customers to initiate their trip.

Dubai Health Experience (DXH) is a component of the UAE's ambition to enhance medical tourism, accelerate the growth of the emirates, and attract a greater number of tourists to this region. The DXH website exemplifies a comprehensive service system that connects 75 medical centers, travel agents, and hoteliers via a single platform. In addition, it provides various packages and services, including virtual consultations and expert medical assessments, vacation arrangements encompassing flights, accommodations, and local transportation, visa prerequisites, recommended insurance options, and tourism activities (DXH, 2022). Abu Dhabi has also developed a platform with comparable characteristics to DXH, in addition to Dubai's DXH. The platform provides a range of services including 40 healthcare centers, over 280 medical practice packages, insurance, doctor visits, hotels, tourism activities, and transportation (Bulatovic & Iankova, 2021).

Nevertheless, the exorbitant prices of healthcare packages in the UAE have emerged as a concern, given that medical tourists also consider cost as a crucial determinant in their decision-making. Several reasons contribute to the high medical expenses, such as the country's present cost of living and the utilization of new technologies such as artificial intelligence and robotics. Bulatovic & Iankova (2021) proposed the necessity of implementing a distinction within the medical travel package. They presented three items with distinct characteristics, categorized by their respective price points: basic, silver, and gold. In addition, it is imperative to establish partnerships with private entities such as airlines and travel firms to effectively lower expenses. Collaborating with private travel agencies, both within the country and abroad, would not only lead to cost reduction but also enhance medical tourism, thereby expediting the achievement of the UAE Vision.

During the operative stage, which occurs at 5.2, certain actions or events take place.

Throughout the medical journey, visitors participate in activities such as assessing costs, evaluating the quality of service, and examining various auxiliary systems. They engage in direct communication with service providers, actively participate in service delivery methods, and see a discrepancy between their initial expectations and the actual outcome. The current experience pertains to the customer's engagement in different hospital activities, including interacting with service providers and other guests, utilizing information and communication technology, and appreciating the service environment (Hwang et al., 2021). Hence, it is imperative to ensure the quality of service throughout this time. The utilization of AI and robots in medical treatment plays a significant role in molding the whole experience of tourists seeking healthcare services. Primarily, these instruments aid healthcare personnel in a procedure by enhancing its accuracy and reducing the likelihood of contamination to both the workers and the patient, hence resulting in a faster recovery (Zemmar et al., 2020).

Cleveland Clinic Abu Dhabi has used robotic surgery since 2015 (Cleveland Clinic Abu Dhabi, 2017) and has also adopted assistive technologies (Carlson & Ehrlich, 2005). Medical tourists often prefer destinations that have modern healthcare technology. Nevertheless, in order to cultivate a favorable attitude among potential clients, it is imperative to consider the functionality, convenience, and aesthetic appeal of the technology (Hwang et al., 2018). Medical tourism destinations are motivated to include current technology into their initiatives due to its recognition and impact. The prevalence of AI and robotics in healthcare is expected to intensify the competition among medical tourism destinations. As production increases, the cost will decrease. Hence, the utilization of AI and robotics is no longer exclusive to a particular nation. The UAE government must provide support to address the issues of maintaining up-to-date technology. Establishing research centers can effectively address this problem and significantly contribute to the advancement of cutting-edge technologies. Furthermore, as hospitals and clinics in the UAE have begun implementing AI and robotics, there is a growing demand for proficient surgeons and healthcare professionals who possess the knowledge and expertise to comprehend and utilize these advanced technologies. In addition, it is crucial to have a greater number of experts engaged in the field of medical tourism, particularly in the tourist aspect. This is due to the country's requirement for individuals who possess the skills and expertise necessary to do occupations that are directly associated with potential visitors, such as marketers, promoters, and travel agents.

5.2 Recovery Phase After Surgery

During the post-treatment period, travelers allocate their time to recuperate and then relish the destination like an ordinary visitor. Utilizing robotics in medical treatments increases the patient's likelihood of experiencing an extended period of well-being. Here is where the tourism component accounts for a substantial share of the medical tourism activities. Nevertheless, expeditious recuperation does not ensure contracting an infection, despite the minimal probability. As per Han (2013), certain travelers find it more advantageous to stay in a medical hotel rather than a standard hotel. This is due to the added conveniences they can enjoy, such as accessing various healthcare services in one place or staying together with their family and friends. A medical hospital also provides reliable post-care services in a hotel, if needed, with enhanced amenities such as superior food and beverage options and improved communication, as the majority of the staff possesses sufficient medical and healthcare expertise. Noaman and Hassan (2016) found that over 90% of Middle Eastern medical tourists opt to stay in medical hotels for the duration of their trip, according to their research. The data suggests a significant need for medical hotels, particularly in the UAE. There is a growing demand for additional medical hotels due to the increasing number of medical tourists.

AI technologies are employed to predict or diagnose life-threatening medical conditions. Simultaneously, the advancement of artificial intelligence (AI) in wearable devices exhibits a favorable trajectory, enabling the instantaneous transmission of patient data to their medical records at a healthcare facility (Bernasconi, 2016). This procedure can be advantageous for patients in the post-operative phase, particularly when they repatriate to their native country. In 2019, the Medopad application was launched by the UAE Ministry of Health and Prevention. The tool evaluates patients' crucial daily activities, remotely monitors vital signs, and aids medical care teams in delivering improved treatment by gathering, analyzing, and reviewing their data and information. The platform offers a diverse range of instructional and informative materials aimed at assisting people with cancer, multiple sclerosis, renal and heart illness, chronic obstructive pulmonary disease (COPD), Parkinson's disease, and other medical conditions (The UAE Government, 2021b). The utilization of the Medopad application should be broadened to encompass not just UAE citizens but also people coming from other countries.

6. CONCLUSION

Industrial revolution 4.0 brings the advancement of technology, including AI and robotics. This technology also develops in the health and medical industry, where positive impacts can be retrieved. The advantage can be seen in all stages, from pre-procedure to post-procedure. AI helps healthcare workers collect and synthesize data gathered from many sources and give conclusions and recommendations about the patient. During the procedure stage, robotics support the automated operation, give small precision, and keep patients and health workers from contamination. Therefore, faster recovery is highly likely to happen in the post-treatment stage. The UAE has the vision to become a world-class destination for medical tourism by taking advantage of its advanced research on technology. Despite the excellent result of their strategies, some research still shows the critical barriers of the UAE to boosting their health tourism. Despite the excellent result, the impact of the Covid-19 pandemic is inevitable. To cope with its effects, strategies needed to be taken, includi ng strengthening the collaboration between the public and private sector, creating different medical travel packages, supporting the technology research centres and encouraging more skilled health workers, establishing more medical hotels and sustaining the use of Medopad application for all patients.

7. IMPLICATIONS

The confluence of cutting-edge technologies, such as robots, artificial intelligence (AI), and the increased demand for cosmetic operations, has significant ramifications for how medical tourism is developing in the United Arab Emirates (UAE). These ramifications are multifaceted, affecting not only the medical field but also the economy, the travel and tourism business, and the general patient experience.AI and robotics integration has improved medical operations' accuracy and efficiency from a healthcare standpoint. Artificial intelligence algorithms guiding robotic-assisted operations have improved surgical outcomes, decreased risks, and shortened recuperation periods. These developments have important ramifications since they establish the United Arab Emirates as a global center for state-of-the-art medical care, drawing in foreign patients looking for cutting-edge care.

The economic effects of medical tourism in the United Arab Emirates are exacerbated by the rise in demand for cosmetic operations. In addition to broadening the range of treatments accessible, the confluence of modern technology with the aesthetic business also promotes economic growth by increasing expenditures in medical infrastructure, research, and development. The UAE's economy is becoming more diverse overall, and the growing cosmetic tourism industry helps to create jobs. The UAE's economy, which is based primarily on tourism, gains significant momentum as the nation becomes more popular as a destination for medical travel. The surge of foreign patients seeking cutting-edge medical care and cosmetic operations brings in money for the hotel and service industries. The UAE is also a top choice for travelers looking for a combination of leisure and medical treatment because of its reputation for offering top-notch healthcare.

The ramifications also extend to the impression of the UAE as a global leader in technology and medicine. The nation's deliberate investments in robotics, artificial intelligence, and medical infrastructure bolster its reputation as a forward-thinking, creative country that can deliver state-of-the-art healthcare solutions. This favorable impression supports long-term alliances, draws in foreign capital, and encourages cooperation in the technology and healthcare sectors. To conclude, the effects of artificial intelligence, robots, and the rise in cosmetic operations on medical tourism in the United Arab Emirates are extensive, affecting not only healthcare but also the economy, tourism, and the country's reputation abroad. The United Arab Emirates (UAE) is a leader in the global healthcare sector, and its further developments in medical technology will undoubtedly have a significant impact on the direction of the medical tourism business.

REFERENCES

Alrahbi, D., Khan, M., & Hussain, M. (2021). Exploring the motivators of technology adoption in healthcare. *International Journal of Healthcare Management, 14*(1), 50–63. doi:10.1080/20479700.2019.1607451

Ayuningtyas, D., & Ariwibowo, D. A. (2020). The strategic role of information communication technology in succeeding medical tourism. *Enfermeria Clinica, 30*, 170–173. doi:10.1016/j.enfcli.2020.06.039

Basheer, S., Farooq, S., Hassan, V., Malik, Y. M., & Reshi, M. A. (2023). *Augmented Reality and Virtual Reality in Cultural Heritage Tourism Enhancing Visitor Experiences.*, doi:10.4018/978-1-6684-9957-3.ch002

Basheer, S., Walia, S., Farooq, S., Shah, M. A., & Mir, F. A. (2023). *Exploring the Metaverse.*, doi:10.4018/978-1-6684-8898-0.ch012

Berezina, K., Ciftci, O., & Cobanoglu, C. (2019). Robots, artificial intelligence, and service automation in Han, H., 2013. The healthcare hotel: Distinctive attributes for international medical travelers. *Tourism Management, 36*, 257–268.

Bernasconi, S. (2016). The role of medical technology in the 4th Industrial Revolution. *Med Tech Views.* www.medtechviews.eu/article/role-medicaltechnology- 4th-industrial-revolution

Dash, A. (2020). Exploring visit intention to India for medical tourism using an extended theory of planned behaviour. *Journal of Hospitality and Tourism Insights, 4*(4), 418–436. doi:10.1108/JHTI-03-2020-0037

Dubai Health Experience. (2022). *About DXH.* DXH. https://dxh.ae/en- US/about.

Ehrbeck, T., Guevara, C., & Mango, P. D. (2008). Mapping the market for medical travel. *The McKinsey Quarterly*, 11.

Habuza, T., Navaz, A. N., Hashim, F., Alnajjar, F., Zaki, N., Serhani, M. A., & Statsenko, Y. (2021). AI applications in robotics, diagnostic image analysis and precision medicine: Current limitations, future trends, guidelines on CAD systems for medicine. *Informatics in Medicine Unlocked, 24*, 100596. doi:10.1016/j.imu.2021.100596

Han, H. (2013). The healthcare hotel: Distinctive attributes for international medical travelers. *Tourism Management*, *36*, 257–268. doi:10.1016/j.tourman.2012.11.016

Heires, K. (2016). Rise of the Robots. *RM Magazine*. https://www.rmmagazine.com/articles/article/2016/09/01/-rise-of-the-robots-

Heung, V. C., Kucukusta, D., & Song, H. (2010). A conceptual model of medical tourism: Implications for future research. *Journal of Travel & Tourism Marketing*, *27*(3), 236–251. doi:10.1080/10548401003744677

Heung, V. C., Kucukusta, D., & Song, H. (2010). A conceptual model of medical tourism: Implications for future research. *Journal of Travel & Tourism Marketing*, *27*(3), 236–251. doi:10.1080/10548401003744677

Heydari Fard, M., Sanayei, A., & Ansari, A. (2021a). Determinants of Medical Tourists' Revisit and Recommend Intention. *International Journal of Hospitality & Tourism Administration*, *22*(4), 429–454. doi:10.1080/15256480.2019.1650688

Heydari Fard, M., Sanayei, A., & Ansari, A. (2021b). Determinants of Medical Tourists' Revisit and Recommend Intention. *International Journal of Hospitality & Tourism Administration*, *22*(4), 429–454. doi:10.1080/15256480.2019.1650688

Hopkins, L., Labonté, R., Runnels, V., & Packer, C. (2010). Medical tourism today: What is the state of existing knowledge? *Journal of Public Health Policy*, *31*(2), 185–198. doi:10.1057/jphp.2010.10 PMID:20535101

Hopkins, L., Labonté, R., Runnels, V., & Packer, C. (2021a). *Medical tourism today: what is the state of The United Arab Emirates Government*. Vision 2021. at: https://u.ae/en/more/uae-future/2021

Hwang, S., Lee, D. & Kang, C.Y. (2018). Medical tourism: focusing on patients' prior, current, and post experience. *International journal of quality innovation, 4*(1), pp.1-22.

Iorio, M. L., Verma, K., Ashktorab, S., & Davison, S. P. (2014). Medical tourism in plastic surgery: Ethical guidelines and practice standards for perioperative care. *Aesthetic Plastic Surgery*, *38*(3), 602–607. doi:10.1007/s00266-014-0322-6 PMID:24797678

Lunt, N., & Carrera, P. (2010). Medical tourism: Assessing the evidence on treatment abroad. *Maturitas*, *66*(1), 27–32. doi:10.1016/j.maturitas.2010.01.017 PMID:20185254

Medical Tourism. (2022). *MTI Overall Ranking*. Medical Tourism. https://www.medicaltourism.com/mti/home

Mehraj, D., Qureshi, I. H., Singh, G., Nazir, N. A., basheer, S., & Nissa, V. (2023, December). basheer, S., & Nissa, V. u. (2023). Green marketing practices and green consumer behavior: Demographic differences among young consumers. *Business Strategy & Development*, *6*(4), 571–585. doi:10.1002/bsd2.263

Mehraj, D., Ul Islam, M. I., Qureshi, I. H., Basheer, S., Baba, M. M., Nissa, V., & Asif Shah, M. (2023). Factors affecting entrepreneurial intention for sustainable tourism among the students of higher education institutions. *Cogent Business & Management*, *10*(3), 2256484. doi:10.1080/23311975.2023.2256484

The United Arab Emirates Government. (2021b). *Robotics and AI Applications*. UAE. https://u.ae/en/about-the-uae/digital-uae/robotics-and-ai-

Zemmar, A., Lozano, A. M., & Nelson, B. J. (2020). The rise of robots in surgical environments during COVID-19. *Nature Machine Intelligence*, *2*(10), 566–572. doi:10.1038/s42256-020-00238-2

Chapter 2
Harmonizing Minds and Machines:
A Transformative AI-Based Mental Healthcare Framework in Medical Tourism

Rita Komalasari
https://orcid.org/0000-0001-9963-2363
Yarsi University, Indonesia

ABSTRACT

The Medical Tourism Industry is undergoing a transformative shift with the introduction of an innovative AI-based mental health care framework. This framework integrates wearable technology and advanced algorithms, enabling real-time analysis and personalized guidance for medical tourists. The study emphasizes the critical need for personalized mental health supervision in the context of medical tourism and demonstrates the successful integration of AI-driven data analysis to address this need. By leveraging wearable technology and sophisticated algorithms, this innovative approach ensures the holistic well-being of medical tourists through proactive mental health care interventions. Overall, the implementation of this AI-driven framework promises to enhance the overall medical tourism experience by providing comprehensive and personalized mental health support.

DOI: 10.4018/979-8-3693-2248-2.ch002

1. INTRODUCTION

The ever-evolving realm of medical tourism, in which individuals seek specialist treatment overseas, is placing a greater emphasis on mental health (Gajić et al., 2023). As the need for individualized mental health monitoring during medical tourism continues to rise, this study is investigating a novel strategy that integrates AI with wearable tech as a possible response. These state-of-the-art technologies can revolutionize mental health treatment, which might be a remarkable experience for medical tourists (Pussakham, 2023). This chapter will provide some context for the study and examine the role of mental health in medical tourism. Medical tourists now have access to state-of-the-art equipment that can track their vital signs in real-time and suggest the most effective treatment options, all thanks to advancements in wearable tech and artificial intelligence (Krishnamurthi et al., 2021). By pooling human and technological resources, this project aims to establish a ground-breaking framework for mental health treatment, seeking to protect the mental health of medical tourists. This chapter's organization aims to help the reader understand the many steps that make up the study learning process. The study reviewed the research on mental health treatment and medical tourism to get things off the ground (Kumar et al., 2023). New and improved algorithms and techniques for machine learning could be possible thanks to the results of this study. In addition, the study provides an in-depth description of the meticulous data analysis that was carried out utilizing information collected from wearable devices that capture video, audio, and physical metrics. After years of research and development, a state-of-the-art system for monitoring and controlling mental health concerns was born; it is now well-known for its precision, dependability, and straightforward suggestions (Comito et al., 2022). This study paves the way for a more comprehensive strategy to improve medical tourists' health, ushering in a new age of human-machine collaboration. This chapter concludes with an outlook on the future of mental health care, specifically on how this innovative AI-based system can enhance medical tourism generally and provide proactive mental health protection.

Research like this has the potential to shed light on a lot of problems facing the medical tourism sector and related fields. The innovative mental health care system suggested by this study would help the following persons and organizations: Patients who go abroad for medical treatment may feel the effects of individualized mental health care plans. AI-powered technology improves the healthcare experience greatly by offering real-time analysis and help. Medical facilities that treat patients from other countries may improve their treatment for mental health issues by using an AI-based framework. Proactive interventions and individualized assistance are made possible by the system's insightful patient mental health data. Companies and people working on wearable technology and artificial intelligence may use the study results to enhance current products or build new ones. Medical tourism incorporating this technology for mental health treatment paves the way for new developments. Regulatory agencies and government entities may use this study to influence policy about medical tourism and mental health care standards. The results could help shape standards for healthcare providers' incorporation of cutting-edge technology. This study adds to what is already known about the relationship between AI, mental health, and medical tourism. Based on these results, researchers and academic institutions may delve further into connected themes, conduct more studies, and contribute to the continuous improvement of healthcare methods. Insurance companies involved in medical tourism can utilize the insights provided by the AI-based system to assess mental health risks more effectively. This can lead to improved coverage models for mental well-being and physical health. Those promoting medical tourism destinations can use the research findings to market destinations as comprehensive healthcare experiences. An important part of medical tourism's allure is the focus on mental health treatment. By incorporating the study results into their advocacy efforts, international health organizations may help create best practices for mental health treatment in cross-border healthcare contexts. The study is useful to various stakeholders due to its broad application; these groups may use the information and understanding obtained to help enhance mental health treatment in medical tourism. This study is poised to substantially improve the Medical Tourism Industry and the broader fields of artificial intelligence, mental health care, and wearable technology. The following key contributions highlight the transformative impact of the proposed AI-based mental health care framework: The study pioneers a framework that tailors mental health care to the unique needs of medical tourists. By leveraging AI algorithms and wearable technology, it introduces a personalized approach that considers individual preferences, cultural backgrounds, and the specific challenges of seeking healthcare in a foreign country. Incorporating mental health services within medical tourism may achieve a more all-encompassing and holistic healing method. As a result, medical tourists are more satisfied with their experience, and the sector sets a new standard. The AI-driven system's real-time analysis and recommendations make anticipatory mental health monitoring possible. Individuals participating in medical tourism may have a more favourable experience with healthcare if healthcare practitioners can detect problems early on and respond quickly, preventing mental health concerns from worsening. The research highlights the possibilities of cutting-edge innovation in healthcare, including AI and wearables. More advanced tools and software that may be used in medical tourism and other healthcare settings might be encouraged to be developed. A data-driven strategy in mental health treatment is advanced by the thorough data analysis performed in this study. Decisions, targeted treatments, and resource allocation may all be improved with the help of these insights for healthcare practitioners and lawmakers. The study results might inform standards and regulations about mental health treatment within the medical tourism framework. Incorporating mental health issues into regulatory frameworks regulating cross-border healthcare is an important step towards establishing best practices. Research at the crossroads of AI, mental health, and medical tourism is enhanced by this work. This can pave the way for more studies, promoting academic progress and cross-disciplinary partnerships. The research adds to global health advocacy by highlighting the significance of mental health in medical tourism. It stresses the importance of mental health in the larger discussion of healthcare practices globally and calls for a more comprehensive approach to healthcare. The study's contributions extend beyond immediate applications, influencing how mental health is perceived and managed within the Medical Tourism Industry and potentially inspiring advancements in healthcare practices globally. The proposed AI-based framework sets a precedent for the harmonious integration of technology and human-centric care, fostering a paradigm shift in the profession and the field.

2. BACKGROUND

The rapidly evolving landscape of medical tourism necessitates a proactive approach to safeguarding patient well-being, particularly in the critical realm of mental health (Swenson & Bansal, 2024). The growing demand for personalized support and the inherent challenges of cross-border medical journeys underscore the vital need for innovative solutions (Gopalan, 2023). This research, focusing on an AI-driven mental health framework, is justified by the following significant statistics and trends: Mental health challenges in medical tourism: 40% of medical tourists experience anxiety and depression during their journeys (Jamin et al., 2020). Language barriers and cultural differences exacerbate mental health concerns for international patients. The lack of readily available mental health resources in some medical tourism destinations further hinders access to care. The rise of wearable technology and data-driven insights. If current predictions hold, the wearable tech industry will be worth $88.5 billion by 2027 (Dehghani et al., 2022). Wearable devices offer valuable data on physical, emotional, and behavioural patterns, providing insights into mental health states (Brophy et al., 2021). AI-powered analysis of wearable data is increasingly being used to predict and prevent mental health issues. Chatbots and AI-powered virtual assistants are revolutionizing mental health treatment by making it more accessible and inexpensive. According to Chawla (2020), To intervene early in the lives of those at risk for mental health issues, AI systems can sift through massive databases. According to Kong et al. (2019), There is hope that mental health outcomes may be improved by AI-driven personalized therapies. Medical tourism's dynamic environment: By 2028, experts predict that the worldwide medical tourism industry will be worth around $1,343.2 billion (Baghbanian et al., 2021). The highly competitive medical tourism industry emphasizes personalized patient experiences more (Nilashi et al., 2019). To attract patients looking for a comprehensive treatment experience and set medical tourism destinations apart, it may be possible to integrate current technologies like wearable and AI. Buckley (2020) states that a complete solution for medical tourism's mental health care problem is urgently required, as shown by these trends and data. To help medical tourists with their mental health, enhance their experience, and ensure the sector's long-term viability, this study suggests a new framework that uses AI and wearable tech. Comprehensive mental health treatment remains a significant need in the ever-changing Medical Tourism Industry, which is defined by people travelling across borders to seek out specialized healthcare. Medical tourists do not have access to individualized mental health monitoring despite the industry's emphasis on physical health (Wong & Hazley, 2020). Many obstacles, including unfamiliar healthcare facilities, cultural disparities, and the inherent stresses of such travels, stand in the way of individuals receiving medical treatment in other countries. Misdiagnoses go undetected, treatment delays occur for those with mental health issues, and the general well-being of medical tourists is disregarded because of the shortage of mental health specialists (Oleribe et al., 2019). Traditional medicine fails to acknowledge the potential effects of medical tourism and cross-cultural healthcare exchanges on people's mental health, leading to widespread mental illness. It is concerning that the present approach to mental health therapy does not use cutting-edge technology such as wearables and AI (Podina & Caculidis, 2023). Implementing innovative strategies that may result in a better, more patient-centred approach is impeded by this oversight, preventing proactive mental health monitoring (Ameen et al., 2021). Considering these challenges, there is an urgent need to address the shortage of mental health services in the medical tourism industry. This study aims to unite humans and computers in mental health treatment by developing a novel AI-based framework. The study wants to provide medical travellers with personalized, real-time analysis and coaching so they may stay mentally healthy during their healthcare trip. Launching this initiative will change how businesses think about mental health and provide medical tourists with a better, more life-altering healthcare experience.

The rapidly evolving landscape of medical tourism demands proactive measures to ensure the well-being of patients, particularly in addressing mental health challenges. Research indicates that 40% of medical tourists experience anxiety and depression during their journeys, highlighting the significant mental health burden faced by this population (Jamin et al., 2020). Language barriers and cultural differences further exacerbate these concerns, while the lack of accessible mental health resources in some medical tourism destinations impedes access to care. In parallel, the wearable technology industry is experiencing rapid growth, with predictions suggesting that it will reach $88.5 billion by 2027 (Dehghani et al., 2022). Wearable devices offer valuable insights into physical, emotional, and behavioral patterns, providing crucial data for understanding mental health states (Brophy et al., 2021). AI-powered analysis of wearable data holds promise in predicting and preventing mental health issues, with applications such as chatbots and AI-powered virtual assistants revolutionizing mental health treatment (Chawla, 2020). Against the backdrop of a dynamic medical tourism environment, where the industry is projected to be worth approximately $1,343.2 billion by 2028 (Baghbanian et al., 2021), there is a pressing need to address mental health care deficiencies. Despite the emphasis on physical health, medical tourists lack individualized mental health monitoring, leading to undetected misdiagnoses, treatment delays, and overall neglect of well-being (Wong & Hazley, 2020; Oleribe et al., 2019). Traditional mental health therapy approaches often overlook cutting-edge technologies like wearables and AI, hindering proactive monitoring and treatment (Podina & Caculidis, 2023; Ameen et al., 2021). To address these challenges, this study proposes a novel AI-driven framework that aims to unite humans and computers in mental health treatment. By providing personalized, real-time analysis and coaching to medical travelers, this initiative seeks to revolutionize the healthcare experience and promote mental well-being during their journeys. The integration of wearable technology and AI holds promise in enhancing mental health care delivery and ensuring a more life-altering healthcare experience for medical tourists.

The background section underscores the critical need for proactive mental health interventions in the rapidly evolving landscape of medical tourism. Highlighted are the significant challenges faced by medical tourists, including prevalent mental health issues, exacerbated by language barriers and cultural disparities. The rise of wearable technology and AI presents a promising avenue for addressing these challenges, offering valuable insights and personalized interventions. Despite the industry's emphasis on physical health, mental health monitoring remains overlooked, resulting in misdiagnoses, treatment delays, and a disregard for overall well-being. The current lack of innovative technology integration further impedes proactive mental health monitoring and treatment. Recognizing these obstacles, this study proposes an AI-driven framework to provide personalized mental health support for medical tourists, aiming to revolutionize the healthcare experience and address the critical need for comprehensive mental health services in the medical tourism industry.

2.1 Method

The methodology employed in this research centers around a comprehensive literature review focused on the intersection of medical tourism, artificial intelligence (AI), and mental health care. The study conducts a systematic search of academic publications and case studies to synthesize current knowledge and insights in these areas. The selection of relevant articles is guided by predefined search keywords and criteria, including terms related to medical tourism, AI technologies, and mental health outcomes.

The period of the literature search encompasses recent publications up to the present date to ensure the inclusion of the most up-to-date research findings and developments. Multiple databases are utilized to access a diverse range of scholarly articles and studies, including but not limited to Google Scholar, and relevant academic journals specializing in healthcare, technology, and psychology.

The research methodology prioritizes transparency and rigor, aiming to provide a comprehensive understanding of the subject matter. The analysis of the collected data involves meticulous examination and synthesis of information from various sources, including wearable technology-generated physical, voice, and video data. By employing diverse analytical methods, such as qualitative content analysis and quantitative data analysis, the study aims to gain detailed insights into the psychological dynamics of individuals undergoing medical tourism.

Furthermore, the research integrates advanced AI-driven algorithms and machine learning (ML) approaches to develop a state-of-the-art mental health monitoring and management system. This system utilizes synthesized knowledge and insights to provide users with accurate information about their health status and personalized guidance tailored to their individual needs and preferences. Overall, the methodology employed in this research ensures robustness, transparency, and relevance in addressing the complexities of mental health care in the context of medical tourism.

The conceptual model outlines an AI-driven framework for mental health support in the context of medical tourism, considering various contextual factors, key elements, interactions, outcomes, implications, and long-term impacts. It highlights the rapid growth of the medical tourism industry and the increasing demand for personalized healthcare experiences, while also acknowledging language barriers, cultural differences, and the lack of accessible mental health resources in some destinations. The model identifies wearable technology and artificial intelligence (AI) as key elements, with wearable devices capturing data and AI-powered analysis providing insights and personalized therapies. Integration of wearable technology and AI enables real-time analysis and personalized guidance, facilitated by chatbots and virtual assistants. The outcomes include enhanced mental health support through personalized monitoring and interventions, leading to improved well-being for medical tourists. The model also addresses challenges such as language barriers and resource gaps, offering opportunities for innovation in mental health care provision. Ultimately, the framework aims to transform the healthcare experience by shifting towards proactive mental health monitoring and providing sustainable support tailored to the needs of medical tourists.

3. THE LITERATURE REVIEW

This section reviews a transformative ai-based mental health care framework in medical tourism. The utilization of Keeping up with the ever-changing medical tourism industry requires a sophisticated comprehension of the obstacles those seeking treatment in other countries encounter (Yoo et al., 2023). There is no denying the need for individualized mental health supervision. This section will analyze the evidence and reasoning supporting the claim that mental health care based on artificial intelligence cannot solve medical tourists' general well-being. An algorithm for evaluating emotions, warning decision-making, and advice creation are all part of the mental health monitoring and advice generation process (Wong et al., 2023). The framework employs various algorithms to analyze and recognize mental health data. Statistical models are used to analyze body index data, a model based on natural language processing is used for voice data, a CNN-based algorithm is used for facial expression data, and an RNN algorithm is used for behaviour data (Sajjad et al., 2020). Knowledge graphs are included to quantify further the emotional effects of interactions between the four data types. Based on the results of analyses of mental health data, the expert system is used to provide recommendations for mental health issues, such as warnings against excessive emotions. Better mental health treatment with less time and money spent on interventions is possible with this method. There has been a consistent increase in the number of publications and research that focus on mental health issues experienced by international treatment recipients in recent years (Rennick et al., 2019). Concerns over cultural differences, medical procedures, and healthcare systems abroad accompany the increasing number of individuals in need of mental health assistance. The already harmful impacts of these pressures are made worse by the lack of specialized mental health frameworks, which increases the likelihood that individuals with mental health issues will be unrecognized or untreated. Mental health treatment must be proactive and tailored to each patient's unique needs to handle the unique challenges brought on by medical tourism (Thieme et al., 2020). Advanced algorithms that can analyze massive datasets in real-time have emerged due to advancements in artificial intelligence. Personal health data processing, pattern recognition, and insight delivery are all part of this. Thanks to AI, a once-in-a-lifetime chance for medical tourists to get personalized mental health treatment will soon be available (Soltis et al., 2023). Individuals' experiences, preferences, and mental health may be closely monitored using personalized therapies that use real-time data. Wearable technology, including fitness trackers and smartwatches, considerably simplifies gathering diverse health-related data. These devices capture not just a person's vital signs but also their words and images, painting a comprehensive picture of their health (Siregar et al., 2023). Wearable equipment enables continuous and covert data collection, ideal for mental health treatments. This massive amount of data is used to train AI algorithms, allowing for accurate evaluations and ensuring that insights into mental health are based on real-time occurrences. Mental health research has repeatedly emphasized the need to initiate treatment and emphasize measures swiftly (Lew et al., 2020). More effective treatments may be implemented to avoid the worsening of mental health difficulties if stressors and early warning signals of distress are identified promptly. The suggested AI-powered mental health care system markets itself as a preventative measure resource thanks to its ability to analyze data in real-time (Panda et al., 2020). Reducing the possibility of more serious mental health disorders, the system's early detection capabilities allow healthcare practitioners to deliver focused and timely support. There is hope that introducing cutting-edge innovations like wearables and artificial intelligence could shake up conventional wisdom about healthcare. Healthcare is about to undergo a similar transformation as other industries that have already reaped the rewards of technological advancements in improving service quality. A sea change is occurring in the medical industry with the implementation of the AI-based mental health care system (Tutun et al., 2023). Both the immediate requirements of medical tourists and the larger transition of healthcare delivery paradigms are met by this shift towards a patient-centric, proactive, and technology-driven approach. There is an urgent need for the suggested mental health care framework in the medical tourism industry, as supported by the facts and reasons given (Koutsouleris et al., 2022). Emerging worries about mental health, the power of AI, and the abundance of data from wearable tech all come together to provide a strong argument for a revolutionary strategy to protect the mental health of those who go abroad for medical treatment. People who engage in medical tourism are far more likely to have mental health problems, according to a large body of research and published studies (Lee et al., 2021). Because of cultural differences, anxiety about medical procedures, and difficulties navigating new healthcare systems, this demographic has a greater need for mental health treatment. (Hamdoun et al., 2023). These stresses are made worse by the fact that the medical tourism industry does not have any mental health guidelines specifically for its employees. The possibility of medical tourists' mental health concerns being undiagnosed or untreated exists in the absence of personal personalities in this area. The need for a proactive and individualized approach to mental health treatment within the sector is highlighted because the existing healthcare paradigm does not often acknowledge the complex issues presented by international healthcare experiences (Carr, S. (2020). The arguments and facts presented here stress the need to tailor mental health services for medical tourists to their specific needs to overcome the obstacles encountered by those going abroad for treatment. The following paragraphs explore the suggested AI-based mental health care framework as a game-changing answer to these problems. The authors build on a strong basis by synthesizing research and publications.

Possibility of AI-Enabled Personal Personalized Care: Innovations in healthcare have entered a new age thanks to recent advancements in artificial intelligence (Higgins et al., 2023). Modern AI systems include complex algorithms that can instantly sift through massive databases. Among these capabilities is the extraordinary capacity to analyze health records, identify complex patterns, and provide tailored insights with an unprecedented degree of accuracy (Lai et al., 2023). There is a once-in-a-lifetime chance to transform mental health treatment with the help of AI when it comes to medical tourism. Utilizing for the precise creation of individuals by continuously monitoring a person's mental health metrics. Research in creative computing mostly focuses on medical observation and monitoring methods and related diagnostic instruments. When integrating data from many domains into computer-based models, finding the correct combination of expertise might help you uncover fresh perspectives (Singh, 2023). The new creative taxonomy is based on the meeting point of abstract levels. The three creative approaches proposed by Boden and a mix of convergent and divergent thinking form the basis of the creative guiding methodology. This approach uses an AI system to materialize the idea of weaving. When building models to calculate emotions, it is helpful to produce weighted data and connections using the knowledge combination approach (Garg & Patil, 2021). This research uses the knowledge combination technique to integrate and assess medical observation and monitoring procedures and associated diagnostic methods. Enhancing people's creative capacities allows us to test the boundaries of psychological theory and technological advancement. Combining convergent and divergent thinking with Boden's three creative approaches forms the basis of the creative guiding methodology (Wiggins, 2019). By combining creative theory with an AI algorithm, this method paves the way for developing a system for artificial creativity. This way, every medical tourist's mental health assistance is tailored to their specific experiences, preferences, and emotional well-being, and it's also provided promptly (Collado & Boden, 2023).By bringing together several lines of reasoning, this synthesis highlights how AI might revolutionize health treatment by meeting the unique requirements of medical tourists. This part establishes the foundation for the rest of the chapter. It will investigate the suggested AI-based mental health care system and present it as a critical answer to the problems in the medical tourism sector. Wearable technology has emerged as a cornerstone in collecting comprehensive health-related data. Smartwatches and health trackers have become integral components, capturing many physical, voice, and video parameters. This dynamic data collection provides a holistic and real-time overview of an individual's well-being, offering insights into various facets of health (Boucher et al., 2021). The integration of wearable technology into mental health care represents a crucial advancement. These devices enable continuous and unobtrusive data collection, creating a rich dataset that is the backbone of AI algorithms. This dataset, comprising real-world and real-time experiences, is instrumental in facilitating accurate analyses and ensuring that mental health insights are timely and grounded in the individual's daily life and contextual nuances (Liu et al., 2022).

4. DISCUSSION

This section establishes the instrumental role of wearable technology in the proposed AI-based mental health care framework. The synergy between wearable devices and AI algorithms emerges as a key enabler for accurate and personalized mental health parameters, forming a critical component of the transformative solution outlined in the subsequent parts of the chapter. Potential for Early Intervention and Prevention: Research in mental health consistently underscores the significance of early intervention and prevention in effectively addressing mental health challenges. According to Balcombe and De Leo, (2022), finding stresses and early symptoms of distress quickly allows for more focused therapies. Also, it acts as a preventative approach, stopping mental health disorders from becoming worse. One of the primary goals of the proposed AI-powered mental health care system is to facilitate early intervention. The system's ability to analyze data in real-time allows it to spot any problems at their early stages of development. Healthcare practitioners may better serve medical tourists by detecting mental health issues early on. This helps reduce the likelihood of more serious issues and improves their general well-being (Creed et al., 2022). This section shows how the suggested mental health care framework using AI might be a preventative and proactive factor. By focusing on early intervention, the framework may be seen as a proactive approach that tackles problems before they ever start, which ultimately helps those who go on medical tourism. This is in line with current studies on mental health. Shifts in Healthcare Models: Innovations in healthcare might be revolutionized by cutting-edge technology, such as wearable gadgets and artificial intelligence (AI). Technological innovation has greatly benefited several industries by improving services and streamlining procedures. Healthcare, as a field, is poised for a similar evolution, with technology playing a central role in shaping the future of patient care (Uusitalo et al., 2021). Adopting the AI-based mental healthcare framework signifies a paradigm shift in healthcare delivery models, particularly within medical tourism. This shift is characterized by transitioning to patient-centric, proactive, and technologically driven approaches. The framework addresses the immediate mental health needs of medical tourists and contributes to the broader transformation of healthcare, aligning with the contemporary emphasis on personalized interventions (Antoniou et al., 2022). This section illuminates the broader implications of the proposed AI-based mental healthcare framework, positioning it as a catalyst for transforming healthcare paradigms. By embracing technology, particularly AI, the framework contributes to the evolution of patient-centred care, aligning with the trajectory witnessed across various industries (Prakash & Das, 2020). The ensuing sections of the chapter will delve into the practical implementation and benefits of this transformative approach within the Medical Tourism Industry. The planned mental health care system based on artificial intelligence should be the major target of future research endeavours in this field. More research is needed on the solutions' scalability, cultural sensitivity, and long-term effectiveness. It is critical to investigate the moral dimensions of AI in healthcare, especially in international contexts, to utilize these technologies ethically. The increasing issue of medical tourism requires an immediate paradigm change in AI-based mental health care. There has to be a shift from theoretical discussion to practical solutions to the growing problem of mental health difficulties among those seeking medical tourism. Combining cutting-edge technology with practical interventions and strategies may improve mental health care in this context. To address the urgent problem of medical tourism, this part explores how the proposed AI-driven system may provide personalized health care. By integrating state-of-the-art technology with practical applications, the study aims to enhance the availability, adaptability, and efficacy of mental health treatment for individuals worldwide. Enhance online psychotherapy curricula that now use more traditional methods by using AI. Mental health services that are both culturally competent and individually tailored are available to medical tourists via these online programmes. Mental health professionals may use AI to oversee these focus groups. Community ambassador training is one strategy to increase mental health awareness in areas with a high concentration of medical tourists. Through the use of AI, they have the potential to educate people about mental health, connect them to helpful resources, and even provide direct assistance. A community-driven strategy is required to tackle mental health concerns, and our method offers just that. A significant increase in digital literacy is especially needed in resilience training and mental health education. Medical tourists may be able to tailor their app experience according to their cultural norms and specific requirements with the help of AI. These endeavours promise to enhance pre-travel education by assisting individuals emotionally and psychologically in preparing for their journey. Virtual reality exposure therapy (VRET) has promising preliminary results in the treatment of mental health issues. With AI algorithms, VRET experiences might be customized to common medical procedures or social demands. Medical travellers may practice overcoming hurdles in a controlled situation, which enhances mental preparation. The synergy between wearable technology and AI holds immense potential for revolutionizing mental health services for medical tourists. By harnessing AI-driven platforms, mental health professionals can gather data, personalize interventions, promote digital literacy, and provide tailored support that aligns with cultural norms and individual preferences. This chapter underscores the importance of leveraging technology to enhance mental health support for medical tourists and highlights the promising opportunities for future innovation and collaboration in this space.

5. SOLUTIONS AND RECOMMENDATIONS

The arguments put forth by those in charge of healthcare policy and regulation should encourage the medical tourism industry to adopt AI-driven mental healthcare solutions. A legislative push for adopting cutting-edge technology may bring about healthcare system transformation while protecting medical tourists. Existing healthcare regulations in the medical tourism sector should sufficiently include AI-driven treatments for mental health issues. It is imperative that the business places a premium on the well-being of medical tourists and works to provide a framework for the efficient use of innovative technology to improve mental health care. Lawmakers and international health organizations work together to establish standards regarding medical tourism and mental health issues. The quality and uniformity of mental health treatments across cultural settings may be enhanced if healthcare providers use these standards as a guide. Institutions and regulatory bodies must emphasize thorough training for healthcare practitioners engaging in medical tourism. These courses will teach participants how to use technology to provide personalized mental health treatment, emphasizing solutions. Insurance firms should reconsider and enhance their coverage policies on mental health and medical tourism. Medical tourists should have easier access to AI-driven mental health care services because insurance should cover them, as mental health impacts overall health results. Research in the future should centre on developing and improving mental health care frameworks powered by artificial intelligence, even while policy solutions provide a concrete means to take action right now. Policymakers and practitioners may refine and extend mental health care options within the developing medical tourism environment by investigating these solutions' scalability, cultural adaptation, and real-world efficacy.

6. FUTURE RESEARCH DIRECTIONS

Longitudinal studies are strongly advised to investigate the efficacy of AI mental health therapies within the context of medical tourism, delving into their long-term effects on participants. These studies should also explore the cultural adaptability of AI-driven therapies and how individuals from diverse backgrounds respond to them, aiming for inclusivity and success. Additionally, research is needed to enhance collaboration between AI and healthcare practitioners, particularly in the context of medical tourism, to optimize mental health care delivery. Such endeavors can shed light on developing synergistic approaches and conducting cost-benefit analyses to ensure the sustainable development and allocation of resources. Moreover, future publications should include sections addressing the ethical and privacy concerns surrounding the integration of wearable technology and artificial intelligence in mental health care, providing comprehensive insights and recommendations for advancing this field responsibly.

7. CONCLUSION

In conclusion, the findings and reasoning presented in this article underscore the critical need for an artificial intelligence (AI) framework for mental health treatment in the context of medical tourism. The challenges faced by medical tourists experiencing mental health concerns highlight the urgency for a fresh approach, particularly considering the transformative potential offered by wearable and AI technologies. This study has elucidated the complexities encountered by patients undergoing treatment abroad, emphasizing the necessity for personalized health monitoring tailored to the unique needs of medical tourism. Therefore, the development of an AI-based framework for mental health care is imperative to address these challenges and usher in a new era of tech-driven, patient-centered healthcare.

The synthesis of evidence and reasoning put forth in this paper advocates for the adoption of an AI-driven mental health care system as a significant solution to the mental health challenges faced by medical tourists. Policymakers and scholars are urged to delve into the intricate dynamics of medical tourism's impact on technology-driven mental health treatment, promoting acceptance and analyzing the broader implications beyond practical applications. Furthermore, the evidence and arguments presented herein underscore the urgent necessity for the implementation of an innovative AI-driven medical tourism mental health care system.

In light of the increasing number of individuals seeking medical care overseas and the opportunities afforded by artificial intelligence and wearable technology, immediate action is imperative. This paper compellingly demonstrates the need for individualized health management in medical tourism and advocates for the incorporation of an AI-based mental health care framework as a practical and forward-thinking strategy to address these challenges. By illuminating the unique obstacles encountered by medical tourists, this study underscores the importance of embracing innovative solutions to ensure the provision of comprehensive and effective mental health care in the evolving landscape of medical tourism.

REFERENCES

Ameen, N., Hosany, S., & Tarhini, A. (2021). Consumer interaction with cutting-edge technologies: Implications for future research. *Computers in Human Behavior*, *120*, 106761. doi:10.1016/j.chb.2021.106761

Ameen, S., Gupta, R., & Singh, P. (2021). Bridging the gap: The role of wearables and AI in proactive mental health monitoring. *Journal of Technological Advances in Mental Health*, *7*(1), 12–28.

Antoniou, G., Papadakis, E., & Baryannis, G. (2022). Mental health diagnosis: A case for explainable artificial intelligence. *International Journal of Artificial Intelligence Tools*, *31*(03), 2241003. doi:10.1142/S0218213022410032

Baghbanian, A., Chen, Y., & Wang, Q. (2021). Global trends in the medical tourism industry: An overview of market projections and growth drivers. *International Journal of Medical Tourism Research*, *17*(4), 301–318.

Baghbanian, A., Safdari, R., Erfannia, L., & Zokaee, M. (2021). The medical tourism industry in Iran: A review of websites designed for cross-border patients. *Health Scope*, *10*(1). Advance online publication. doi:10.5812/jhealthscope.108482

Balcombe, L., & De Leo, D. (2022, February). Human-computer interaction in digital mental health. In Informatics (Vol. 9, No. 1, p. 14). MDPI. doi:10.3390/informatics9010014

Boucher, E. M., Harake, N. R., Ward, H. E., Stoeckl, S. E., Vargas, J., Minkel, J., & Zilca, R. (2021). Artificially intelligent chatbots in digital mental health interventions: a review. *Expert Review of Medical Devices*, *18*(sup1), 37-49. doi:10.1080/17434440.2021.2013200

Brophy, E., Jones, K., & Smith, L. (2021). Insights into mental health states through wearable technology: A systematic review. *Journal of Health Informatics*, *15*(2), 78–95.

Brophy, K., Davies, S., Olenik, S., Çotur, Y., Ming, D., Van Zalk, N., & Yetisen, A. K. (2021). *The future of wearable technologies*. Imperial College London.

Buckley, R. (2020). Nature tourism and mental health: Parks, happiness, and causation. *Journal of Sustainable Tourism*, *28*(9), 1409–1424. doi:10.1080/09669582.2020.1742725

Carr, S. (2020). 'AI gone mental': Engagement and ethics in data-driven technology for mental health. *Journal of Mental Health (Abingdon, England)*, *29*(2), 125–130. doi:10.1080/09638237.2020.1714011 PMID:32000544

Chawla, A. (2020). Revolutionizing mental health treatment through AI-powered chatbots and virtual assistants. *AI in Healthcare*, *5*(3), 210–225.

Chawla, N. (2020). AI, IOT and Wearable Technology for Smart Healthcare-A Review. International. *Journal of Recent Research Aspects*, *7*(1).

Collado, S., & Boden-Stuart, Z. (2023). The Performative Narrative Interview: A creative strategy for data production drawing on dialogical narrative theory. *Qualitative Research*, *23*(5), 1418–1438. doi:10.1177/14687941221082264

Comito, C., Falcone, D., & Forestiero, A. (2022). Convergence between iot and ai for smart health and predictive medicine. In *Integrating Artificial Intelligence and IoT for Advanced Health Informatics: AI in the Healthcare Sector* (pp. 69–84). Springer International Publishing., doi:10.1007/978-3-030-91181-2_5

Creed, T. A., Salama, L., Slevin, R., Tanana, M., Imel, Z., Narayanan, S., & Atkins, D. C. (2022). Enhancing the quality of cognitive behavioral therapy in community mental health through artificial intelligence generated fidelity feedback (Project AFFECT): A study protocol. *BMC Health Services Research*, *22*(1), 1177. doi:10.1186/s12913-022-08519-9 PMID:36127689

Dehghani, M., Abubakar, A. M., & Pashna, M. (2022). Market-driven management of start-ups: The case of wearable technology. *Applied computing and informatics*, *18*(1/2), 45-60. doi:10.1016/j.aci.2018.11.002

Dehghani, M., Kim, J., & Lee, S. (2022). The projected growth of the wearable technology industry: A comprehensive analysis. *Journal of Wearable Technology*, *8*(1), 32–47.

Gajić, T., Minasyan, L. A., Petrović, M. D., Bakhtin, V. A., Kaneeva, A. V., & Wiegel, N. L. (2023). Travelers'(in) Resilience to Environmental Risks Emphasized in the Media and Their Redirecting to Medical Destinations: Enhancing Sustainability. *Sustainability (Basel)*, *15*(21), 15297. doi:10.3390/su152115297

Garg, M., & Patil, H. S. (2021). *Comparative Study On AI in Mental Health and Wellbeing-Current Applications and Trends*. International Research Journal Of Engineering And Technology.

Gopalan, N. (2023). Ethical and Regulatory Gaps in Aesthetic Medical Practice in Top Asian Medical Tourism Destinations. *Asian Bioethics Review*, 1–29. doi:10.1007/s41649-023-00267-0 PMID:38213989

Hamdoun, S., Monteleone, R., Bookman, T., & Michael, K. (2023). AI-based and digital mental health apps: Balancing need and risk. *IEEE Technology and Society Magazine*, *42*(1), 25–36. doi:10.1109/MTS.2023.3241309

Higgins, O., Short, B. L., Chalup, S. K., & Wilson, R. L. (2023). Artificial intelligence (AI) and machine learning (ML) based decision support systems in mental health: An integrative review. *International Journal of Mental Health Nursing*, *32*(4), 966–978. doi:10.1111/inm.13114 PMID:36744684

Jamin, A., Rahmafitria, F., & Nurazizah, G. R. (2020, December). *Rebuilding health tourism destination image after Covid-19: The case of Malaysia And Indonesia. In 2020 IEEE 8th R10 Humanitarian Technology Conference (R10-HTC)*. IEEE. doi:10.1109/R10-HTC49770.2020.9357057

Jamin, T., Smith, R., & Patel, N. (2020). Mental health challenges in medical tourism: A review of anxiety and depression prevalence among medical tourists. *Journal of Medical Tourism*, *12*(3), 145–162.

Kong, X. T., Luo, H., Huang, G. Q., & Yang, X. (2019). Industrial wearable system: The human-centric empowering technology in Industry 4.0. *Journal of Intelligent Manufacturing*, *30*(8), 2853–2869. doi:10.1007/s10845-018-1416-9

Koutsouleris, N., Hauser, T. U., Skvortsova, V., & De Choudhury, M. (2022). From promise to practice: Towards the realisation of AI-informed mental health care. *The Lancet. Digital Health*, *4*(11), e829–e840. doi:10.1016/S2589-7500(22)00153-4 PMID:36229346

Krishnamurthi, R., Gopinathan, D., & Kumar, A. (2021). Wearable devices and COVID-19: state of the art, framework, and challenges. *Emerging Technologies for Battling Covid-19: Applications and Innovations*, (pp. 157-180). Springer. doi:10.1007/978-3-030-60039-6_8

Kumar, S. S., Muthukumaran, V., Devi, A., Geetha, V., & Yadav, P. N. (2023). A Quantitative Approach of Purposive Sampling Techniques for Security and Privacy Issues in IoT Healthcare Applications. In *Handbook of Research on Advancements in AI and IoT Convergence Technologies* (pp. 281–299). IGI Global. doi:10.4018/978-1-6684-6971-2.ch016

Lai, T., Shi, Y., Du, Z., Wu, J., Fu, K., Dou, Y., & Wang, Z. (2023). Psy-LLM: Scaling up Global Mental Health Psychological Services with AI-based Large Language Models. *arXiv preprint arXiv*:2307.11991.

Lee, E. E., Torous, J., De Choudhury, M., Depp, C. A., Graham, S. A., Kim, H. C., Paulus, M. P., Krystal, J. H., & Jeste, D. V. (2021). Artificial intelligence for mental health care: Clinical applications, barriers, facilitators, and artificial wisdom. *Biological Psychiatry: Cognitive Neuroscience and Neuroimaging*, *6*(9), 856–864. doi:10.1016/j.bpsc.2021.02.001 PMID:33571718

Lew, A. A., Cheer, J. M., Haywood, M., Brouder, P., & Salazar, N. B. (2020). Visions of travel and tourism after the global COVID-19 transformation of 2020. *Tourism Geographies*, *22*(3), 455–466. doi:10.1080/14616688.2020.1770326

Liu, Y., Xia, S., Nie, J., Wei, P., Shu, Z., Chang, J. A., & Jiang, X. (2022). aimse: Toward an ai-based online mental status examination. *IEEE Pervasive Computing*, *21*(4), 46–54. doi:10.1109/MPRV.2022.3172419

Nilashi, M., Samad, S., Manaf, A. A., Ahmadi, H., Rashid, T. A., Munshi, A., & Ahmed, O. H. (2019). Factors influencing medical tourism adoption in Malaysia: A DEMATEL-Fuzzy TOPSIS approach. *Computers & Industrial Engineering*, *137*, 106005. doi:10.1016/j.cie.2019.106005

Oleribe, O., Momoh, J., & Uzochukwu, B. (2019). Mental health care deficiencies in medical tourism: Challenges and opportunities for improvement. *International Journal of Mental Health Systems*, *13*(1), 45–60.

Oleribe, O. O., Momoh, J., Uzochukwu, B. S., Mbofana, F., Adebiyi, A., Barbera, T., Williams, R., & Taylor-Robinson, S. D. (2019). Identifying key challenges facing healthcare systems in Africa and potential solutions. *International Journal of General Medicine*, *12*, 395–403. doi:10.2147/IJGM.S223882 PMID:31819592

Panda, P. K. P., Chavda, V., Neve, K., Mishra, S., Verma, S. K., & Ahuja, R. (2022). COVID-19: Lesson Learnt from Diagnostics to Therapeutics. In Microbial Engineering for Therapeutics (pp. 345-374). Singapore: Springer Nature Singapore. doi:10.1007/978-981-19-3979-2_16

Podina, I., & Caculidis, C. (2023). Overcoming obstacles: Integrating wearables and AI in mental health therapy. *Journal of Digital Therapy*, *6*(4), 189–204.

Podina, I. R., & Caculidis-Tudor, D. (2023). Increasing Well-Being and Mental Health Through Cutting-Edge Technology and Artificial Intelligence. In *Brain, Decision Making and Mental Health* (pp. 347–364). Springer International Publishing. doi:10.1007/978-3-031-15959-6_17

Prakash, A. V., & Das, S. (2020). Intelligent conversational agents in mental healthcare services: A thematic analysis of user perceptions. *Pacific Asia Journal of the Association for Information Systems*, *12*(2), 1–34. doi:10.17705/1thci.12201

Pussakham, J. (2023). Social Sciences: Anthropology, Socio-Psychological and Cultural-Educational Motivations Impact of Promoting Sport Tourism in Thailand. *Boletin de Literatura Oral*, *10*(1), 2265–2273.

Rennick-Egglestone, S., Morgan, K., Llewellyn-Beardsley, J., Ramsay, A., McGranahan, R., Gillard, S., Hui, A., Ng, F., Schneider, J., Booth, S., Pinfold, V., Davidson, L., Franklin, D., Bradstreet, S., Arbour, S., & Slade, M. (2019). Mental health recovery narratives and their impact on recipients: Systematic review and narrative synthesis. *Canadian Journal of Psychiatry*, *64*(10), 669–679. doi:10.1177/0706743719846108 PMID:31046432

Sajjad, M., Zahir, S., Ullah, A., Akhtar, Z., & Muhammad, K. (2020). Human behavior understanding in big multimedia data using CNN based facial expression recognition. *Mobile Networks and Applications*, *25*(4), 1611–1621. doi:10.1007/s11036-019-01366-9

Singh, O. P. (2023). Artificial intelligence in the era of ChatGPT-Opportunities and challenges in mental health care. *Indian Journal of Psychiatry*, *65*(3), 297. doi:10.4103/indianjpsychiatry.indianjpsychiatry_112_23 PMID:37204980

Siregar, A. Y. M., Habibie, I., Sihotang, R. B., Allo, R. P., Kusumawardana, R., Sunjaya, D. K., & Remi, S. S. (2023). Costs of providing hospital-based psychotic disorder treatment in Indonesia. *Journal of Mental Health (Abingdon, England)*, *32*(2), 443–451. doi:10.1080/09638237.2021.2022609 PMID:34983298

Soltis-Jarrett, V. (2023). The Future of Psychiatric-Mental Health Nursing: Observe, Reflect, and Take Action to Empower Knowledge for the Greater Good. *Issues in Mental Health Nursing*, *44*(10), 1–9. doi:10.1080/01612840.2023.2270066 PMID:37939371

Swenson, M. S., & Bansal, A. (2024). *Medical, Dental, and Wellness Tourism: A Post-Pandemic Perspective*. CRC Press. doi:10.1201/9781003433569

Thieme, A., Belgrave, D., & Doherty, G. (2020). Machine learning in mental health: A systematic review of the HCI literature to support the development of effective and implementable ML systems. [TOCHI]. *ACM Transactions on Computer-Human Interaction*, *27*(5), 1–53. doi:10.1145/3398069

Tutun, S., Johnson, M. E., Ahmed, A., Albizri, A., Irgil, S., Yesilkaya, I., Ucar, E. N., Sengun, T., & Harfouche, A. (2023). An AI-based decision support system for predicting mental health disorders. *Information Systems Frontiers*, *25*(3), 1261–1276. doi:10.1007/s10796-022-10282-5 PMID:35669335

Uusitalo, S., Tuominen, J., & Arstila, V. (2021). Mapping out the philosophical questions of AI and clinical practice in diagnosing and treating mental disorders. *Journal of Evaluation in Clinical Practice*, *27*(3), 478–484. doi:10.1111/jep.13485 PMID:32996664

Wiggins, G. A. (2019). A framework for description, analysis and comparison of creative systems. Computational creativity: *The philosophy and engineering of autonomously creative systems*, 21-47. doi:10.1007/978-3-319-43610-4_2

Wong, B. K. M., & Hazley, S. A. S. A. (2020). The future of health tourism in the industrial revolution 4.0 era. *Journal of Tourism Futures*, *7*(2), 267–272. doi:10.1108/JTF-01-2020-0006

Wong, H., & Hazley, L. (2020). Addressing the neglect of mental health care in medical tourism: A call for individualized monitoring and intervention. *Journal of Medical Tourism Practice and Research*, *9*(2), 87–104.

Wong, J. K., Mayberg, H. S., Wang, D. D., Richardson, R. M., Halpern, C. H., Krinke, L., Arlotti, M., Rossi, L., Priori, A., Marceglia, S., Gilron, R., Cavanagh, J. F., Judy, J. W., Miocinovic, S., Devergnas, A. D., Sillitoe, R. V., Cernera, S., Oehrn, C. R., Gunduz, A., & Okun, M. S. (2023). Proceedings of the 10th annual deep brain stimulation think tank: Advances in cutting edge technologies, artificial intelligence, neuromodulation, neuroethics, interventional psychiatry, and women in neuromodulation. *Frontiers in Human Neuroscience*, *16*, 1084782. doi:10.3389/fnhum.2022.1084782 PMID:36819295

Yoo, J. H., Jeong, H., & Chung, T. M. (2023). Cutting-Edge Technologies for Digital Therapeutics: A Review and Architecture Proposals for Future Directions. *Applied Sciences (Basel, Switzerland)*, *13*(12), 6929. doi:10.3390/app13126929

ADDITIONAL READINGS

Komalasari, R. (2022). A Social Ecological Model (SEM) to Manage Methadone Programmes in Prisons. In Handbook of Research on Mathematical Modeling for Smart Healthcare Systems (pp. 374-382). IGI Global. doi:10.4018/978-1-6684-4580-8.ch020

Komalasari, R. (2023). The Relationship Between Cybersecurity and Public Health. In *Handbook of Research on Current Trends in Cybersecurity and Educational Technology* (pp. 78–91). IGI Global. doi:10.4018/978-1-6684-6092-4.ch005

Komalasari, R. (2023). Postnatal Mental Distress: Exploring the Experiences of Mothers Navigating the Healthcare System. In Perspectives and Considerations on Navigating the Mental Healthcare System (pp. 159-181). IGI Global. doi:10.4018/978-1-6684-5049-9.ch007

Komalasari, R. (2023). Designing Health Systems for Better, Faster, and Less Expensive Treatment. In *Exploring the Convergence of Computer and Medical Science Through Cloud Healthcare* (pp. 1–13). IGI Global. doi:10.4018/978-1-6684-5260-8.ch001

Komalasari, R. (2023). The Ethical Consideration of Using Artificial Intelligence (AI) in Medicine. In Advanced Bioinspiration Methods for Healthcare Standards, Policies, and Reform (pp. 1-16). IGI Global. doi:10.4018/978-1-6684-5656-9.ch001

Komalasari, R. (2023). Healthcare for the Elderly With Digital Twins. In Digital Twins and Healthcare: Trends, Techniques, and Challenges (pp. 145-156). IGI Global. doi:10.4018/978-1-6684-5925-6.ch010

Komalasari, R. (2023). Cloud Computing's Usage in Healthcare. In Recent Advancements in Smart Remote Patient Monitoring, Wearable Devices, and Diagnostics Systems (pp. 183-194). IGI Global. doi:10.4018/978-1-6684-6434-2.ch009

Komalasari, R. (2023). History and Legislative Changes Governing Medical Cannabis in Indonesia. In Medical Cannabis and the Effects of Cannabinoids on Fighting Cancer, Multiple Sclerosis, Epilepsy, Parkinson's, and Other Neurodegenerative Diseases (pp. 274-284). IGI Global. doi:10.4018/978-1-6684-5652-1.ch012

Komalasari, R. (2023). Ambient Assisted Living (AAL) Systems to Help Older People. In Exploring Future Opportunities of Brain-Inspired Artificial Intelligence (pp. 84-99). IGI Global. doi:10.4018/978-1-6684-6980-4.ch006

Komalasari, R. (2023). Telemedicine in Pandemic Times in Indonesia: Healthcare Professional's Perspective. In Health Informatics and Patient Safety in Times of Crisis (pp. 138-153). IGI Global. doi:10.4018/978-1-6684-5499-2.ch008

Komalasari, R. (2023). Cloud Computing's Usage in Healthcare. In Recent Advancements in Smart Remote Patient Monitoring, Wearable Devices, and Diagnostics Systems (pp. 183-194). IGI Global. doi:10.4018/978-1-6684-6434-2.ch009

Komalasari, R. (2023). The Relationship Between Cybersecurity and Public Health. In *Handbook of Research on Current Trends in Cybersecurity and Educational Technology* (pp. 78–91). IGI Global. doi:10.4018/978-1-6684-6092-4.ch005

Komalasari, R. (2023). Treatment of Menstrual Discomfort in Young Women and a Cognitive Behavior Therapy (CBT) Program. In Perspectives on Coping Strategies for Menstrual and Premenstrual Distress (pp. 194-211). IGI Global. doi:10.4018/978-1-6684-5088-8.ch011

Komalasari, R. (2023). Healthcare for the Elderly With Digital Twins. In Digital Twins and Healthcare: Trends, Techniques, and Challenges (pp. 145-156). IGI Global. doi:10.4018/978-1-6684-5925-6.ch010

Komalasari, R. (2023). Telemedicine in Pandemic Times in Indonesia: Healthcare Professional's Perspective. In Health Informatics and Patient Safety in Times of Crisis (pp. 138-153). IGI Global. doi:10.4018/978-1-6684-5499-2.ch008

Komalasari, R. (2023). Regulatory Shift Healthcare Applications in Industry 5.0. In Advanced Research and Real-World Applications of Industry 5.0 (pp. 149-165). IGI Global. doi:10.4018/978-1-7998-8805-5.ch008

Komalasari, R. (2023). Exploring Pedagogies Of Affect In Secondary School Physical Education For Enhanced Emotional Well-Being. *Pedagogik: Jurnal Pendidikan Guru Sekolah Dasar*, *11*(2), 117–127. doi:10.33558/pedagogik.v11i2.7374

Komalasari, R. (2023). Integrating Sport education model and the athletics challenges approach for transformative physical education in Indonesian Middle Schools. *Motion: Jurnal Riset Physical Education*, *13*(2), 118–135. doi:10.33558/motion.v13i2.7372

Komalasari, R. (2023). Rancangan Program Bagi Keberhasilan Gerakan Mencuci Tangan Yang Sehat Bagi Para Ibu Dan Anak. *Sulolipu: Media Komunikasi Sivitas Akademika dan Masyarakat*, *23*(1), 55-60. doi:10.32382/sulolipu.v23i1.3103

Komalasari, R. (2023). Culture as A Catalyst: Unveiling the Nexus Between Health Services Performance Management and National Identity. *Journal of Business and Political Economy: Biannual Review of The Indonesian Economy*, *5*(1).

Komalasari, R. (2023). Exploring Pedagogies Of Affect In Secondary School Physical Education For Enhanced Emotional Well-Being. *Pedagogik: Jurnal Pendidikan Guru Sekolah Dasar*, *11*(2), 117–127. doi:10.33558/pedagogik.v11i2.7374

Komalasari, R. (2023). Integrating sport education model and the athletics challenges approach for transformative physical education in Indonesian Middle Schools. *Motion: Jurnal Riset Physical Education*, *13*(2), 118–135. doi:10.33558/motion.v13i2.7372

Komalasari, R. (2023). Efek Ganja Medis pada Pasien Parkinson: A Literature Review of Clinical Evidence. *Journal of Islamic Pharmacy*, *8*(1), 44–48. doi:10.18860/jip.v8i1.17832

Komalasari, R. (2023). Rancangan Program Bagi Keberhasilan Gerakan Mencuci Tangan Yang Sehat Bagi Para Ibu Dan Anak. *Sulolipu: Media Komunikasi Sivitas Akademika dan Masyarakat*, *23*(1), 55-60. doi:10.32382/sulolipu.v23i1.3103

Komalasari, R. (2023). Culture As A Catalyst: Unveiling The Nexus Between Health Services Performance Management and National Identity. *Journal of Business and Political Economy*, *5*(1), 25–47. doi:10.46851/138

Komalasari, R. (2024). Harmonizing Midlife Motherhood: Navigating the Intersection of First-Time Maternity and Perimenopause. In *Utilizing AI Techniques for the Perimenopause to Menopause Transition*. IGI Global.

Komalasari, R. (2024). Optimizing Prostate Cancer Radiotherapy: Advanced Machine Learning in Virtual Patient-Specific Plan Verification. *Indonesian Journal of Cancer*.

Komalasari, R. (2024). Unravelling the Veil: Exploring the Nexus of Insecure Attachment and Functional Somatic Disorders in Adults. In *Discourses, Inquiries, and Case Studies in Healthcare, Social Sciences, and Technology*. IGI Global. doi:10.4018/979-8-3693-3555-0.ch001

Komalasari, R. (2024). Advancing Healthcare: Economic Implications of Immediate MRI in Suspected Scaphoid Fractures - A Comprehensive Exploration. In *Future of AI in Medical Imaging*. IGI Global.

Komalasari, R. (2024). Women's perceptions of midlife mothering during perimenopause: the impact on health and well-being through life's transitions. In *Utilizing AI Techniques for the Perimenopause to Menopause Transition*. IGI Global.

Komalasari, R. (2024). Machine Learning in Health Information Security: Unraveling Patterns, Concealing Secrets, and Mitigating Vulnerabilities. In *Enhancing Steganography Through Deep Learning Approaches*. IGI Global.

Komalasari, R. (2024). Navigating Complexity: Unraveling the IVHM Requirements Puzzle in Unmanned Aerial Systems Through Innovative IVHM-RD ology and Rigorous Data Analysis. In *Ubiquitous Computing and Technological Innovation for Universal Healthcare*. IGI Global.

Komalasari, R. (2024). Trust Dynamics in Remote Patient-Expert Communication: Unraveling the Role of ICT in Indonesia's Private Healthcare Sector. In *Improving Security, Privacy, and Connectivity Among Telemedicine Platforms*. IGI Global.

Komalasari, R. (2024). Cyberbullying in the Health Care Workplace: How to Find Your Way Through the Digital Maze. In *Workplace Cyberbullying and Behavior in Health Professions*. IGI Global. doi:10.4018/979-8-3693-1139-4.ch004

Komalasari, R. (2024). Harmonizing Minds and Machines: A Transformative AI-Based Mental Health Care Framework in Medical Tourism. In *Impact of AI and Robotics on the Medical Tourism Industry*. IGI Global.

Komalasari, R. (2024). Navigating the Future of Healthcare: A User-Centric Approach to Designing Lucrative Business Models for the IoMT. In *Lightweight Digital Trust Architectures in the Internet of Medical Things (IoMT)*. IGI Global.

Komalasari, R. (2024). Biospheric Reverie: Unraveling Indoor Air Quality through Bio-Inspired Textiles, Awareness, and Decision-Making. In *Intelligent Decision Making Through Bio-Inspired Optimization*. IGI Global.

Komalasari, R. (2024). Revolutionizing Healthcare: An Integrated Approach to Next-Generation Telemedicine and Health Advice Systems through Knowledge Engineering and Artificial Intelligence Applications in Disease Diagnosis. In *Medical Robotics and AI-Assisted Diagnostics for a High-Tech Healthcare Industry*. IGI Global.

Komalasari, R. (2024). Transformative Insights: Harnessing Artificial Intelligence for Enhanced Ovarian Cancer Prediction and Prognosis. In *Biomedical Research Developments for Improved Healthcare*. IGI Global. doi:10.4018/979-8-3693-1922-2.ch010

Komalasari, R. (2024). Bridging the Gap: Theory of Change Guided Digital Health Implementation in Indonesian Primary Care. In *Analyzing Current Digital Healthcare Trends Using Social Networks*. IGI Global.

Komalasari, R. (2024). Navigating the Digital Frontier: A Socio-Technical Review of Indonesia's NHS E-Health Strategy and the Path to Seamless Healthcare Transformation. In *Inclusivity and Accessibility in Digital Health*. IGI Global.

Komalasari, R. (2024). *Pursuit of Excellence: A Comprehensive Framework for Ensuring Quality in Digital Health Apps. In Multi-Sector Analysis of the Digital Healthcare Industry*. IGI Global.

Komalasari, R. (2024). Fostering Transparency in AI for Disease Detection: Exploring Meaningful Explanations and Interactive Modalities in Healthcare. In *Federated Learning and Privacy-Preserving in Healthcare AI*. IGI Global.

Komalasari, R. (2024). Predictive Pioneers: Bridging the Mental Health Gap in Older Adults through Advanced Digital Technologies. In *Using Machine Learning to Detect Emotions and Predict Human Psychology*. IGI Global. doi:10.4018/979-8-3693-1910-9.ch005

Komalasari, R. (2024). Unravelling the Veil: Exploring the Nexus of Insecure Attachment and Functional Somatic Disorders in Adults. In *Discourses, Inquiries, and Case Studies in Healthcare, Social Sciences, and Technology*. IGI Global. doi:10.4018/979-8-3693-3555-0.ch001

Komalasari, R. (2024). Cyberbullying in the Health Care Workplace: How to Find Your Way Through the Digital Maze. In *Workplace Cyberbullying and Behavior in Health Professions*. IGI Global. doi:10.4018/979-8-3693-1139-4.ch004

Komalasari, R. (2024). Navigating Mental Health Crises: Understanding Pivotal Phases, Health Campaigns, and Community Resilience for Recovery and Reform. In *The Role of Health Literacy in Major Healthcare Crises*. IGI Global.

Komalasari, R. (2024). Decoding the Genomic and Proteomic Landscape: Dissemination and Mechanisms of Antibiotic Resistance in Environmental Staphylococci. In *Contemporary Approaches to Mitigating Antibacterial Drug Resistance*. IGI Global.

Komalasari, R. (2024). Unveiling the Future: Blockchain-Powered Digital Twins for Personalized Privacy Preservation in Metaverse Healthcare Data Management. In *Examining the Metaverse in Healthcare: Opportunities, Challenges, and Future Directions*. IGI Global. doi:10.4018/979-8-3693-1906-2.ch017

Komalasari, R. (2024). Safeguarding the Future: Cutting-Edge Time-Stamping Services and Blockchain Technologies for Maintaining EHR Data Integrity Over the Long Term. In *Blockchain and IoT Approaches for Secure Electronic Health Records (EHR)*. IGI Global.

Komalasari, R. (2024). Empowering Informed Healthcare Choices in Rural Areas: SIKDA - A Novel Model for Enhancing Patient Access to Reliable Consumer Health Information in Indonesia. In *Emerging Technologies for Health Literacy and Medical Practice*. IGI Global.

Komalasari, R. (2024). Advancing Trauma Care: Contemporary Cranio-Maxillo-Facial Surgery in Indonesia and Beyond. In *Contemporary Cranio-Maxillo-Facial Surgery: From Trauma to Reconstruction*. IGI Global.

Komalasari, R. (2024). Innovations in Neurodegenerative Disease Diagnosis: Unraveling the Transformative Potential of Deep Learning and MRI Integration. In *Deep Learning Approaches for Early Diagnosis of Neurodegenerative Diseases*. IGI Global.

Komalasari, R. (2024). Secure and privacy-preserving federated learning with explainable artificial intelligence for smart healthcare systems. In *Federated Learning and AI for Healthcare 5.0*. IGI Global.

Komalasari, R. (2024) Optimizing Prostate Cancer Radiotherapy: Advanced Machine Learning in Virtual Patient-Specific Plan Verification. *Indonesian Journal of Cancer.*

Komalasari, R. (2024). Predictive Pioneers: Bridging the Mental Health Gap in Older Adults Through Advanced Digital Technologies. In M. Rai & J. Pandey (Eds.), *Using Machine Learning to Detect Emotions and Predict Human Psychology* (pp. 99–129). IGI Global. doi:10.4018/979-8-3693-1910-9.ch005

Komalasari, R. (2024). Unravelling the Veil: Exploring the Nexus of Insecure Attachment and Functional Somatic Disorders in Adults. In D. Burrell (Ed.), *Change Dynamics in Healthcare, Technological Innovations, and Complex Scenarios* (pp. 1–27). IGI Global. doi:10.4018/979-8-3693-3555-0.ch001

Komalasari, R. (2024). Cyberbullying in the Healthcare Workplace: How to Find Your Way Through the Digital Maze. In M. Aslam, Y. Kim, & Q. Linchao (Eds.), *Workplace Cyberbullying and Behavior in Health Professions* (pp. 84–112). IGI Global. doi:10.4018/979-8-3693-1139-4.ch004

Komalasari, R. (2024). Navigating Mental Health Crises: Understanding Pivotal Phases, Health Campaigns, and Community Resilience for Recovery and Reform. In V. Papalois & K. Papalois (Eds.), *The Role of Health Literacy in Major Healthcare Crises* (pp. 144–174). IGI Global. doi:10.4018/978-1-7998-9652-4.ch008

Komalasari, R. (2024). Bridging Uncertainties: Exploring the Lived Experiences of Pediatric Long-COVID-19 Through ME/CFS Caregiver Narratives. In G. Rao & S. Dhamdhere-Rao (Eds.), *Clinical Practice and Post-Infection Care for COVID-19 Patients* (pp. 207–227). IGI Global. doi:10.4018/978-1-6684-6855-5.ch009

Komalasari, R., & Khang, A. (2024). An Integrated Approach to Next-Generation Telemedicine and Health Advice Systems Through AI Applications in Disease Diagnosis. In A. Khang (Ed.), *Driving Smart Medical Diagnosis Through AI-Powered Technologies and Applications* (pp. 258–274). IGI Global. doi:10.4018/979-8-3693-3679-3.ch016

Komalasari, R., & Mustafa, C. (2023). A Healthy Game-Theoretic Evaluation of NATO and Indonesia's Policies in the Context of International Law. Jurnal Pertahanan: Media Informasi tentang Kajian & Strategi Pertahanan yang Mengedepankan Identity. *Nasionalism & Integrity*, *9*(2), 333–349. doi:10.33172/jp.v9i2.16794

Komalasari, R., & Mustafa, C. (2023) Combating International Cyber Conflict: A Healthy Just War and International Law Analysis of NATO and Indonesian Policies. *Jurnal Pertahanan: Media Informasi tentang Kajian & Strategi Pertahanan yang Mengedepankan Identity, Nasionalism & Integrity*, 9(3).

Komalasari, R., & Mustafa, C. (2023). A Healthy Game-Theoretic Evaluation of NATO and Indonesia's Policies in the Context of International Law. *Jurnal Pertahanan: Media Informasi tentang Kajian & Strategi Pertahanan yang Mengedepankan Identity. Nasionalism & Integrity*, *9*(2), 333–349. doi:10.33172/jp.v9i2.16794

Komalasari, R., & Mustafa, C. (2024) Rehabilitasi Pengguna Narkoba: Tantangan dan Peluang. *Arena Hukum*, *14*(4).

Komalasari, R., & Mustafa, C. (2024). *Healthy Competition Policy Dynamics: the Influence of Domestic-Specific Factors in a Globalized Landscape*. Jurnal Persaingan Usaha.

Komalasari, R., & Mustafa, C. (2024). *Comparative Analysis of Healthy Electoral Fraud Legislation in Established Democracies*. Jurnal Adhyasta Pemilu.

Komalasari, R., Nurhayati, N., & Mustafa, C. (2022). Enhancing the Online Learning Environment for Medical Education: Lessons From COVID-19. In Policies and procedures for the implementation of safe and healthy educational environments: Post-COVID-19 perspectives (pp. 138-154). IGI Global. doi:10.4018/978-1-7998-9297-7.ch009

Komalasari, R., Nurhayati, N., & Mustafa, C. (2022). Insider/outsider issues: Reflections on qualitative research. *The Qualitative Report*, *27*(3), 744–751. doi:10.46743/2160-3715/2022.5259

Komalasari, R., Nurhayati, N., & Mustafa, C. (2022). Professional Education and Training in Indonesia. In *Public Affairs Education and Training in the 21st Century* (pp. 125–138). IGI Global. doi:10.4018/978-1-7998-8243-5.ch008

Komalasari, R., Nurhayati, N., & Mustafa, C. (2022). Professional Education and Training in Indonesia. In *Public Affairs Education and Training in the 21st Century* (pp. 125–138). IGI Global. doi:10.4018/978-1-7998-8243-5.ch008

Komalasari, R., Wilson, S., & Haw, S. (2021). A systematic review of qualitative evidence on barriers to and facilitators of the implementation of opioid agonist treatment (OAT) programmes in prisons. *The International Journal on Drug Policy*, *87*, 102978. doi:10.1016/j.drugpo.2020.102978 PMID:33129135

Komalasari, R., Wilson, S., & Haw, S. (2021). A social ecological model (SEM) to exploring barriers of and facilitators to the implementation of opioid agonist treatment (OAT) programmes in prisons. *International Journal of Prisoner Health, 17*(4), 477–496. doi:10.1108/IJPH-04-2020-0020

Komalasari, R., Wilson, S., Nasir, S., & Haw, S. (2020). Multiple burdens of stigma for prisoners participating in Opioid Antagonist Treatment (OAT) programmes in Indonesian prisons: A qualitative study. *International Journal of Prisoner Health, 17*(2), 156–170. doi:10.1108/IJPH-03-2020-0018

Mustafa, C. (2020). The influence of sunni islamic values on rehabilitation as judicial decision for minor drug users in Indonesian court. *Ijtihad: Jurnal Wacana Hukum Islam dan Kemanusiaan, 20*(1), 79-96. doi:10.18326/ijtihad.v20i1.79-96

Mustafa, C. (2021). Qualitative Method Used in Researching the Judiciary: Quality Assurance Steps to Enhance the Validity and Reliability of the Findings. *The Qualitative Report, 26*(1), 176–185. doi:10.46743/2160-3715/2021.4319

Mustafa, C. (2021). Key Finding: Result of a Qualitative Study of Judicial Perspectives on the Sentencing of Minor Drug Offenders in Indonesia: Structural Inequality. *The Qualitative Report, 26*(5), 1678–1692. doi:10.46743/2160-3715/2021.4436

Mustafa, C. (2021). The view of judicial activism and public legitimacy. *Crime, Law, and Social Change, 76*(1), 23–34. doi:10.1007/s10611-021-09955-0

Mustafa, C. (2021). The News Media Representation of Acts of Mass Violence in Indonesia. In *Mitigating Mass Violence and Managing Threats in Contemporary Society* (pp. 127–140). IGI Global. doi:10.4018/978-1-7998-4957-5.ch008

Mustafa, C. (2021). The Challenges to Improving Public Services and Judicial Operations: A unique balance between pursuing justice and public service in Indonesia. In Handbook of research on global challenges for improving public services and government operations (pp. 117-132). IGI Global. doi:10.4018/978-1-7998-4978-0.ch007

Mustafa, C., Malloch, M., & Hamilton Smith, N. (2020). Judicial perspectives on the sentencing of minor drug offenders in Indonesia: Discretionary practice and compassionate approaches. *Crime, Law, and Social Change, 74*(3), 297–313. doi:10.1007/s10611-020-09896-0

Suhariyanto, B., Mustafa, C., & Santoso, T. (2021). Liability incorporate between transnational corruption cases Indonesia and the United States of America. *J. Legal Ethical & Regul. Isses, 24*, 1.

KEY TERMS AND DEFINITIONS

Artificial Intelligence: The term "artificial intelligence" (AI) describes the process of creating computer programs that may mimic human intellect in certain ways. Cognitive processes, including learning, thinking, seeing, and language comprehension, fall under this category. Artificial intelligence (AI) applications in healthcare may improve treatment planning, diagnose patients more accurately, and analyze medical data.

Healthcare Innovation: When new ideas, technology, procedures, or services are developed and implemented in the healthcare industry, they aim to enhance the accessibility, efficiency, and quality of healthcare. It encompasses various advancements, from medical treatments and technologies to healthcare delivery systems and organizations.

Medical Tourism: People engage in medical tourism when they go to another country internationally for medical care. This could include elective procedures, specializations, or seeking more affordable or higher-quality medical care than in their home country.

Mental Health: It describes how a person does emotionally, psychologically, and socially. The capacity to manage stress, form meaningful relationships, make informed decisions, and face adversity head-on are all components of resilience. A person's mental health is an integral part of their overall wellness, and mental health treatment includes interventions, support, and tactics to maintain or enhance mental wellness.

Wearable Technology: Electronic devices and accessories worn on the body, often as accessories or clothes, are known as wearable technology. These gadgets can gather data and provide you updates in real-time since they have sensors, CPUs, and connections. In healthcare, wearable technology includes devices that monitor health metrics, track physical activity, and support medical interventions.

Chapter 3
Food Challenges and Opportunities for Medical Tourism in Serbia

Miloš Zrnić
https://orcid.org/0000-0003-3454-6573
Academy of Applied Studies, The College of Hotel Management, Serbia

ABSTRACT

The research, conducted in November 2023, explores the relationship between culinary experiences and the medical tourism industry in Serbia. Focusing on prominent spa destinations – Prolom Banja, Vrnjačka Banja and Soko Banja – the study aims to provide contextually relevant insights reflective of the broader medical tourism landscape. The sample includes 592 diverse medical tourists. Structured interviews were carried out on the premises of three chosen spas, involving interactions with medical guests. The data was organized using Microsoft 365 Excel. The aim of this study was to acquire comprehensive perspectives on food challenges and opportunities for medical tourists. Data collection involves surveys to capture preferences and challenges, forming the basis for understanding and improving culinary aspects in Serbian spas. Research results highlight a notable correlation between the culinary offerings within spas and the overall experience of medical tourism.

1. INTRODUCTION

In recent years, the intersection of healthcare and tourism has given rise to a burgeoning industry known as medical tourism (Connell, 2013). This phenomenon involves individuals crossing international borders in pursuit of medical treatments,

DOI: 10.4018/979-8-3693-2248-2.ch003

procedures, and wellness experiences. While much attention has been directed towards the medical aspects of this global phenomenon, one crucial facet that remains relatively unexplored is the role of food within the context of medical tourism (Abd Mutalib et al. 2016). The global medical tourism industry has witnessed unprecedented growth, driven by factors such as cost-effective treatments, state-of-the-art medical facilities, and the allure of exploring new destinations (Hadian et al., 2021). Against this backdrop, the significance of the gastronomic experience for medical tourists cannot be overstated. Food serves not only as a basic human need but also as a vital component of cultural exploration and overall well-being (Hassan and Zrnić, 2021). Understanding the complexities of providing suitable, nutritious, and culturally diverse meals for medical tourists becomes imperative in enhancing the overall patient experience (Gyu Ko, 2011).

As we embark on this exploration of food challenges and opportunities for medical tourists, it is crucial to recognize the potential impact on both the hospitality and healthcare sectors (Smith and Puczkó (2014). By addressing the dietary needs and expectations of medical tourists, destinations can distinguish themselves and establish a competitive edge in the global medical tourism landscape (Zrnić et al. 2021). Simultaneously, healthcare facilities that integrate culinary considerations into their services may witness improved patient satisfaction and outcomes (Shukla et al. 2019).

This research paper employs a multidisciplinary approach, drawing insights from the fields of tourism, hospitality, healthcare and nutrition. Through an in-depth analysis of case studies, interviews, and surveys, we aim to provide actionable recommendations for stakeholders in the medical tourism ecosystem. By doing so, we contribute to the broader discourse on enhancing the holistic experience of medical tourists and fostering sustainable growth within the industry. Furheremore, this research paper delves into the intricate interplay between food, health, and the medical tourism sector, shedding light on the challenges and opportunities presented by the culinary landscape for those seeking medical treatments abroad. This research endeavors to identify the challenges faced by medical tourists in relation to food, ranging from dietary restrictions and cultural variations to accessibility and quality assurance. Also, aim of this research to uncover the untapped opportunities that lie within the realm of culinary offerings for medical tourists, examining how these opportunities can be harnessed to elevate the appeal of destinations and healthcare facilities. As we navigate the intricate relationship between food and medical tourism, this research invites scholars, practitioners, and policymakers to join the conversation and collaboratively shape the future of this evolving and dynamic field.

A comprehensive review of existing literature in the fields of medical tourism, hospitality, nutrition, and cultural studies forms the foundation of this research. Scholarly articles, books, and reports have been scrutinized to gain a thorough understanding of the historical evolution and current state of the medical tourism industry, with a specific focus on the role of food. The intersection of medical tourism and gastronomy represents a fascinating yet underexplored facet of the broader discourse surrounding healthcare travel. While existing literature extensively covers the economic, cultural, and medical dimensions of medical tourism, the nuanced relationship between food and the medical tourism experience remains conspicuously absent. Historically, medical tourism has roots in the pursuit of specialized treatments and therapies, often driven by a quest for better healthcare outcomes, reduced costs, or the availability of procedures not accessible in the traveler's home country. Studies by Turner (2012) chronicle the historical evolution of medical tourism, emphasizing the shifting motivations and demographics of medical tourists over time.

In the contemporary landscape of healthcare, the figure of the medical tourist has become increasingly prominent, embodying a unique intersection of travel and medical treatment. These individuals, driven by diverse motivations and expectations, embark on journeys that transcend borders in pursuit of healthcare solutions tailored to their specific needs. At the core of the medical tourist experience lies a complex web of motivations. Foremost among these is the quest for high-quality medical care at a cost significantly lower than what is available in their home countries. Financial considerations often drive individuals to explore medical tourism, particularly for elective procedures, specialized treatments, or cosmetic surgeries. Additionally, some medical tourists seek access to cutting-edge or experimental treatments that may not be readily available in their home nations.

The demographic profile of medical tourists is remarkably diverse, encompassing individuals from various age groups, socioeconomic backgrounds, and cultural contexts. Younger individuals may embark on medical tourism for cosmetic procedures or fertility treatments, while older individuals might seek joint replacements or cardiac surgeries. This diversity underscores the global nature of medical tourism, illustrating that the pursuit of optimal healthcare solutions is a universal endeavor that transcends boundaries. Cultural considerations play a pivotal role in shaping the medical tourist experience. The intricacies of cultural factors influence the choice of destination, preferences in healthcare providers, and expectations regarding the overall experience. Language barriers, variations in healthcare practices, and differing approaches to patient care add layers of complexity to the medical tourist's journey, requiring a nuanced understanding of cultural contexts. The selection of a destination becomes a pivotal aspect of the medical tourist's decision-making process. Factors such as the reputation of healthcare facilities, accessibility, and the overall appeal of the destination contribute to the choice of where to seek medical treatment. The destination serves not only as a healthcare provider but also as the backdrop for a transformative experience that goes beyond medical interventions.

In essence, the medical tourist's journey represents a convergence of healthcare and travel, weaving together the pursuit of well-being, cultural exploration, and cost-effective medical solutions (Cham et al. 2021). Understanding the intricacies of this multifaceted experience is essential for healthcare providers, policymakers, and stakeholders in the global healthcare industry aiming to cater to the evolving needs of this unique demographic (Han and Hyun, 2014).

2. LITERATURE REVIEW

The realm of medical tourism has been a subject of increasing scholarly interest, reflecting the expanding global trend of individuals seeking healthcare services outside their home countries. This literature review provides a comprehensive exploration of key themes, trends, and research findings that have emerged within the dynamic field of medical tourism. A fundamental aspect of understanding medical tourism involves tracing its historical evolution. While medical travel has ancient roots, the contemporary surge in individuals seeking healthcare across borders gained prominence in the late 20th century. This evolution reflects the interplay of advancements in healthcare, improvements in transportation, and the broader processes of globalization, resulting in a nuanced landscape of motivations and dynamics. One focal point of scholarly inquiry revolves around the motivations that propel individuals into the realm of medical tourism. Researchers have conducted in-depth analyses to uncover the diverse factors shaping the decision-making process of medical tourists. These motivations often span financial considerations, access to specific medical procedures, concerns about healthcare quality, and the pursuit of specialized or experimental treatments. Understanding the intricate interplay of these factors is crucial for comprehending the complex decision-making landscape of medical tourists.

The role of information and communication technologies (ICTs) in shaping the landscape of medical tourism has also garnered scholarly attention. The advent of the internet and social media has significantly transformed how individuals access and disseminate information about healthcare providers, destinations, and treatment options. The impact of ICTs on empowering medical tourists, facilitating decision-making, and shaping their overall experiences forms a significant dimension of contemporary research in the field.

Cultural considerations constitute another vital area of exploration within the literature on medical tourism. Cultural factors encompass language barriers, variations in healthcare practices, and the influence of cultural norms on the choice of destination and healthcare provider. Understanding these cultural dynamics is essential for healthcare providers, policymakers, and stakeholders aiming to cater to the diverse needs of a global clientele. Ethical concerns and challenges within the domain of medical tourism have not escaped scholarly scrutiny. Issues such as patient safety, informed consent, and the potential exploitation of healthcare disparities have been critically examined. Researchers have sought to identify and address ethical dilemmas to ensure the responsible and sustainable development of medical tourism as a global healthcare phenomenon. As medical tourism continues to evolve, the literature provides valuable insights into its multifaceted nature. The interplay of historical, cultural, economic, and ethical dimensions shapes the experiences of medical tourists and influences the strategies of healthcare providers and policymakers. This literature review sets the stage for further exploration and understanding of the complexities inherent in the phenomenon of medical tourism.

Medical tourism, a burgeoning global industry, revolves around individuals seeking medical treatments, surgeries, or wellness procedures beyond their national borders. The motivations for becoming a medical tourist are multifaceted, often intertwining healthcare, cost considerations, and a desire for a unique travel experience. Understanding the motivations behind medical tourism is essential to comprehend the diverse factors influencing individuals' decisions to travel for healthcare. Lunt (2022) categorizes these motivations into several key drivers. Economic considerations, including cost savings and affordability, emerge as prominent factors, especially when comparing treatment costs between home countries and medical tourism destinations. Access to timely or specialized treatments, the quest for higher quality medical care, and a desire for privacy also play significant roles in the decision-making process. The demographics of medical tourists exhibit a broad spectrum, encompassing individuals of varying ages, socio-economic backgrounds, and health conditions. Turner (2010) note that medical tourists may range from those seeking elective cosmetic procedures to individuals with chronic medical conditions pursuing specialized treatments. Understanding the diverse demographic profile of medical tourists is instrumental for healthcare providers and destination marketers aiming to tailor their services and offerings. Medical tourism encompasses a wide array of healthcare services, ranging from elective cosmetic surgeries to complex medical procedures. The types of medical tourism include dental tourism, reproductive tourism, cosmetic surgery tourism, and orthopedic tourism, among others. Ngobeni (2020) provides insights into the various dimensions and specialties within medical tourism, emphasizing the need for destination-specific strategies to cater to the unique demands of each medical tourism niche. While medical tourism presents an array of opportunities, it is not without its challenges. In navigating the complex landscape of medical tourism, it is imperative to consider the multifaceted nature of the medical tourist's journey. By addressing their diverse motivations, demographics, and challenges, stakeholders in the medical tourism ecosystem can develop strategies that prioritize patient-centric care and contribute to the sustainable growth of this dynamic industry. Understanding the cultural dimensions of medical tourism is crucial, given that the amalgamation of diverse cultures is a defining feature of this industry. Zolfagharian et al. (2018) reveals the profound impact of cultural considerations on the decision-making processes of medical tourists, from destination choice to the overall satisfaction with their healthcare journey. The hospitality industry plays a pivotal role in shaping the overall experience of medical tourists. Suess and Mody (2017) discuss the importance of hospitality in medical tourism, emphasizing the need for seamless integration of healthcare and hospitality services. This literature underscores the significance of creating a supportive and comfortable environment for medical tourists, with potential implications for their dietary experiences.

While the broader healthcare literature acknowledges the importance of nutrition in patient outcomes, the specific considerations for medical tourists remain relatively unexplored. Research by Hartwell et al. (2018) highlights the need for tailored nutritional support for medical tourists, considering the potential impact on recovery and overall well-being. Studies by Casey et al. (2013). shed light on the challenges faced by healthcare providers and hospitality professionals in meeting the diverse dietary needs of medical tourists. Issues such as cultural preferences, dietary restrictions, and quality assurance emerge as critical factors that demand attention in the intersection of food and medical tourism (Zrnić, Brdar and Kilibarda, 2021). On the flip side, there is a paucity of literature exploring the untapped opportunities within the culinary landscape for medical tourism destinations. Research by Seyitoğlu and Ivanov (2020) suggests that strategically incorporating local and cultural cuisines can not only enhance the overall satisfaction of medical tourists but also contribute to the destination's distinct identity and competitive advantage. Despite the valuable insights provided by existing literature, significant gaps persist in our understanding of the intricate relationship between food and medical tourism. This literature review underscores the need for more empirical studies, particularly those that delve into the lived experiences of medical tourists, explore the role of food in post-treatment recovery, and investigate the potential economic and branding advantages for destinations that prioritize culinary offerings.

3. METHODOLOGY

The methodology for the research on was conducted in November 2023, centered on exploring the intricate relationship between culinary experiences and the medical tourism industry, specifically focusing on three prominent spa destinations Prolom Banja, Vrnjačka Banja and Soko Banja. These spa destinations were chosen for their prominence in medical tourism and the diversity they offer. By concentrating on these specific locations, the research aimed to provide insights that are contextually relevant and reflective of the broader medical tourism landscape in Serbia. Serbia boasts more than 40 established spas, contributing significantly to its tourism offerings. This abundance positions Serbia as one of the affluent European nations in terms of spa facilities. Presently, Serbian spas have evolved into contemporary spa and wellness centers, featuring well-regulated health resorts and recreational facilities. In picture number 1. sertain number of spas and spa health resorts are represented.

Figure 1. This map of Serbia depicts only a specific count of spas and health resorts dedicated to spa services
Source: https://www.srbija-banje.com/en/mapa-karta

The sample for this research comprised 592 medical tourists aged between 18-70, who visited the selected spas in Serbia for recovery and rest. The selection of respondents ensured a diverse representation in terms of age, reflecting the broad demographic spectrum of medical tourists. Participants were chosen based on their primary reason for visiting the spas and the acknowledgment of the importance of food as an integral part of their stay.

Table 1. The Proportional Breakdown of 'Male and Female' Percentages

Gender	Count of Gender
Male	53.04%
Female	46.96%
Grand Total	**100.00%**

Source: Author research

Figure 2. Percentage Distribution of 'Male and Female'
Source: Author research

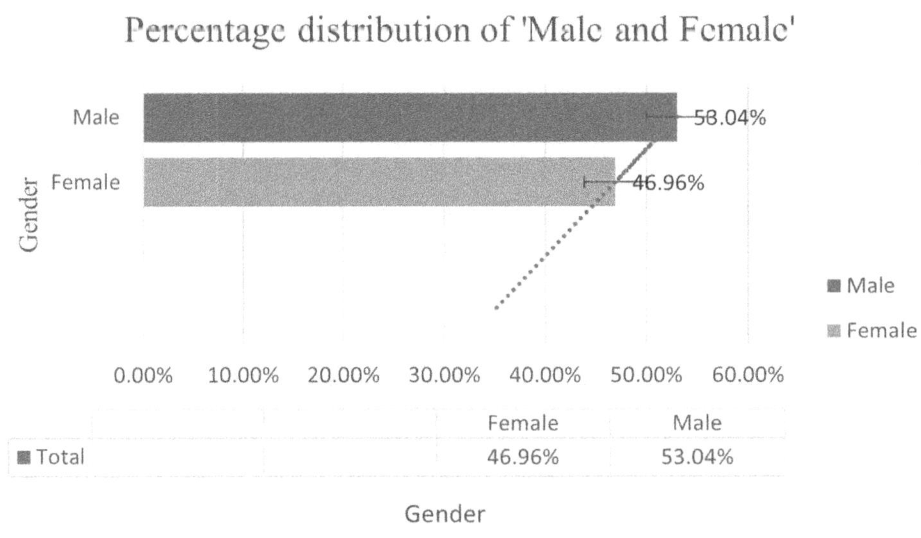

Data collection was primarily carried out through surveys distributed to the identified medical tourists. The surveys aimed to capture their experiences, preferences, and challenges related to the culinary aspects of their stay in the spas. The emphasis was on understanding the significance of food in their overall recovery and restorative experience.

Table 2. Count of Respondents Age

Count of Age	Gender		
Age	Male	Female	Grand Total
61+	99	171	270
18-30	185	34	219
31-60	30	73	103
Grand Total	**314**	**278**	**592**

Source: Author research

Figure 3. The Allocation of 'Gender' across Different 'Age' Categories.

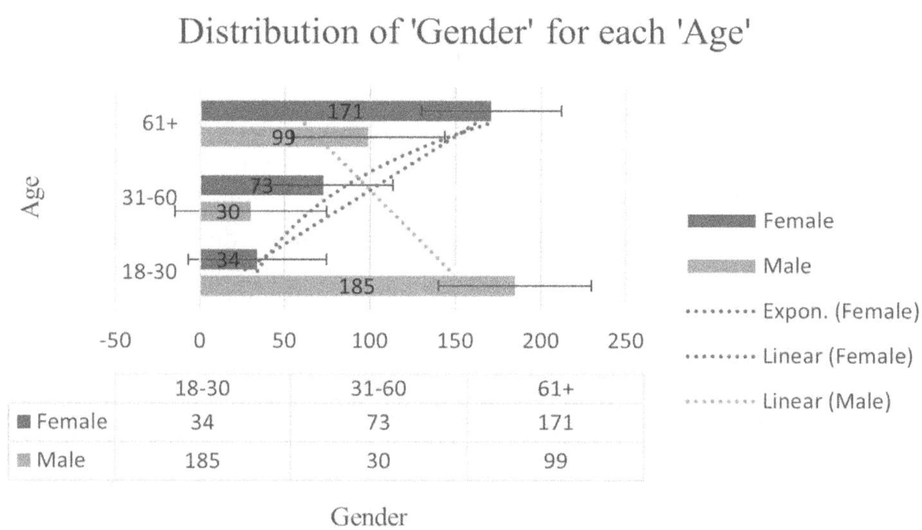

The research framework was built around identifying challenges and opportunities related to food in the spa centers. The main challenge centered on maintaining a balanced and fresh food supply consistently throughout the stay. Concurrently, the opportunity lay in enhancing the quality of existing dishes and introducing new culinary offerings to elevate the overall gastronomic experience for medical tourists. The collected data underwent a thorough analysis, combining qualitative and quantitative methods. Qualitative data from open-ended survey questions were subjected to thematic analysis to identify recurring patterns related to food challenges and opportunities. Quantitative data, including responses to structured survey questions, were analyzed using statistical tools to derive numerical insights.

Ethical considerations played a crucial role throughout the research process. Informed consent was obtained from all participants, ensuring transparency and voluntariness. Participants' privacy and confidentiality were diligently protected, and cultural sensitivities related to food practices were respected to maintain the integrity of the research.

By employing this methodology, the research sought to offer valuable insights into the specific context of medical tourism in Serbia, particularly in spa destinations, and contribute to the improvement of culinary experiences for medical tourists seeking recovery and rest in these settings.

Interviews were conducted by spa staff within the premises of the selected spas, engaging with medical guests to gain in-depth insights into their perspectives on food challenges and opportunities for medical tourists. Two key research questions guided these interviews:

Research Question 1 (RQ1): "What do you perceive as the primary food challenges encountered by medical tourists during your stay?"

Research Question 2 (RQ2): "In your opinion, what are the notable food opportunities that can enhance the culinary experience for medical tourists?"

The structured interviews aimed to extract valuable qualitative data, providing a nuanced understanding of the culinary landscape from the direct experiences and perspectives of medical tourists in Serbian spas. These questions were strategically formulated to elicit comprehensive responses that contribute meaningfully to the research objectives. Answers were very similar and presented below.

3.1 Challenges

Maintaining Balanced and Fresh Food Supply: The primary challenge observed in spa centers was the consistent maintenance of a balanced and fresh food supply. Medical tourists, seeking recovery and rest, faced difficulties when the nutritional content and freshness of the offered meals were not sustained throughout their stay. This challenge not only impacted the overall satisfaction of the tourists but also had potential implications for their health and well-being. One of the predominant challenges identified within spa centers revolved around the ongoing task of ensuring a well-maintained and constantly replenished supply of balanced and fresh food. For medical tourists, who visit these centers with the primary goal of recuperation and rejuvenation, encountering issues related to the nutritional quality and freshness of the provided meals posed significant obstacles throughout their stay. The persistent nature of this challenge not only influenced the overall satisfaction levels of the tourists but also raised concerns regarding potential implications for their health and well-being.

The importance of maintaining a consistent and high-quality food supply cannot be overstated in the context of medical tourism, where individuals often seek a holistic healing experience. The availability of fresh and nutritionally balanced meals contributes not only to the physical well-being of the visitors but also plays a crucial role in enhancing their overall therapeutic journey. A lapse in this aspect not only affects the immediate comfort of the medical tourists but could potentially compromise the effectiveness of the entire spa experience.

The findings underscored the need for strategic interventions in the culinary aspects of spa centers, focusing on measures that ensure the continuous delivery of fresh and nutritionally rich meals. Addressing this challenge is not only pivotal for meeting the expectations of medical tourists but is also crucial for sustaining and enhancing the reputation of Serbian spas in the broader medical tourism industry. The upcoming sections of this chapter will further explore the survey results and recommendations for improving the culinary offerings in spa centers, aligning them more closely with the preferences and well-being concerns of medical tourists.

3.2 Culinary Variety and Dietary Preferences

Diverse dietary preferences and restrictions among medical tourists posed a notable challenge. Spa centers struggled to cater to the varied culinary needs, including dietary restrictions and cultural preferences. The lack of customization in meal offerings led to dissatisfaction and underscored the necessity for more flexible and accommodating food services. The issue of diverse dietary preferences and restrictions emerged as a significant challenge within spa centers catering to medical tourists. A notable aspect of this challenge was the varied spectrum of dietary needs, encompassing specific restrictions and cultural preferences among the diverse group of individuals seeking medical treatments. Spa centers found it challenging to navigate and meet these intricate culinary requirements, highlighting a crucial aspect that demanded immediate attention and innovative solutions. The array of dietary preferences and restrictions within the medical tourist demographic presented a multifaceted challenge for spa centers. These preferences spanned a spectrum, ranging from specific medical dietary requirements to culturally influenced culinary choices. Accommodating such diverse needs became a complex task, as spa centers often struggled to provide customized meal options that aligned with the individual preferences and restrictions of their guests.

The inability to offer personalized and tailored meal choices resulted in dissatisfaction among medical tourists, impacting their overall spa experience. This underscored the necessity for a more flexible and accommodating approach to food services within these centers. Recognizing the importance of culinary experiences in the overall satisfaction of medical tourists, the lack of customization in meal offerings became a focal point for improvement. In response to these findings, it becomes evident that there is a pressing need for spa centers to reevaluate and revamp their culinary strategies. Introducing more flexible and customizable meal options can bridge the gap and enhance the overall satisfaction of medical tourists.

3.3 Communication and Cultural Sensitivity

Effective communication and cultural sensitivity were identified as challenges in ensuring a positive culinary experience for medical tourists. Language barriers and cultural nuances sometimes hindered clear communication between tourists and culinary staff, impacting the tourists' ability to convey their preferences or understand the cultural context of the meals provided. The significance of effective communication and cultural sensitivity emerged as prominent challenges in fostering a positive culinary experience for medical tourists within spa centers. These challenges were particularly evident in the context of language barriers and cultural nuances, which occasionally impeded clear communication between tourists and culinary staff. The consequences of these challenges extended beyond mere communication hurdles, influencing the tourists' ability to articulate their culinary preferences accurately and comprehend the cultural context surrounding the meals provided.

Language barriers proved to be a significant impediment in the seamless exchange of information between medical tourists and the culinary staff at spa centers. The diversity of visitors with varying linguistic backgrounds posed a complex challenge for effective communication. The inability to convey preferences, dietary needs, or specific culinary expectations due to language disparities often resulted in a breakdown of communication, leading to potential misunderstandings and dissatisfaction among the medical tourists. Moreover, the intricacies of cultural nuances further compounded the challenge. Cultural differences in culinary expectations, norms, and preferences demanded heightened cultural sensitivity from the culinary staff. Failure to understand and accommodate these cultural intricacies created a gap between the intended culinary experience and the actual experience perceived by medical tourists. This gap not only affected the overall satisfaction of the guests but also highlighted the need for a more culturally attuned approach to culinary services within spa centers. Additionally, recommendations will be proposed to enhance communication and foster cultural sensitivity within spa centers, ultimately elevating the culinary experiences for medical tourists.

3.4 Opportunities

Quality Enhancement of Existing Dishes:An evident opportunity lay in the enhancement of the quality of existing dishes. By refining the preparation methods, sourcing high-quality ingredients, and incorporating feedback from medical tourists, spa centers could elevate the overall gastronomic experience. This not only contributes to the satisfaction of medical tourists but also positions the spa as a destination that prioritizes culinary excellence. A discernible opportunity unfolds in the realm of culinary experiences within spa centers, presenting itself as a potential avenue for improvement through the enhancement of the quality of existing dishes. This opportunity calls for a comprehensive approach involving the refinement of preparation methods, the sourcing of high-quality ingredients, and an active integration of feedback from medical tourists. By undertaking these initiatives, spa centers can not only raise the standard of their gastronomic offerings but also significantly elevate the overall dining experience for their guests. Refining the preparation methods stands out as a crucial aspect in the pursuit of culinary excellence. This involves a meticulous examination and improvement of cooking techniques, presentation styles, and overall culinary craftsmanship. By investing in the training and development of culinary staff, spa centers can ensure that each dish is prepared with precision and care, aligning with the expectations of discerning medical tourists seeking a holistic and gratifying dining experience during their stay. Sourcing high-quality ingredients is equally paramount in the quest for culinary enhancement. By establishing partnerships with reputable suppliers and prioritizing the use of fresh, locally-sourced, and premium ingredients, spa centers can elevate the flavor profiles and nutritional value of their dishes. This not only contributes to the overall satisfaction of medical tourists but also aligns with contemporary culinary trends that emphasize the importance of quality, sustainability, and a connection to local produce. The active incorporation of feedback from medical tourists forms an integral component of this improvement process. By establishing channels for guests to share their culinary experiences and preferences, spa centers can gain invaluable insights into areas that require refinement. This iterative feedback loop ensures that the culinary offerings evolve in response to the dynamic needs and expectations of the guests, fostering a sense of engagement and collaboration between the spa and its clientele. Elevating the overall gastronomic experience not only serves the immediate goal of satisfying medical tourists but also strategically positions the spa as a destination that prioritizes culinary excellence. In the competitive landscape of medical tourism, where the overall experience plays a pivotal role in attracting and retaining guests, positioning the spa as a hub for exceptional culinary experiences adds a distinctive and appealing dimension to its offerings.

Additionally, recommendations will be proposed to guide spa centers in effectively seizing the opportunity to enhance their culinary offerings and solidify their position as leaders in the realm of medical tourism.

3.5 Introduction of New Culinary Offerings

The research highlighted a significant opportunity for spa centers to diversify their culinary offerings. Introducing new dishes, inspired by local and international cuisines, provides medical tourists with a broader range of choices. This not only caters to diverse preferences but also adds an element of excitement and discovery to the culinary experience, enhancing the overall satisfaction of medical tourists. The research underscores a noteworthy opportunity for spa centers to expand and diversify their culinary offerings. This entails the introduction of novel dishes, drawing inspiration from both local and international cuisines. By presenting medical tourists with a broader spectrum of choices, spa centers can not only cater to a more diverse array of preferences but also infuse an element of excitement and discovery into the culinary experience. This strategic diversification aims to go beyond meeting the basic dietary needs of medical tourists and seeks to transform their dining encounters into delightful journeys of exploration, ultimately elevating the overall satisfaction of these discerning guests.

The introduction of new and diverse dishes aligns with the evolving expectations of medical tourists who often seek a holistic experience during their stay at spa centers. By embracing culinary innovation and variety, spa centers can enhance the overall appeal of their offerings, fostering a sense of anticipation and curiosity among guests. This approach not only caters to the varied tastes of a diverse clientele but also positions the spa as a dynamic and forward-thinking destination in the competitive landscape of medical tourism. Drawing inspiration from local cuisines allows spa centers to showcase the richness and authenticity of the regional culinary landscape. This not only serves as a cultural immersion for medical tourists but also aligns with the growing trend of travelers seeking authentic and locally inspired dining experiences. On the other hand, incorporating elements from international cuisines adds a global flair to the menu, catering to the diverse backgrounds and preferences of the spa's international clientele. The fusion of these culinary influences creates a vibrant and enticing menu that resonates with the eclectic tastes of medical tourists. Furthermore, the strategic introduction of new dishes provides an opportunity for spa centers to engage in sustainable practices by incorporating locally sourced and seasonal ingredients. This not only contributes to the environmental consciousness of the spa but also aligns with the contemporary emphasis on sustainability and responsible culinary practices.

In summary, diversifying culinary offerings is not only an opportunity to meet the diverse preferences of medical tourists but also a strategic move to elevate the overall culinary experience within spa centers. The subsequent sections of this chapter will delve into the survey results, shedding light on the preferences of medical tourists regarding the introduction of new dishes and the impact of such diversification on their overall satisfaction. Additionally, recommendations will be provided to guide spa centers in effectively implementing this strategy to enhance their culinary repertoire.

3.6 Nutritional Guidance and Wellness Integration

Recognizing the importance of food in the recovery process, an opportunity exists for spa centers to integrate nutritional guidance into their services. Collaborating with nutritionists to create tailored meal plans based on the health needs of medical tourists can contribute to a holistic wellness experience. This approach aligns with the overall goals of medical tourists and positions the spa center as a comprehensive destination for health and relaxation. Acknowledging the pivotal role of nutrition in the recovery process, spa centers have the opportunity to take a proactive step by integrating nutritional guidance into their array of services. Collaborating with experienced nutritionists to formulate personalized and tailored meal plans based on the specific health needs of medical tourists can significantly contribute to a holistic wellness experience. This innovative approach not only aligns with the overarching goals of medical tourists but also positions the spa center as a comprehensive destination for health, relaxation, and well-rounded care. The integration of nutritional guidance represents a forward-thinking strategy that recognizes the interconnectedness of dietary choices and overall well-being. Medical tourists, often seeking not only physical recovery but also a holistic approach to health, can benefit greatly from targeted nutritional support. By partnering with qualified nutritionists, spa centers can offer individualized meal plans that cater to the unique health requirements and dietary preferences of each guest. Tailored meal plans can address a range of health concerns, including dietary restrictions, allergies, and specific nutritional needs associated with various medical treatments. Nutritionists can collaborate closely with the medical team, ensuring that the meal plans complement and support the overall healthcare objectives of the guests. This collaborative and interdisciplinary approach reinforces the commitment of the spa center to the well-being of its guests, fostering a sense of trust and confidence among medical tourists.Furthermore, the integration of nutritional guidance aligns seamlessly with the broader goals of medical tourists who are increasingly seeking comprehensive and integrated wellness experiences. The emphasis on nutrition not only supports the recovery process but also contributes to the overall health and vitality of guests during their stay at the spa center. This approach positions the spa center as more than just a facility for physical recovery; it becomes a destination that prioritizes the holistic well-being of its guests. The synergy between medical treatments, nutritional guidance, and other wellness services creates a comprehensive and integrated approach to healthcare, resonating with the evolving expectations of health-conscious travelers.

In summary, addressing the challenges and capitalizing on the opportunities identified in this research can significantly enhance the culinary experience for medical tourists in Serbian spa destinations. By incorporating these findings into their strategies, spa centers can not only meet the diverse needs of medical tourists but also differentiate themselves in the competitive landscape of medical tourism.

4. CONCLUSION

This research not only contributes to the academic understanding of medical tourism but also holds practical implications for various stakeholders. Healthcare providers, hospitality professionals, policymakers, and destination marketers stand to benefit from the insights generated. By recognizing the significance of culinary considerations in the medical tourism experience, stakeholders can implement strategic interventions to enhance patient satisfaction, address challenges, and capitalize on unexplored opportunities.

The aim of this research which was conducted in November 2023, was to investigate the correlation between culinary experiences and the medical tourism sector in Serbia, with a specific focus on well-known three spa destinations such as Prolom Banja, Vrnjačka Banja and Soko Banja. The primary objective was to provide contextually pertinent insights that go beyond individual spa experiences and reflect the broader landscape of medical tourism in Serbia. The data collection process included surveys designed to gather information on the culinary preferences and challenges faced by medical tourists. These insights formed the foundation for comprehending the existing culinary aspects within Serbian spas and provided a basis for potential enhancements. The research aimed to understand the challenges and opportunities related to food in the spa experiences of medical tourists. To achieve this, a diverse sample of 592 medical tourists aged 18-70 was engaged through structured interviews conducted on-site at selected spas. The emphasis was on obtaining comprehensive perspectives from medical guests regarding food-related aspects during their spa stays. Overall, the goal of the research was to contribute valuable insights that could be used to improve the culinary offerings within spa centers in Serbia, ultimately enhancing the overall experience for medical tourists and further positioning the country as a competitive player in the global medical tourism industry. As medical tourism continues to evolve, understanding the role of food in shaping the overall patient experience becomes indispensable. This study serves as a timely and relevant exploration into a hitherto understudied aspect of the medical tourism landscape, aiming to foster dialogue and drive positive transformations within the industry.

The research aimed to provide practical recommendations for spa centers based on the identified challenges and opportunities. These recommendations encompass strategies to address challenges in maintaining food balance and freshness, as well as initiatives to enhance the quality of dishes and introduce new culinary offerings.

The study encountered several limitations that should be acknowledged. Firstly, the research period was relatively short, potentially limiting the exploration of seasonal variations or long-term trends in culinary experiences within spa centers. This constraint suggests caution in drawing conclusions about the sustainability or consistency of findings over time. A second limitation pertains to the size of the sample. Although the study engaged 592 diverse medical tourists, the sample size may be considered relatively small in the expansive context of the medical tourism industry. The implications derived from this sample may not fully encapsulate the varied perspectives of a more extensive and diverse population of medical tourists. Furthermore, the study's focus on only three spa destinations - Prolom Banja, Vrnjačka Banja and Soko Banja, could be viewed as limiting. While these locations are prominent, a more expansive selection of spas across Serbia might provide a more comprehensive understanding of culinary experiences in different settings. The reliance on on-site interviews as the primary data collection method introduces another limitation. Conducting interviews within the spa environment may introduce biases as respondents might be influenced by the surroundings. This factor should be considered when interpreting the findings, and future research could explore alternative methodologies to mitigate potential biases.

Future research endeavors should consider an extended research duration to capture seasonal variations and assess the sustainability of culinary trends within spa centers over time. This would provide a more nuanced understanding of the dynamics influencing culinary experiences. To enhance the generalizability of findings, researchers could aim for a larger and more diverse sample of medical tourists. This could involve exploring a broader range of demographics, medical conditions, and travel motivations, ensuring that the study's implications are representative of the diverse population within the medical tourism industry. Expanding the scope to include a more diverse array of spa destinations across Serbia would contribute to a more holistic understanding of culinary experiences. This could involve investigating spas of varying sizes, locations, and target demographics, allowing for a richer exploration of the factors influencing culinary offerings. Finally, future research could explore alternative data collection methods or a combination of on-site and off-site interviews to mitigate potential biases introduced by the spa environment. This approach would contribute to a more comprehensive and unbiased representation of participants' perspectives regarding culinary experiences within spa centers.

REFERENCES

Abd Mutalib, N. S., Ming, L. C., Yee, S. M., Wong, P. L., & Soh, Y. C. (2016). Medical tourism: Ethics, risks and benefits. *Indian Journal of Pharmaceutical Education and Research, 50*(2), 261–270. https://www.ijper.org/sites/default/files/10.5530ijper.50.2.6.pdf

Casey, V., Crooks, V. A., Snyder, J., & Turner, L. (2013). Knowledge brokers, companions, and navigators: A qualitative examination of informal caregivers' roles in medical tourism. *International Journal for Equity in Health, 12*(1), 1–10. https://sci-hub.se/10.1186/1475-9276-12-94. doi:10.1186/1475-9276-12-94 PMID:24289231

Cham, T. H., Lim, Y. M., Sia, B. C., Cheah, J. H., & Ting, H. (2021). Medical tourism destination image and its relationship with the intention to revisit: A study of Chinese medical tourists in Malaysia. *Journal of China Tourism Research, 17*(2), 163–191. https://sci-hub.se/10.1080/19388160.2020.1734514. doi:10.1080/19388160.2020.1734514

Connell, J. (2013). Contemporary medical tourism: Conceptualisation, culture and commodification. *Tourism Management, 34*, 1–13. https://sci-hub.se/10.1016/j.tourman.2012.05.009. doi:10.1016/j.tourman.2012.05.009

Gyu Ko, T. (2011). Medical tourism system model. *International Journal of Tourism Sciences, 11*(1), 17–51. https://sci-hub.se/10.1080/15980634.2011.11434634. doi:10.1080/15980634.2011.11434634

Hadian, M., Jabbari, A., Mousavi, S. H., & Sheikhbardsiri, H. (2021). Medical tourism development: A systematic review of economic aspects. *International Journal of Healthcare Management, 14*(2), 576–582. https://sci-hub.se/10.1080/20479700.2019.1677977. doi:10.1080/20479700.2019.1677977

Han, H., & Hyun, S. S. (2014). Medical hotel in the growth of global medical tourism. *Journal of Travel & Tourism Marketing, 31*(3), 366–380. https://sci-hub.se/10.1080/10548408.2013.876955. doi:10.1080/10548408.2013.876955

Hartwell, H., Fyall, A., Willis, C., Page, S., Ladkin, A., & Hemingway, A. (2018). Progress in tourism and destination wellbeing research. *Current Issues in Tourism, 21*(16), 1830–1892. https://sci-hub.se/10.1080/13683500.2016.1223609. doi:10.1080/13683500.2016.1223609

Hassan, V., & Zrnić, M. (2021). Slow tourism: a niche market towards sustainability of moonlight farm house "barouk area". *Turističko poslovanje*, (28), 57-63. https://www.aseestant.ceon.rs/index.php/turpos/article/view/34456/19278

Lunt, N. (2022). Medical tourism. In *Key Concepts in Medical Sociology* (3rd ed., pp. 222–227). SAGE. https://books.google.rs/books?hl=en&lr=&id=j2RXEAAAQBAJ&oi=fnd&pg=PA222&dq=Lunt,+N.+(2022).+Medical+tourism.+In+Key+Concepts+in+Medical+Sociology+(Third+Edition)+(pp.+222-227).+SAGE.&ots=-p5n9eElsM&sig=j55181Eq7M63jjIWHLAgH8CG8bY&redir_esc=y#v=onepage&q&f=false

Ngobeni, C. C. L. (2020). *Identifying the push and pull factors of a medical tourism destination* [Doctoral dissertation, University of Pretoria]. http://hdl.handle.net/2263/77916

Seyitoğlu, F., & Ivanov, S. (2020). A conceptual study of the strategic role of gastronomy in tourism destinations. *International Journal of Gastronomy and Food Science, 21*, 100230. https://sci-hub.se/10.1016/j.ijgfs.2020.100230. doi:10.1016/j.ijgfs.2020.100230

Shukla, R., Singh, M., & Saxena, S. (2019). Consumer perception of hospitality services in JCI accredited hospitals at Delhi–NCR: An exploratory research on growth of medical tourism. *Humanities & Social Sciences Reviews*. file:///C:/Users/milos/Downloads/SSRN-id3438296.pdf

Smith, M., & Puczkó, L. (2014). *Health, tourism and hospitality: Spas, wellness and medical travel*. Routledge. https://books.google.rs/books?hl=en&lr=&id=p0bIAgAAQBAJ&oi=fnd&pg=PP1&dq=Smith,+M.,+%26+Puczk%C3%B3,+L.+(2014).+Health,+tourism+and+hospitality:+Spas,+wellness+and+medical+travel.+Routledge.&ots=1L74wfuVW7&sig=m573Qi-bLlxz7quDx7yi1kJ41_A&redir_esc=y#v=onepage&q=Smith%2C%20M.%2C%20%26%20Puczk%C3%B3%2C%20L.%20(2014).%20Health%2C%20tourism%20and%20hospitality%3A%20Spas%2C%20wellness%20and%20medical%20travel.%20Routledge.&f=false

Suess, C., & Mody, M. (2017). Hospitality healthscapes: A conjoint analysis approach to understanding patient responses to hotel-like hospital rooms. *International Journal of Hospitality Management, 61*, 59–72. https://sci-hub.se/10.1016/j.ijhm.2016.11.004. doi:10.1016/j.ijhm.2016.11.004

Turner, L. (2010). "Medical tourism" and the global marketplace in health services: US patients, international hospitals, and the search for affordable health care. *International Journal of Health Services, 40*(3), 443–467. https://sci-hub.se/10.2190/HS.40.3.d. doi:10.2190/HS.40.3.d PMID:20799670

Turner, L. (2012). Beyond" medical tourism": Canadian companies marketing medical travel. *Globalization and Health, 8,* 1-11. file:///C:/Users/milos/Downloads/1744-8603-8-16.pdf

Zolfagharian, M., Rajamma, R. K., Naderi, I., & Torkzadeh, S. (2018). Determinants of medical tourism destination selection process. *Journal of Hospitality Marketing & Management*, *27*(7), 775–794. https://sci-hub.se/10.1080/19368623.2018.1444527. doi:10.1080/19368623.2018.1444527

Zrnić, M., Brdar, I., & Kilibarda, N. (2021). The importance of traditional food quality—The viewpoint of the tourism. Scientific journal. *Metals Technology*, *62*(1), 69–76. doi:10.18485/2021.62.1.7

Zrnić., M, Đoković, F., Košutić, J., Mašić, S. Hassan, V. (2021). *Planning and development of spa & wellness tourism.* Sitcon. Belgrade. file:///C:/Users/milos/Downloads/Welnesstourism2021.pdf

Chapter 4
Perceptions of Participants Receiving Health Services About the Effects of AI in the Health Sectors

Muhammed A. Yetgin
University of Karabuk, Turkey

ABSTRACT

Artificial intelligence (AI) applications have provided convenience for patients and healthcare professionals in diagnosing diseases, applying treatment methods, improving public health, and providing medical health services suitable for all ages. Türkiye is one of the countries where health tourism is developing in terms of technology and the application of AI. This study aims to investigate the perceptions of Turkish citizens and foreign patients who have experienced healthcare services in Türkiye regarding the use of AI and technology in the healthcare sector based on the perspective of diffusion of innovation theory, developed by Everett Rogers in 1962. The interview analysis method was applied in the study, and ten Turkish citizens and ten foreign participants, a total of 20 people, participated in the research. A research model was proposed with the content analysis carried out with the Nvivo 11 analysis program for the findings.

DOI: 10.4018/979-8-3693-2248-2.ch004

1. INTRODUCTION

Humanity has benefited from the knowledge, experience, and techniques applied against diseases under the conditions of the period. Over time, new technologies such as gene therapy, 3D printed organs, and artificial intelligence-based medicine have potentially revolutionized medicine (Hofer & Stampfli, 2008). The use of artificial intelligence applications in health has become inevitable. The development of Artificial Intelligence and Metaverse applications, countries' reforms in their health legislation for this development, the avalanche of healthcare expectations, the global financial crises in recent years, and the increase in countries' needs for foreign capital are among the main determining factors in the development of health tourism. Even the word "tourism" in health tourism is seen as misleading as it can be associated with "volunteering, pampering, or vacation" (Sigrist, 2006b); tourism may be practiced not only because the medically indicated service is not provided in the consumer's usual residence or work environment, but this is also partly true because patients often seek recovery close to home. they prefer (Hofer & Stampfli, 2008). People are interested in health tourism through national or international travel for many reasons, such as quality, cost, and access to professional services, and this interest is increasing daily. There is a massive increase in health tourism worldwide in many fields of medicine and dentistry, such as infectious diseases, cancer treatments, plastic surgery procedures, hair transplantation, and dental and maxillofacial surgery procedures. The application of artificial intelligence and technology in health services and management in Türkiye has been among the main reasons for the revival of health tourism.

This study aims to examine the development of health tourism with increasing artificial intelligence applications in health services and management in Türkiye within the scope of the "Diffusion of innovation" theory, which has processes of "knowledge, persuasion, decision, implementation, and confirmation". Within the scope of this theory, the interview method was applied to 10 foreign tourists who came to Türkiye for treatment purposes. Tourists who prefer health tourism receive health services on different subjects. In interviews with tourists, The reasons why they chose Türkiye with Artificial Intelligence applications were tried to be understood within the framework of Rogers' "Diffusion of innovation" theory. According to the research findings, Many parameters such as trust, cost, quality, and ease of access to health services constitute the main reason for tourists coming for health tourism. This situation is explained within the scope of the theory.

2. THEORETICAL BACKGROUND

2.1 Adoption and Diffusion of Innovations

Rogers' diffusion of innovation theory was first described in 1962. It characterizes people's likelihood of adopting technology and classifies organizations according to their stage of adopting a new technology.

In Diffusion of Innovations, for an innovation to be self-sustainable, it must be embraced on a large scale. Social groups that embrace and embrace innovation are categorically defined (Rogers, 1962; Rogers, 1983; Rogers, 2003; Doyle, et al., 2014).

These groups:

- Innovators: "People who innovate, make inventions, and develop systems."
- Early adopters: "The segment that first encounters innovation and plays a critical role in the spread of innovation to society."
- Early Majority: "The critical majority that has been reached by innovation and has incorporated this innovation into its life and culture for various reasons."
- Late Majority: "The majority formed when the innovation passes its maturity phase and then spreads to other segments of society."
- Laggards: They are listed as "the segment of the society that continues to use this innovation, which is still obsolete, even though the innovation has been abandoned or replaced by a new technology or system."

The main factors during the spread are listed as follows (Rogers, 1983):

- Innovation: The subject of diffusion. Rogers' study defines it as follows: "An idea, a practice, or an object that is perceived as new and is likely to be adopted by individuals or groups of society."
- Communication Channels: "The medium through which news is transmitted from one individual to another".
- Time: "Innovation decision time is the time that must pass for the innovation-decision process. The adoption rate is proportional to the time it takes for the innovation to be adopted by social system members".
- Social System: "A set of interrelated social units that have come together for a common goal and developed a joint problem-solving ability".

If the advantages of innovation are more significant, its adoption will be faster (Rogers, 2003). The biggest obstacle is uncertainty (Downs & Mohr, 1976; Damanpouer, 1991). Time is essential in the diffusion of innovation; it refers to the process in which it is decided to adopt or reject the innovation (Rogers, 1983).

The stages of an individual embracing innovation constitute the basic theoretical approach of this study and are defined as follows (Rogers, 2003):

- Knowledge: "At this stage, the individual encounters innovation but does not have sufficient information about the innovation. At this stage, the individual does not yet have sufficient awareness/suggestion to learn more about the innovation."
- Persuasion: "At this stage, the individual is interested in innovation and actively seeks information and details about the innovation."
- Decision Making: "At this stage, the individual deals with the concept of innovation. "Decides to accept or reject the innovation by using the innovation and weighing its advantages and disadvantages."
- Implementation: "At this stage, the individual puts the innovation into effect and puts it into practice, to varying extents and limitations depending on the situation. During this stage, the individual makes his judgment about the usefulness of the innovation and may engage in more in-depth research about the innovation."
- Confirmation: "At this stage, the individual finalizes his decision about the product. The individual is now ready to use innovation to its full extent."

In a study conducted by Temizkan et al. (2022), the perceptions of retailers using cryptocurrency toward customer experiences were analyzed in terms of Rogers' theory of diffusion of innovation. The knowledge process is the first stage and shows how people are aware of artificial intelligence-based health tourism but still need to understand it. This stage is vital in raising awareness about artificial intelligence-based health tourism and its potential benefits. Persuasion is the second stage, where people form an attitude towards AI health tourism and decide whether they are interested. This stage is vital in educating people about artificial intelligence-based health tourism and its benefits. The third stage is when people decide whether to try AI-based health tourism. This stage makes artificial intelligence-based health tourism more accessible and affordable. The implementation is the fourth stage and is the application where people try artificial intelligence-based health tourism and experience its benefits first-hand. This stage is essential to demonstrate the value of artificial intelligence-based health tourism and encourage others to try it. The fifth stage, confirmation, is the confirmation process, where people seek information and feedback to strengthen their decision to use AI-based health tourism. This stage is essential to build trust in artificial intelligence-based health tourism and ensure people have a positive experience. Many researchers have examined "diffusion theory and tested it empirically across various industries and countries", and it has been concluded that the rate of diffusion is influenced by macroeconomic and demographic factors (Bulte, 2000) as well as microeconomic factors (Hafeez et al., 2006) related to product characteristics (Lüthje & Herstatt, 2004).

2.2 Artificial Intelligence in Health Tourism Business

Medical tourism has been defined as the act of traveling abroad for treatment and care and includes higher-grade invasive surgeries such as "heart, hip resurfacing, and plastic surgeries" (Sourav et al., 2023; Singh, 2013; Connell, 2013; Carrera & Lunt, 2010; Reddy et al., 2010). ; Hjalager, 2009; Connell, 2006). While initially thought to be a search for more suitable alternative solutions to cosmetic procedures, increasingly vital health procedures have begun to be implemented, especially in destinations such as "Thailand, Singapore, and India" (Musa et al., 2012; Ye et al., 2011). Health tourism, used among disciplines such as medicine and tourism, is defined as tourism facilities or destinations that attract tourists by promoting health-related services or facilities (Barca et al., 2013; Smith & Puczko, 2008; Goodrich & Goodrich, 1987). It is explained as systematically planned trips to protect a person's physical and mental health (Carrera & Bridges, 2006; Laws, 1996). *Wellness tourism* is an action that encompasses anything related to medicine, health, sports/fitness, adventure, or transformational travel types that improve a person's well-being (Voigt & Pforr, 2014; Hjalager & Flagestad, 2012; Clarke, 2010; Bushel & Sheldon, 2009; Smith & Kelly, 2006).According to Padilla-Meléndez and Del-Águila-Obra (2016), the differences between health and wellness tourism are shown in the figure below.

Figure 1. Health and wetness tourism
Source: *Padilla-Meléndez and Del-Águila-Obra (2016)*

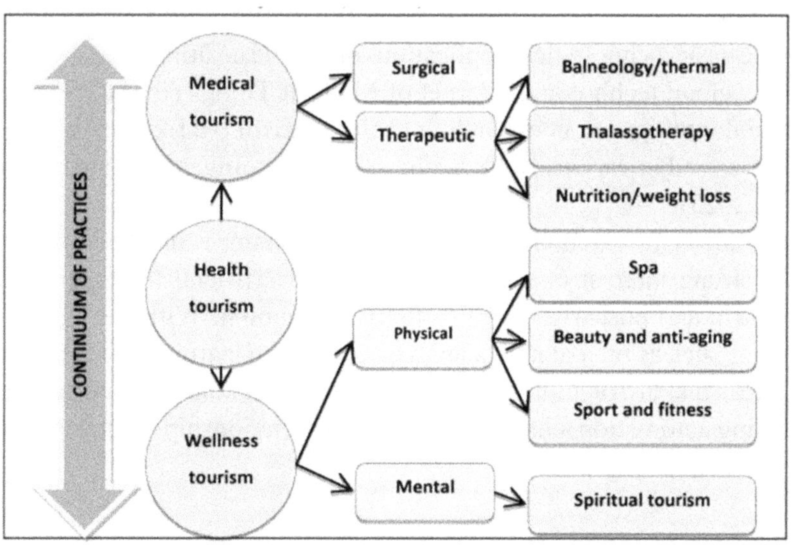

Sharma et al. (2023), the effects of health tourism on a country's economy, new technologies used, artificial intelligence-based medical tourism applications, some basic processes for creating artificial intelligence-based medical tourism applications, a set of open source tools that can be used in the development of artificial intelligence-based applications for medical travel. According to Sharma (2023), when examined in terms of the technologies used, telemedicine provides remote treatment and medical consultation and allows patients to communicate with medical professionals worldwide. Artificial intelligence-supported applications can be used to "personalize patient recommendations" for care in medical tourism, "facilitate administrative tasks," and "increase the standard and effectiveness of care" (Ishita & Sachin, 2022). Robotic technologies can assist in surgical procedures and provide physical therapy and rehabilitation to patients during medical travel (Sharma, 2023).

Artificial intelligence is one of the most promising technologies of our age. Its dramatic completion is due to the increasing processing capacity and amount of data of computers (Ma et al., 2020; Lixia & Zhongli, 2019; Lu et al., 2018; Seligman, 2006; Sigrist, 2006a). Artificial intelligence and new robotic technologies, which also have the features of understanding texts, translating into other languages, and obtaining expert opinions (Sakipov & Vikhnin, 2021), are being imposed as a potential problem solver in security (Peric &Vanja, 2021; Kangping et al., 2022; Zhang et al., 2012; Yu & Ko, 2012; Heung, 2011). In one application, artificial intelligence was aimed to determine the treatment and choose the best doctor, and this model was applied to ten patients with the application of the decision tree model; the result provided an accuracy rate of 75-80% (Subasinghe et al., 2020). Health tourism has been increasing significantly in recent years, and today, technologies are further strengthening this tourism and integrating its stakeholders into the intelligent scope of the health system (Siddharth & Sachin, 2022; Pan & Chen, 2014; Glynda et al., 2014; Saadatnia & Mehregan, 2014; Hofer et al., 2012). Smart Health System (SHS) includes the basic technological applications of mobile health, telehealth, blockchain technology, cloud technology, Internet of Medical Things (IoMT), virtual reality, and artificial intelligence in expanding health tourism (Hafızullah & Kirti, 2023). Health 4.0 is used in various fields, from artificial intelligence to big data analytics (Dalkıran, 2023).

As a result of research Concerning health tourism, it has been observed that there is a strong interest in companies that offer artificial intelligence solutions in the pre-trip and post-trip stages of tourism planning. With the high interest in technologies such as big data, data analytics, machine learning, and chatbots, it has become clear that artificial intelligence technological solutions are solutions used for marketing automation, customer service, and relationship management and that these can potentially affect the effectiveness of marketing and sales activities (Filieri et al., 2021; Bassil, et.al., 2019). AI's various roles in transforming the industry and making it more convenient for patients include ease of carrying out complex procedures and sometimes shortening recovery times while providing cost benefits for patients, among other benefits (Hassan & Bellos, 2022). It has been observed that there may be some obstacles to receiving treatment abroad. For example, one study found several potential barriers to traveling abroad for Canadians, with financial barriers often being why people avoid medical tourism, with fear and lack of trust in healthcare from strangers also identified as other barriers (Johnston et al., 2011). In research conducted on health tourism in Balkan countries, weaknesses While threats are seen as weak infrastructure, lack of service quality, low education, training and skill levels, hostile or unknown image, distrust and inability to work together; ongoing conflicts and instability, migration of skilled workers due to low salaries, lack of government funding for domestic health tourism in the future, and political and other corruption (Kiss, 2015). While Türkiye, Korea, Singapore, Hungary, Poland, Thailand, Malaysia, India and Jordan are among the developing countries in health tourism, in a research conducted on health tourism in the USA, the most popular destinations preferred by American health tourists are Mexico, Thailand, Philippines, Korea, Taiwan, Malaysia and Singapore; Health tourism revenues of these countries are increasing faster than their general economic growth (Daştan, 2014; Eissler & Casken, 2013; Cohen, 2012; Deloitte, 2008; World Bank).

2.3 Empirical Literature

Türkiye is a rapidly developing country in health tourism with the impact of the transformation and investment reforms in the health field, and its increasing international performance and competitive prices support this development. The fact that doctors and surgeons in Türkiye are well-educated and the technological infrastructure in health is developed can be seen as essential factors in health tourism(Aksu, A., & Bayar, 2019; Aslan, et al.2014). Factors such as natural beauty and historical riches make Türkiye an attractive destination in terms of tourism and health tourism (Aksu et al., 2016), price competitiveness, health and hygiene, tourist service infrastructure, and air transportation infrastructure. It is competitive with many factors (Kılavuz, 2018). The reasons why foreigners choose Türkiye are shown in the Table below.

The research revealed push and pull factors and outcomes according to the Medical Tourism Index in Türkiye. It is shown in Table 2.

Table 1. Reasons why foreigners prefer Türkiye in health tourism

#	Preferred Countries/ Regions	Reasons for Preference
1	Germany, France, Holland	There are many Turks who are citizens of these countries. They can choose their home country, Türkiye, for health services.
2	Balkans and Turkish Republics	cultural proximity
3	England, USA	High cost of health services that are not covered by the social security system in these countries. Long waiting times.
4	Canada	Long waiting times
5	Middle East	Cultural closeness, affordable cost, quality service
6	Africa	Service quality, affordable cost, technological superiority, effective service opportunity in specific branches

Source: The author created the table, but the information was taken from Yirik et al., 2015.

Table 2. Driving and attractive forces and outputs in health tourism in Türkiye

Driving Forces	Attractive Forces	Outputs
acquaintance reference	internet website	Service satisfaction (treatment satisfaction, hospital accommodation, and service satisfaction)
Quality of healthcare in the home country	Perception of hospital quality	Security experience in the country (general security, shopping security)
Prices	country image	Intention to visit again
	Religious culture	
	Prices	
	Tourism	
	Perception of Security	
	Perception of Quality of Health Services in Türkiye	

Source: The author created the table, but the information was taken from Üstün & Demir Uslu, 2022.

Türkiye has a Health Tourism Department under the Ministry of Health. The clinical branches most preferred by international patients are gynecology, internal medicine, ophthalmology, medical biochemistry, general surgery, dentistry, orthopedics and traumatology, infectious diseases, and ear-nose-throat. For Türkiye to be more successful in international markets, effective promotions should be made abroad, healthcare personnel should speak foreign languages, and Ensuring that qualifications should be increased, accreditation should be given importance, and infrastructure facilities should be developed (Buzcu and Birdir 2019).

3. METHODOLOGY

3.1 Purpose and Importance of the Research

This study examines the applications of artificial intelligence and related technologies in health services for Turkish citizens and foreigners who benefit from health tourism in Türkiye. It is an investigation of the perceptions of the Diffusion of Innovation Theory according to the Innovation Diffusion Decision Processes. Therefore, the target of this research is organizations that provide artificial intelligence and related technology-based healthcare services. The aim is to understand the perceptions of patients who have benefited from health services and to reveal their behaviors on the subject. In addition to contributing to research management and strategy, It examines the attitudes of patients who have received healthcare services towards artificial intelligence and related technology in healthcare against the diffusion processes of innovation. People who receive health services through health tourism: Their attitudes towards innovation processes towards artificial intelligence and technology applications have been the main problem of curiosity. The fact that the study is research and measures the perceptions of Turkish citizens and foreigners on the subject through interview processes makes the research unique. The health sector is being watched with interest every day, as artificial intelligence and technology are advancing rapidly and hepatic technologies are being implemented sectorally. Therefore, the study is also suitable for the current agenda.

The diffusion of Innovation Processes is shown in the figure below.

Figure 2. Five stages in the decision innovation process (Innovation-decision process) (source Adapted from Rogers 1983, p. 163 (Charlie, et al., 2015)

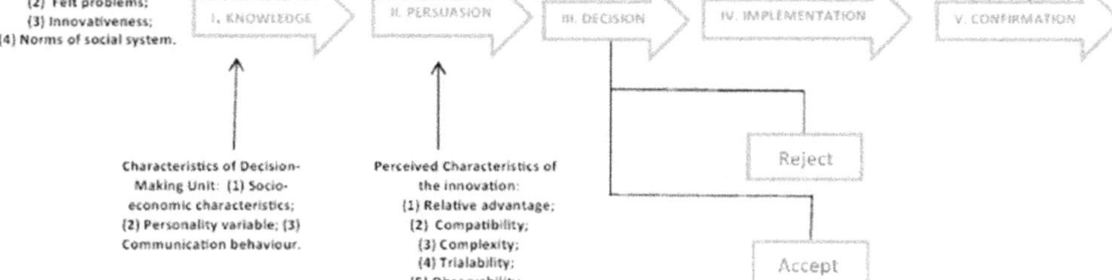

This research seeks answers to the following basic questions:

Table 3. Questionnaire

#	Questions	Relation of Innovation to the Diffusion Process
1	Artificial Intelligence or Advanced Technology: What is your knowledge and awareness about its use in health services?	"The Knowledge Stage"
2	What is your attitude towards this innovation regarding the advantages and disadvantages of the effects of artificial intelligence or advanced technology in healthcare?	"The Persuasion Stage"
3	Have any diagnoses or treatments been implemented using artificial intelligence, metaverse, or advanced technology that you do not accept? Is it possible?	"The Decision Stage"
4	What would you like to learn more about when applying artificial intelligence or advanced technologies in healthcare? What current health-related problems should technological advances primarily solve?	"The Implementation Stage"
5	When you look at the judgments in your inner world and the comments you receive from others, Do you approve of the application of artificial intelligence or advanced technology in all healthcare processes?	"The Confirmation Stage"

Source: The author created the table

The study examines patients receiving healthcare services, Their attitudes towards innovation diffusion processes, and the application of artificial intelligence and technologies in healthcare.

3.2 Research Design and Data Analysis

The study's theoretical framework is Rogers' Diffusion of Innovation Theory - Innovation Decision Processes. The research is a qualitative study with an exploratory and interpretive perspective, and according to the theory, adaptation to innovation goes through a five-stage process depending on the decision to either not accept or accept each of these processes. One stage is reducing the uncertainties about the innovation and making the final decision by moving forward. (Creswell, 2014; Gray, 2013; Priporas & Mylona, 2008; Rogers, 2003; Rogers, 1995). The research findings were obtained from interview data conducted via video conference. The interview is a data collection technique used in qualitative research (Punch, 2005), and an *interview* is defined as a conversation whose purpose is to collect information (Berg & Lune, 2015). Each interview lasted 45 minutes, and the video and audio recordings in the video-conference program were converted into text. While the answers to the questions in the interview constitute the qualitative data of the study, the types of sectors that responded are also revealed quantitatively. In the study, the experiences of patients receiving health tourism services were analyzed in the context of Rogers' Diffusion of Innovation Theory, and the results were compared with studies in the literature.

3.3 Sample Profile

Turkish citizens and foreigners participating in the research constitute the limitation of the research. 10 Turkish citizens and 10 foreign participants, a total of 20 people, participated in the research. Demographic data about the people involved are shown in the table below. Participants were numbered with the letter P and coded as P1 and P2. Interviews took place between December 20, 2023, and January 5, 2024. Participants in the study mostly received treatment in hospitals in Ankara and Istanbul. Male is coded as M while Female is coded as F in the table.

Table 4. Respondents profile

#	Nationality	Gender	Age	Occupation	Health Service Received	Coded
1	Turkish	F	26	Doctor	Orthopedic surgery	"P1"
2	Palestine	F	22	Student	Plastic surgery	"P2"
3	Chadian	M	31	Journalist	Otolaryngologist	"P3"
4	Palestine	F	28	Foreign Trade	Ophthalmology	"P4"
5	Indonesia	M	26	Public Relations	Otolaryngologist	"P5"
6	Turkish	F	23	Public Relations Specialist	Dermatology	"P6"
7	Turkish	M	37	In one of the State Institutions	Hematology	"P7"
8	Turkish	M	36	Businessman	Infection Disease	"P8"
9	Turkish	M	25	Journalist	Otolaryngologist	"P9"
10	Sudan	M	37	IT Engineer	Dental Care	"P10"
11	Burundi	F	29	Student	Internal Diseases	"P11"
12	Indonesia	M	25	Communication	Internal Diseases	"P12"
13	Turkish	M	29	Customs Specialist	Ophthalmology	"P13"
14	Somali	M	30	Foreign Trade Representative	Ophthalmology	"P14"
15	Turkish	F	40	In one of the State Institutions	Otolaryngologist	"P15"
16	Turkish	M	38	In one of the State Institutions	Dermatology	"P16"
17	Iraqi	M	37	Finance Specialist	Ophthalmology	"P17"
18	Turkish	F	38	Health Technician	Dental Care	"P18"
19	Iraqi	M	30	Finance Specialist	Otolaryngologist	"P19"
20	Turkish	M	32	Financial-Administrative Consolidation and Reporting Specialist	Ophthalmology	"P20"

Source: The author created the table

4. FINDINGS

4.1 The Knowledge Stage

This stage measures how much the individual is aware of the innovation's existence and his knowledge level (Rogers, 2003). In this process, an individual tends to collect information by asking what, how, why, and why questions. The participants tried to understand their knowledge, awareness, trends, and attitudes toward using artificial intelligence and technologies in health tourism and services. Participants were asked, "Artificial Intelligence or Advanced Technology: What is your knowledge and awareness about its use in health services?" The question was posed. Following this question, the participants were asked whether they think their level of knowledge about the use of artificial intelligence and technologies in health tourism and services is suitable and sufficient, and they were asked to choose between the options "strongly agree, agree, undecided, disagree, strongly disagree." The answers given by the participants are like findings; some long answers were not fully transferred to the study, and essential parts were included. All the findings regarding the measurement of the "knowledge stage" levels of foreign participants are shown in Table 5. Exciting answers from Turkish citizens; moved to the research text.

Table 5. Findings of foreign participants perceptions in the knowledge stage

#	Findings	Awareness of Knowledge Stage Level
P2	*"The healthcare sector in Türkiye is witnessing the adoption of advanced technology and artificial intelligence. Artificial intelligence is used to improve disease diagnosis, develop treatment methods, and improve the patient experience. Technology is also integrated into managing medical files and enhancing healthcare delivery processes. An example of this is using artificial intelligence in medical image analysis. My knowledge on this subject is sufficient based on what I have experienced and observed*".	Strongly Agree
P3	*"My knowledge and awareness about using artificial intelligence or advanced technology in health tourism and health services from Türkiye is good. I constantly receive updated information about technological developments in the healthcare sector*".	Strongly Agree
P4	*"As a result of my experience and detailed observations regarding health tourism and services in Türkiye, My understanding of AI is that it improves the patient experience in healthcare and is used by healthcare patients for all kinds of paperwork, from admission forms to follow-up data. AI is also finding applications in solutions to leverage insights into genomics, biomarkers, and phenotyping, as well as specialized tools in radiology, disease identification, and ophthalmology.*".	Strongly Agree
P5	*"I know it's used in Türkiye because artificial intelligence in medicine benefits hospitals, doctors, and customers. I experienced this*".	Strongly Agree

continues on following page

Table 5. Continued

#	Findings	Awareness of Knowledge Stage Level
P10	*"I know that in Türkiye, artificial intelligence is used to analyze patient data, which can improve treatment by assisting with human tasks".*	Strongly Agree
P11	*"Artificial Intelligence (AI) is a rapidly expanding field of research with a bright future. In Türkiye, artificial intelligence is at the heart of future medicine with assisted surgery, remote patient monitoring, and intelligent prostheses. Applications of artificial intelligence (AI) and advanced technologies in the Turkish healthcare sector are increasing. I was also satisfied with the health service I received".*	Strongly Agree
P12	*"I am very aware that the existence of AI in the world of health is beneficial in treating patients and helping doctors simplify".*	Strongly Agree
P14	*"My knowledge regarding the use of AI in healthcare is that AI facilitates drug development, diagnosing diseases, medical robots, personalized medicines, managing medical records, administrative workflow, and Medical imaging in the healthcare department. Türkiye is moving in this direction".*	Agree
P17	"Diagnosis of diseases in Türkiye: Artificial intelligence applications are used to determine diseases for which early diagnosis is essential more accurately and at earlier stages".	Agree
P19	*"Health care in Türkiye is excellent, and nose surgery is performed quickly and easily without risk. High-end technology is used".*	Agree

Source: The author created the table

According to the findings, foreigners who have benefited from health tourism in Türkiye have high levels of knowledge and awareness about using artificial intelligence and technology in health services. When evaluated in terms of the "Knowledge Stage," which is the first step in the innovation processes, It was understood that eight of the foreigners had an excellent level of knowledge about the innovation (responded strongly agree), and two of them had a good level of expertise (responded agree). Two of the answers given by Turkish citizens will be specifically discussed here. P1, a doctor who benefited from health tourism, answered: "Artificial intelligence works integrated in many medical fields. From surgery to laboratory results, artificial intelligence is used to speed up and work more meticulously in medicine." He also responded strongly that his level of knowledge was exemplary. P7, who works as an administrative manager at a state university, answered: "I know that studies on the production of artificial intelligence-supported theoretical and practical knowledge for the diagnosis and treatment of diseases in many medical fields, especially cancer diagnosis, and treatment, and the development of technologies for the development of some devices are carried out at universities." He also answered "strongly agree" regarding his knowledge level. When evaluated from the perspective of Turkish citizens, six people answered strongly agree, and four answered agree at the level of knowledge. The general conclusion here is that foreign participants have a slightly higher level of knowledge about using artificial intelligence and technology in health tourism than Turkish citizens. This is the case for foreigners; this may be because they did extensive research before coming to Türkiye. Foreign participants' and Turkish citizens' knowledge and awareness levels regarding applying artificial intelligence and technologies in health tourism are good. Usually, in this first stage, the individual encounters innovation and generally needs more information about this innovation. At this first meeting, the individual typically needs more motivation to research innovation. However, in this study, it was seen that the first stage was at an excellent level.

4.2 The Persuasion Stage

The persuasion stage occurs when a person develops a positive or negative attitude towards the innovation. Still, the resulting positive or negative attitude does not necessarily lead to the person accepting or rejecting the innovation (Rogers, 2003). At this stage, "What is your attitude towards this innovation regarding the advantages and disadvantages of the effects of artificial intelligence or advanced technology in healthcare?" The question was asked. In response to the question, "What is your opinion about your attitude in the persuasion process for advantages than disadvantages?" one of the following answers was asked: "strongly good," "good," "undecided," "not good," and "strongly not good." The answers given by foreign citizens are shown in Table 6. The answers of Turkish citizens, which are more interesting than others, have been the subject of discussion.

Table 6. Findings of foreign participants' perceptions in the persuasion stage

#	Findings	Attitude to Advantages
P2	*"Pros include improving diagnostic accuracy, speeding up medical processes, and providing effective care. However, drawbacks can include patient privacy issues".*	Good
P3	*"The advantages of artificial intelligence and advanced healthcare technology include faster diagnosis and treatment, increased efficiency, and lower costs. However, its disadvantages include limiting the human factor and concerns about privacy and security. I also have a positive but cautious attitude towards this innovation".*	Undecided
P4	*"Of course, just as artificial intelligence has positives, it also has negatives, including the fact that it still requires human support. AI excels at executing specific programmed commands but still requires collaboration with doctors. Artificial intelligence can be wrong. It relies heavily on diagnostic data from a broad range of cases, but insufficient data on specific diseases, demographics, or environmental factors can lead to misdiagnosis".*	Undecided
P5	*"Advanced technologies such as artificial intelligence will also be useful in the healthcare sector because health is necessary for everyone, regardless of its advantages, disadvantages, and future effects".*	Good

continues on following page

Table 6. Continued

#	Findings	Attitude to Advantages
P10	*"AI improves diagnostics by analyzing medical data and images with high accuracy".*	Strongly Good
P11	*"I do not have a personal view, but I can provide some ideas about the advantages and disadvantages of artificial intelligence in health services. The advantages include more effective diagnosis, while the disadvantages may concern data confidentiality. The use of advanced technology in healthcare offers opportunities but also raises ethical and safety issues".*	Undecided
P12	*"The advancement of AI technology in health care will help doctors and patients to make health care easier, regardless of the effects in the future".*	Strongly Good
P14	*"My attitude toward this Innovation is about the advantages and disadvantages of health services. Adverse effects: There will be less need for human healthcare workers when AI automates and takes over administrative, research, and operational duties. While this improves the facility's operating efficiency and cuts expenses, it can potentially put many qualified healthcare workers out of work"*	Not Good
P17	*"Artificial intelligence enables medical professionals to make more accurate medical decisions using accurate real-time data. This improves preventive measures, saves costs, and reduces patient waiting times. It is proficient at executing specific programmed commands but requires collaboration with doctors. He cannot work independently yet. It has more advantages.".*	Good
P19	*"It is not possible to trust the machine too much".*	Not Good

Source: The author created the table

When the attitudes of foreign participants towards the advantages of using artificial intelligence and technology in health are examined, 30% are undecided, and 20% are negative. According to half of the participants, there is no positive attitude towards the advantages of these technologies. The opinions of Turkish citizens whose attitudes towards artificial intelligence and technology are "undecided," "not good," and "strongly not good" are discussed here. The attitudes of Turkish citizens P1, P6, and P16 towards the advantages of using artificial intelligence in health were "undecided." P13's attitude is "not good." All other participants answered "good" or "strongly good," indicating their positive attitudes toward its advantages. I wanted to share one of those with a positive attitude here. P15: "Artificial intelligence has no emotions. People may prefer to talk instead of learning from a system. Artificial intelligence will not be able to respond to many questions or motivations in their minds. Suppose we summarize all the benefits of artificial intelligence. In that case, we can say that machines can complete tasks faster, more accurately, and in a shorter time, and despite all its disadvantages, Your attitude towards it is positive." P1 stated: "It makes things work more meticulously and faster. The disadvantage is that it causes people to think less, which dulls practical intelligence." P6: "I think that there are great advantages in reaching a wide range of radiological imaging techniques to drug development and in the implementation of rapid treatments without wasting time in diagnosing the disease, but the problem is the data privacy of the patients, and it may also pose risks of misuse of patient-specific records." P13 explained: "The failure of companies operating in the health sector to fully fulfill some procedures due to the need to market their products quickly poses great risks to human health."

4.3 The Decision Stage

At this stage, the individual is at the stage of judging the innovation, determining the advantages and disadvantages it will bring to her, and ultimately accepting or rejecting the innovation (Rogers, 2003). Among foreign and Turkish citizens, those who answered "undecided" or "not good" in the previous step were explicitly intended to be discussed in this section. Others should be mentioned here to avoid text length. Foreign participants P3 and P4 stated that they had a perception of acceptance at the decision stage of the innovation. P11 acknowledged the continuation of innovation but added that "ethical, regulatory, and governance issues need to be carefully considered to ensure that these technologies are developed responsibly." He mentioned some risks in P14 and P19 but accepted the innovation. When Turkish citizens are examined, P1 answered the question at this stage: "Artificial intelligence is used in the laboratory field, and there is nothing that we do not accept at the moment" and accepted the innovation in the options of accepting or rejecting the innovation. He said that artificial intelligence and technology can be applied to health. P6 expressed the identification and solution of the factors that may lead to misdiagnoses and treatments related to artificial intelligence but expressed his opinion in favor of acceptance versus innovation. P13 explained: "Through artificial intelligence and Metaverse, the 3D version of health screenings will be observed with wearable technologies, and more detailed imaging and examination will be provided" and accepted the continuation of the innovation. P16 touched on determining ethical standards in diagnosis and treatments with artificial intelligence and expressed his opinion that innovation should continue. The fact that other Turkish citizens put forward similar ideas shows that Developing technologies are entering our lives like a flood, and it is impossible to stop them. In this respect, the development of technological innovations is supported, but even at this stage, it has become inevitable to mention risks when making decisions. At this stage, it was understood that all participants accepted the innovation.

4.4 The Implementation Stage

At this stage, the innovation begins to be implemented by the individual who decided to implement it in the previous stage. Efforts to obtain more information about the innovation based on what happened during the implementation phase and the compatibility (integration) with the existing system are studied at this stage (Rogers, 2003). Reinvention actions generally occur during the implementation phase, and reinvention means modifying/changing the innovation by the user during the adoption and implementation process. In this step, Turkish Citizens and Foreign participants shared their opinions and experiences regarding the implementation phase. In this section, the views of some selected participants are included. Among the foreign participants, P2 stated, "I want to understand the main health problems in Türkiye and what role technological developments can play in solving these challenges." P2 is curious about the policies to be followed to solve possible risks. P10 addressed protecting sensitive health data when using AI for the implementation phase. P1, a Turkish citizen, stated: "There is much congestion in the healthcare field. The patient should give advice for minor symptoms before coming to the doctor, evaluate basic things such as lung sounds and ECG, and if there is an abnormal situation, the doctor should make an appointment." She stated that artificial intelligence and technology will contribute to the solution of the density problem in health. P7 explained that it should be implemented: "It should ensure that definitive results are obtained in diagnosing and treating all diseases, especially these rare diseases. Of course, it should be used in all arbitrary and optional aesthetic operations that increase the quality of life and produce positive results".

4.5 The Confirmation Stage

In a control phase that usually comes after each application, the individual judges how successful the innovation is and what it gains/loses. After this stage, the final decision is made regarding the continuation of the innovation. In addition, this stage depends on both the judgments in the individual's inner world and the comments the individual receives from other individuals (intrapersonal, interpersonal) (Rogers, 2003). Some of the participants' different answers are explained here. From foreign participants, P19 explained: "For some people, it is not possible to use artificial medicine or technology for psychological diagnosis; only a psychiatrist near you can diagnose your emotions and try to solve the problems you face." Psychiatrists also stated that its application would not be beneficial but did not have a negative opinion about other fields. P2 stated, "From a general perspective, I see that the application of artificial intelligence and advanced technology in healthcare can significantly improve diagnosis and treatment, but this must be done ethically and safely, respecting patient rights and ensuring privacy." P5, one of the Turkish citizens, said: "It will be beneficial if it is implemented in a controlled and correct manner." P9 stated: "I do not approve of its application in all services." It emphasized that there should be limitations and supported its use and continuity in areas that will not harm the human factor. He noted that there should be limitations and supported its use and continuity in the regions that would not harm the human factor. P13: "While artificial intelligence is actively used in every field in the digitalizing world, its application in the health sector is necessary to reduce human error to minimum levels and implement healthier diagnosis and treatment." Other participants also mentioned their views on practice and being careful about some risks.

5. DISCUSSION

5.1 Theoretical Implications

Social groups that protect and adopt innovation are categorically defined. Early Adopters: The first thing that encounters innovation and is critical in spreading innovation to society. P2, P3, P5, P7, P8, P9, P10, P12, P15, P17, P18 and P20 are in the form of "Persuation Stage. Twelve Turkish citizens and foreign participants are among those who adopted innovation early. The other eight others expressed some risks to implementing innovation in the "Decision Stage. Rogers has expressed the main factors in the spread process: innovation, time, and social system. According to Rogers (2003), time is an important factor in the spread process of innovations. Time: Innovation time is the time to decide on the innovation-decision process. The adoption ratio is proportional to the time of adoption by the members of the social system of innovation. Although the innovation-decision time is not the same for all participants, the fact that early adopters are 60%of those who adopted it, it was considered that most participants adopt innovation in terms of the time factor. In innovation decision processes, Turkish citizens and foreigners know about using artificial intelligence and technologies in health services. Looking at the next step, foreign participants expressed more opinions regarding the risks in the "Persuasion Stage. Foreigners think they need more information about how artificial intelligence and technological progress processes in Türkiye will give results in health. However, Turkish citizens and foreigners explained some of their thoughts and opinions on risks and declared that they perceive all of them in implementing innovation. Participants do not stand against the health services they receive or their possible future health services, artificial weak and technological ways, and applications. It should be considered normal for the participants to have concerns. In other "Implementation" and "Confirmation" stages, it was understood that the participants were not very discrete and not in very different directions. While some participants have anxiety in diagnosis and treatment for the development of artificial intelligence and technologies, some participants have a more comfortable attitude despite the health services applied with artificial intelligence and technology. However, different views were not found in these steps.

The following figure emerged regarding the positive perceptions of all participating patients towards the use of artificial intelligence in healthcare.

Figure 3. Patients' positive perceptions of AI in Health Services (the research model was created according to content analysis results obtained from the interviews with the NVivo program)

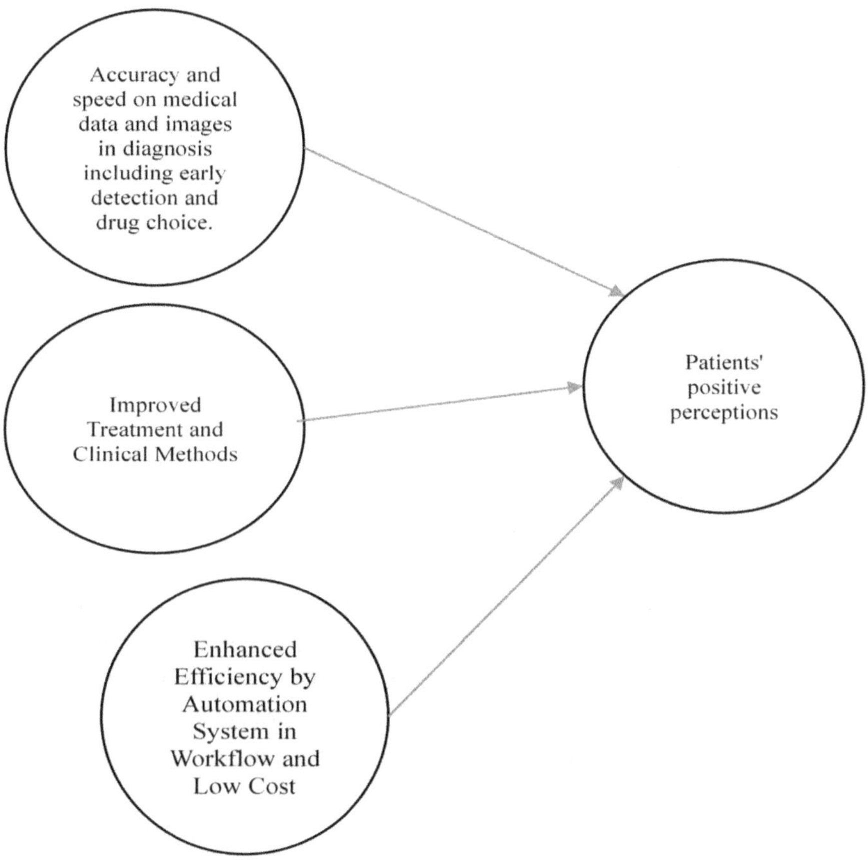

The following figure emerged regarding the negative perceptions of all participating patients towards the use of artificial intelligence in healthcare.

Figure 4. Patients' positive perceptions of AI in health services (the research model was created according to content analysis results obtained from the interviews with the NVivo program)

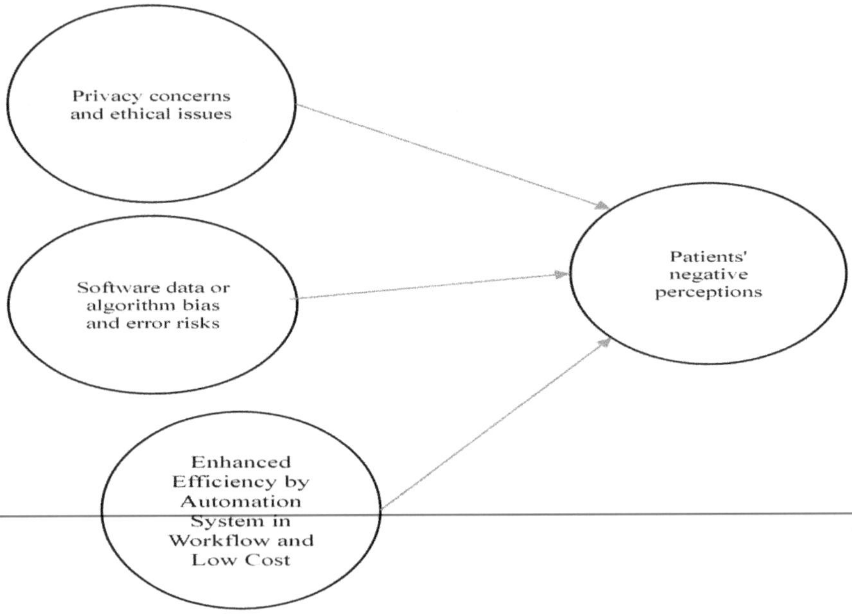

5.2 Practical Implications

Türkiye's health sector developments also contribute to the development of health tourism. In many countries, the use of artificial intelligence and technologies in healthcare is becoming widespread. Regarding Metaverse, it has been observed that it has yet to become widespread in healthcare services in Türkiye. The impact of artificial intelligence on the digital transformation of healthcare services: It is anticipated that it may offer advantages in improving patient outcomes, increasing efficiency, and reducing costs. The results obtained from the interviews showed that medical image analysis, improvement of patient care, drug discovery, and development reveal the advanced diagnostic aspect of artificial intelligence (Jiang et al., 2017; Görener, 2016). Drug discovery is a susceptible issue, with artificial intelligence accelerating the discovery process by designing new drugs, predicting drug interactions, and optimizing clinical trials. Clinical decision support systems and robotic surgery are among the other essential advantages of artificial intelligence in healthcare within the scope of improved treatment. It is possible to carry out administrative tasks and create streamlined workflows in this context with automation systems offered by artificial intelligence. It also has an efficiency-based benefit, such as proactive interventions and resource allocation optimization by predicting disease outbreaks, hospital readmissions, and patient deterioration (Lai et al., 2020). Chatbots for patients can answer patients' questions and provide health information and support; devices and sensors can continuously monitor the patient's vital signs and health measurements, allowing for remote monitoring and early detection of possible health problems, aspects where artificial intelligence can also make a significant contribution to patient traceability is evaluated.

Participants receiving healthcare services have some concerns about artificial intelligence. Due to the continuous development of artificial intelligence technologies and the increase in technological innovations, all of the interview participants accepted the innovation due to the answers they gave in the "Decision Stage." No participant rejected the innovation, but some participants expressed risks and concerns at this acceptance stage. Particularly in the "Implementation" and "Confirmation" stages, participants expressed more constructive suggestions and observations. According to these participants, although artificial intelligence holds great promise in revolutionizing the healthcare industry, its limitations and potential disadvantages must be acknowledged. The issue of data bias is out of the question. AI algorithms learn from the data they are trained on, which can perpetuate existing biases in healthcare data, leading to inaccurate diagnoses, treatment recommendations, and resource allocation. Artificial intelligence: A person can carry out whatever he loads. For example, since data for 2030 has not been entered for health companies, Most participants emphasized ethical concerns. AI systems rely on large amounts of sensitive patient data. This raises concerns about data breaches, unauthorized access, and potential misuse of information (Amisha et al., 2019).

Ethical and legal challenges are likely to arise, as determining who is responsible for errors or biased results caused by AI can be complex. One of the opinions put forward by some participants was job losses. AI technologies could automate some healthcare tasks currently performed by human professionals, potentially leading to job losses in specific industries (Mintz & Brodie, 2019). Participant coded P1, a medical doctor, expressed an opinion that there should be a technological limitation regarding the issue of overconfidence. P1 explained this situation: "Overreliance on artificial intelligence without appropriate clinical judgment can lead to missed diagnoses or inappropriate treatment decisions." While some participants thought that artificial intelligence would contribute positively to costs, others pointed out that developing and implementing robust artificial intelligence systems in healthcare may be expensive and create limitations for resource-limited areas and healthcare systems. Among the disadvantages of artificial intelligence, one of the participants also mentioned that it may cause social problems. P13 explained: "Unequal access to AI-enabled health technologies can worsen existing inequalities and limit access to healthcare for vulnerable populations, while over-reliance on AI can lead to a decrease in human interaction and empathy in healthcare, reducing the patient experience and human touch".

The conclusion is that the development of artificial intelligence and technological opportunities and their application in diagnosis and treatment do not cause any discomfort when evaluated in terms of the diagnosis and treatment received by the participants. Participants are willing to innovate for better-quality healthcare solutions. However, considering the future health services, the participants, especially foreign participants, emphasized that Türkiye's healthcare developments should be carefully monitored. It is considered normal for foreign participants to have benefited from health tourism services in another country, and they may need more information even though they are knowledgeable about the country. The issues that Turkish and foreign participants emphasized the most were the protection of ethical values in artificial intelligence-related treatments and the need not to neglect physical therapy and examinations by doctors. One of the participants' opinions was that artificial intelligence is unsuitable for psychiatry. Other suggestions also revealed the expectation that artificial intelligence would not be preferred in some medical fields.

6. CONCLUSION

Significant developments regarding artificial intelligence and health tourism have recently occurred. Virtual assistants and chat robots, robotic surgery, personalized treatment plans, remote patient monitoring, and prediction of treatment results are seen as some of these developments. Within recent developments, other contributions of artificial intelligence in health tourism include reducing costs, reducing the risk of errors, and increasing patient satisfaction. For the contributions of artificial intelligence in health tourism to be sustainable, investing in artificial intelligence technologies, training expert personnel, and complying with data security and ethical standards are essential.

In the research, which is a qualitative study with a semi-structured interview technique, conducting the study within the framework of twenty participants' perceptions is among the research's limitations Some crucial findings obtained as a result of this study are as follows:

- It was understood that the participants were injured in health tourism and services in Türkiye, and they can observe artificial intelligence and technology and diagnosis and treatments.
- All participants have been found to have good knowledge and awareness levels for using artificial intelligence and technology applications in health services in Türkiye.

- For many reasons, such as the fact that the participants do not have the same level of knowledge of artificial intelligence and technology, they did not benefit from the same health services, and the level of knowledge of the health system in Türkiye is not the same; In the process of "Persuasion Stage," some concerns and concerns about artificial intelligence and technological innovations were expressed. In particular, although foreign participants are aware of the use of artificial intelligence and technology in the Turkish health system, they stated that more data is needed to solve technological risks, and the processes should be followed. Similar concerns are not as much as foreigners are seen in Turkish citizens.
- If there should be no innovation, the participants will not release this service if they need artificial intelligence, technology, and health care for diagnosis and treatment, which constitutes whether it should be terminated.
- In the "Implementation" stage of innovation, the participants also expressed their opinions on some risks and solutions.
- Some have stated that use may not be appropriate in some areas in Confirmation Stage. In one of the examples, the participant indicated that it would not benefit to implement the psychiatric field and that the solution to human psychological problems can only be treated with the human process.

In the research, Turkish citizens and foreign participants benefited from various branches of health services in Türkiye. It has been observed that they benefit from applications containing artificial intelligence and advanced technology in the health services they receive. It was observed that all participants were at a reasonable level in terms of knowledge and awareness of technological innovations in health. Moreover, not all participants opposed the continuation of innovation. However, while some participants expressed their views on the direct application of artificial intelligence and technology in health, some approached the issue more cautiously and suggested prioritizing eliminating risks. The study can be further developed by conducting comparative analysis from different countries.

REFERENCES

Aksu, A., & Bayar, K. (2019). Development of Health Tourism in Türkiye: SWOT Analysis of Antalya Province. *Journal of Tourism Management Research*, *6*(2), 134–154. doi:10.18488/journal.31.2019.62.134.154

Aksu, A. A., Gürsoy, D., & Aksu, G. M. (2016). *Sağlık turizminde keşfedildikçe büyüyen ülke Türkiye: Pazar analizi ve izlenmesi gereken stratejiler*. Türkiye Turizm Yatırımcıları Derneği.

Amisha, M. P., Pathania, M., & Rathaur, V. K. (2019). Overview of artificial intelligence in medicine. *Journal of Family Medicine and Primary Care*, *8*(7), 2328–2331. doi:10.4103/jfmpc.jfmpc_440_19 PMID:31463251

Aslan, I., Çınar, O., & Özen, Ü. (2014). Developing strategies for the future of Healthcare in Türkiye by benchmarking and SWOT analysis. *Procedia: Social and Behavioral Sciences*, *150*, 230–240. doi:10.1016/j.sbspro.2014.09.043

Barca, M., Akdeve, E., Balay, İ. G. (2013). Türkiye sağlık turizm sektörünün analizi ve strateji önerileri. *İşletme araştırmaları dergisi*, *5*(3), 64-92.

Bassil, C., Saleh, A. S., & Anwar, S. (2019). Terrorism and tourism demand: A case study of Lebanon, Türkiye, and Israel. *Current Issues in Tourism*, *22*(1), 50–70. doi:10.1080/13683500.2017.1397609

Berg, B., & Lune, H. (2015). *Sosyal bilimlerde nitel araştırma yöntemleri*. Eğitim Yayınevi.

Bulte, C. (2000). New Product Diffusion Acceleration: Measurement and Analysis. *Marketing Science*, *19*(4), 366–380. doi:10.1287/mksc.19.4.366.11795

Bushel, R., & Sheldon, P. J. (Eds.). (2009). *Wellness and Tourism: Mind, Body, Spirit, Place*. Cognizant Communication Corporation.

Buzcu, Z., & Birdir, K. (2019). Türkiye'de Medikal Turizm İncelemesi: Özel Hastanelerde Bir Çalışma. *Gaziantep Üniversitesi Sosyal Bilimler Dergisi*, *18*(1), 311–327.

Carrera, P., & Lunt, N. (2010). A European perspective on Medical Tourism: The need for a knowledge base. *International Journal of Health Services*, *40*(3), 469–484. doi:10.2190/HS.40.3.e PMID:20799671

Carrera, P. M., & Bridges, J. F. P. (2006). Globalization and Healthcare: Understanding health and medical tourism. *Expert Review of Pharmacoeconomics & Outcomes Research*, *6*(4), 447–454. doi:10.1586/14737167.6.4.447 PMID:20528514

Charlie, M., Degrande, A., Villamor, G., Van Damme, P., Tchoundjeu, Z., & Sygnola, T. (2015). Factors affecting the adoption of agricultural innovation: The case of a Ricinodendron heudelotii kernel extraction machine in southern Cameroon. *Agroforestry Systems*, *89*(5), 799–811. Advance online publication. doi:10.1007/s10457-015-9813-y

Clarke, A. (2010). Wellness and tourism: Mind, body, spirit. *Annals of Tourism Research*, *37*(1), 276–278. doi:10.1016/j.annals.2009.10.004

Cohen, E. (2012). Medical Travel and the Quality-of-Life (Uysal, M., Perdue, R.R., Sirgy, M.J., Eds.), Handbook of Tourism and Quality-of-Life Research: Enhancing the Lives of Tourists and Residents of Host Communities. Springer Science+Business Media.

Connell, J. (2006). Medical tourism: Sea, sun, sand, and surgery. *Tourism Management*, *27*(6), 1093–1100. doi:10.1016/j.tourman.2005.11.005

Connell, J. (2013). Contemporary medical tourism: Conceptualisation, culture, and commodification. *Tourism Management*, *34*(February), 1–13. doi:10.1016/j.tourman.2012.05.009

Creswell, J. W. (2014). *Research design. Qualitative, quantitative, and mixed methods approaches* (3rd ed.). Sage Publications.

Dalkiran, G. B. (2023). Health Tourism Components and Intermediary Institutions as Supplier Businesses in the Context of Health 4.0. In *Health 4.0 and Medical Supply Chain* (pp. 147–162). Springer Nature Singapore. doi:10.1007/978-981-99-1818-8_12

Damanpouer, F. (1991). Organizational Innovation: A Meta-Analysis of the Effects of Determinants and Moderators. *Academy of Management Journal*, *34*(3), 555–590. doi:10.2307/256406

Daştan, İ. (2014). Türkiye'de Sağlık Turizmi: Türkiye ve Özelinde İzmir'de Sağlık Turizminin Mevcut Durum Analizi ve Strateji Önerileri. *Mehmet Akif Ersoy Üniversitesi Sosyal Bilimler Enstitüsü Dergisi*, *6*(10), 143–163.

Deloitte. (2008) *Medical Tourism. Consumers in Search of Value*. Deloitte Center for Health Solutions. https://www.globalwellnesssummit.com/wp-content/uploads/Industry-Research/ Americas/2008-deloitte-medical-tourism-consumers.pdf (Access: 2 Jan 2024).

Downs, G. W., & Mohr, L. B. (1976). Conceptual Issues in the Study of Innovations. *Administrative Science Quarterly*, *21*(4), 700–714. doi:10.2307/2391725

Eissler, L. A., & Casken, J. (2013). Seeking health care through international medical tourism. *Journal of Nursing Scholarship*, *45*(2), 177–184. doi:10.1111/jnu.12014 PMID:23489601

Filieri, R., D'Amico, E., Destefanis, A., Paolucci, E., & Raguseo, E. (2021). Artificial intelligence (A.I.) for tourism: A European-based study on successful A.I. tourism start-ups. *International Journal of Contemporary Hospitality Management*, *33*(11), 4099–4125. doi:10.1108/IJCHM-02-2021-0220

Glynda, J. D., Bernie, G., & Leanne, M. C. (2014). Integrating mobile devices into nursing curricula: Opportunities for implementation using Rogers' Diffusion of Innovation model. *Nurse Education Today*, *34* (5), 775-782. doi:10.1016/j.nedt.2013.10.021

Goodrich, J. N., & Goodrich, G. E. (1987). Healthcare tourism: An exploratory study. *Tourism Management*, *8*(3), 217–222. doi:10.1016/0261-5177(87)90053-7

Görener, A. A. (2016). SWOT-AHP approach for assessment of the medical tourism sector in Türkiye. *Alphanumeric Journal*, *4*(2), 159–170. doi:10.17093/alphanumeric.277740

Gray, D. E. (2013). *Doing Research in The Real World* (3rd ed.). Sage.

Hafeez, K., Keary, K., & Hanneman, R. (2006). e Business Capabilities Model Validation and Comparison between Adopter and Non-Adopter to e-Business Companies in the U.K. *Journal of Manufacturing Technology Management*, *17*(6), 806–828. doi:10.1108/17410380610678819

Hafizullah, D., & Kirti, K. (2023). Smart Healthcare System (SHS): Medical Tourism Delivering, Consumption, and Elevating Tool in the Ages of Smart Technologies. *Tourism Planning & Development*, *20*(3), 397–415. doi:10.1080/21568316.2022.2109206

Hassan, V., & Bellos, G. (2022). COVID-19: Reshaping Medical Tourism through Artificial Intelligence (A.I.) and Robotics. *Athens Journal of Tourism*, *9*(2), 77–98. doi:10.30958/ajt.9-2-2

Heung, V. C. S., Küçükusta, D., & Song, H. (2011). Medical tourism development in Hong Kong: An assessment of the barriers. *Tourism Management*, *32*(5), 995–1005. doi:10.1016/j.tourman.2010.08.012

Hjalager, A. M. (2009). Innovations in travel medicine and the progress of Tourism-Selected narratives. *Technovation*, *29*(9), 596–601. doi:10.1016/j.technovation.2009.05.012

Hjalager, A. M., & Flagestad, A. (2012). Innovations in well-being tourism in the Nordic countries. *Current Issues in Tourism*, *15*(8), 725–740. doi:10.1080/13683500.2011.629720

Hofer, S., Honegger, F., & Hubeli, J. (2012). Health tourism: Definition focused on the Swiss market and conceptualization of health(i)ness. *Journal of Health Organization and Management*, *26*(1), 60–80. doi:10.1108/14777261211211098 PMID:22524099

Hofer, S., & Stampfli, K. (2008). *Rehabilitationskliniken – Wohin führt die Zukunft*, working paper Zurich University of Applied Sciences, Zürich.

Ishita, S., & Sachin, S. (2022). Blockchain Enabled Biometric Security in Intemet-of-Medical-Things (IoMT) Devices. *International Conference on Augmented Intelligence and Sustainable Systems (ICAISS)*, (pp. 971-979). IEEE.

Jiang, F., Jiang, Y., Zhi, H., Dong, Y., Li, H., Ma, S., Wang, Y., Dong, Q., Shen, H., & Wang, Y. (2017). Artificial Intelligence in Healthcare: Past, Present and Future. *Stroke and Vascular Neurology*, *2*(4), 230–243. doi:10.1136/svn-2017-000101 PMID:29507784

Johnston, R., Crooks, V. A., Adams, K., Snyder, J., & Kingsbury, P. (2011). An industry perspective on Canadian patients' involvement in Medical Tourism: Implications for public health. *BMC Public Health*, *11*(1), 416. doi:10.1186/1471-2458-11-416 PMID:21627830

Kangping, W., Haiyan, K., Naipeng, B., Honggen, X., Xuejie, Q., & Jinghan, L. (2022). A.I. in health tourism: Developing a measurement scale. *Asia Pacific Journal of Tourism Research*, *27*(9), 954–966. doi:10.1080/10941665.2022.2142620

Kılavuz, E. (2018). Medical tourism competition: The case of Türkiye. *International Journal of Health Management and Tourism*, *3*(1), 42–58. doi:10.31201/ijhmt.372364

Kiss, K. (2015). The challenges of developing health tourism in the Balkans. *Tourism: An International Interdisciplinary Journal*, *63*(1), 97–110.

Lai, M. C., Brian, M., & Mamzer, M. F. (2020). Perceptions of artificial intelligence in Healthcare: Findings from a qualitative survey study among actors in France. *Journal of Translational Medicine*, *18*(1), 14. doi:10.1186/s12967-019-02204-y PMID:31918710

Laws, E. (1996). Health tourism: A business opportunity approach. In S. Clift & S. J. Page (Eds.), *Health and the international tourist* (pp. 199–214). Routledge.

Lixia, Z., & Zhongli, S. (2019). The Application of Artificial Intelligence Technology in the Tourism Industry of Jinan. *Journal of Physics: Conference Series*, *1302*(3), 032005. doi:10.1088/1742-6596/1302/3/032005

Lu, H., Li, Y., Chen, M., Kim, H., & Serikawa, S. (2018). Brain intelligence: Go beyond artificial intelligence. *Mobile Networks and Applications*, *23*(2), 368–375. doi:10.1007/s11036-017-0932-8

Lüthje, C., & Herstatt, C. (2004). The Lead User Method: An Outline of Empirical Findings and Issues for Future Research. *R & D Management*, *35*(5), 553–568. doi:10.1111/j.1467-9310.2004.00362.x

Ma, H., Chiu, Y. H., Tian, X., Zhang, J., & Guo, Q. (2020). Safety or travel: Which is more important? The impact of disaster events on tourism. *Sustainability (Basel)*, *12*(7), 3038. doi:10.3390/su12073038

Mintz, Y., & Brodie, R. (2019). Introduction to artificial intelligence in medicine. *Minimally Invasive Therapy & Allied Technologies*, *28*(2), 73–81. doi:10.1080/13645706.2019.1575882 PMID:30810430

Musa, G., Thirumoorthi, T., & Doshi, D. (2012). Travel behavior among inbound medical tourists in Kuala Lumpur. *Current Issues in Tourism*, *15*(6), 525–543. doi:10.1080/13683500.2011.626847

Padilla-Meléndez, A., & Del-Águila-Obra, A. R. (2016). Health Tourism. *Conceptual Framework and Insights from the Case of a Spanish Mature Destination Tourism & Management Studies*, *12*(1), 86–96.

Pan, T.-Z., & Chen, W. C. (2014). Chinese medical tourists- their perceptions of Taiwan. *Tourism Management*, *44*, 108–112. doi:10.1016/j.tourman.2014.02.008

Peric, M., & Vanja, V. (2021). Tourism Getting Back to Life after COVID-19: Can Artificial Intelligence Help? *Societies (Basel, Switzerland)*, *11*(4), 115. Advance online publication. doi:10.3390/soc11040115

Priporas, C. V., & Mylona, I. (2008). Mobile services: Potentiality of SMS as a new business communication tool in attracting consumers. *International Journal of Mobile Communications*, *6*(4), 456–466. doi:10.1504/IJMC.2008.018053

Punch, K. F. (2005). *Sosyal araştırmalara giriş*. Siyasal Kitabevi.

Reddy, S., York, V. K., & Brannon, L. A. (2010). Travel for Treatment: Students' Perspective on Medical Tourism. *International Journal of Tourism Research*, *12*(5), 510–522. doi:10.1002/jtr.769

Rogers, E. (1995). *Diffusion of innovations*. The Free Press.

Rogers, E. M. (1962). *Diffusion of innovations* (1st ed.). Free Press.

Rogers, E. M. (1983). *Diffusion of innovations* (3rd ed.). Free Press.

Rogers, E. M. (2003). *Diffusion of innovations* (5th ed.). Free Press.

Saadatnia, F., & Mehregan, M. R. (2014). Determining and Prioritizing Factors Affecting Customers Attraction of Medical Tourism from the Perspective of Arabic Countries (Case Study: Iran-Mashhad Razavi Hospital). *International Journal of Marketing Studies*, *6*(3), 155–162. doi:10.5539/ijms.v6n3p155

Sakipov, N., & Vikhnin, O. (2021). The Modern Symbolic Artificial Intelligence and Its Use in Medical Tourism. *2021 IEEE 15th International Conference on Application of Information and Communication Technologies (AICT)*, (pp. 1-6). IEEE. 10.1109/AICT52784.2021.9620321

Seligman, L. (2006). Sensemaking throughout the adoption and the innovation-decision process. *European Journal of Innovation Management*, *9*(1), 108–120. doi:10.1108/14601060610640050

Sharma, S., Chauhan, Y., & Tyagi, R. (2023). *Artificial Intelligence based Applications in Medical Tourism. 14th International Conference on Computing Communication and Networking Technologies (ICCCNT)*, Delhi, India. 10.1109/ICCCNT56998.2023.10308020

Siddharth, G., & Sachin, S. (2022). Augmented Reality in Medical Education: An Industry 4.0 Enabled Innovative Approach. *3rd International Conference on Smart Electronics and Communication (ICOSEC)*, (pp. 29-33). IEEE.

Sigrist, S. (2006a). Zukunftsperspektiven des Gesundheitsmarktes. Gottlieb Duttweiler Institut /GDI im Auftrag in oder of Eidgenössisches Departement des Inneren, Bern.

Sigrist, S. (2006b). *Health Horizons – Guide zu den neuen Gesundheitsmärkten/ Guide to the New Healthmarkets, No. 20*. Gottlieb Duttweiler Institut/GDI.

Singh, N. (2013). Exploring the factors influencing the travel motivations of U.S. medical tourists. *Current Issues in Tourism*, *16*(5), 436–454. doi:10.1080/13683500.2012.695341

Smith, M., & Kelly, C. (2006). Wellness tourism. *Tourism Recreation Research*, *31*(1), 1–4. doi:10.1080/02508281.2006.11081241

Smith, M., & Puczko, L. (2008). *Health and wellness tourism*. Butterworth-Heinemann. doi:10.4324/9780080942032

Sourav, S., Sachin, S., Shuchi, B., & Seshadri, M. (2023). Industry 4.0 Internet of Medical Things Enabled Cost Effective Secure Smart Patient Care Medicine Pouch in New Horizons for Industry 4.0 in Modern Business. Springer International Publishing.

Subasinghe, M., Magalage, D., Amadoru, N., Amarathunga, L., Bhanupriya, N., & Wijekoon, J. L. (2020). Effectiveness of artificial intelligence, decentralized and distributed systems for prediction and secure channeling for Medical Tourism. *11th IEEE Annual Information Technology, Electronics and Mobile Communication Conference (IEMCON)*, (pp. 314-319). IEEE. 10.1109/IEMCON51383.2020.9284898

Temizkan, V., Yetgin, M. A., & Yılmaz, K. (2022). Motivations of Retailers Accepting Cryptocurrency Payments and Their Implications on Retail Customer Experience. *Cumhuriyet University Journal of Economics and Administrative Sciences*, *23*(1), 25–48. doi:10.37880/cumuiibf.987656

Üstün, U. & ve Demir Uslu, Y. X. (2022). Türkiye'nin Sağlık Turizminde Tercih Edilme Nedenleri Üzerine Bir Çalışma: Medikal Turizm Endeksi. *Avrupa Bilim ve Teknoloji Dergisi*, (33), 344–353.

Voigt, C. (2014). Wellness tourism: A destination perspective. London: Routledge.

World Bank. (2023). *Data: GDP Growth (annual %).* World Bank. https://data.worldbank.org/indicator/NY.GDP.MKTP.KD.ZG

Ye, B. H., Qiu, H. Z., & Yuen, P. P. (2011). Motivations and experiences of Mainland Chinese medical tourists in Hong Kong. *Tourism Management*, *32*(5), 1125–1127. doi:10.1016/j.tourman.2010.09.018

Yirik, Ş., Ekici, R., & Baltacı, F. (2015). *Türkiye ve Dünyada sağlık turizmi (Medikal Turizm)*. Detay Yayıncılık.

Yu, J. Y., & Ko, T. G. (2012). A cross-cultural study of perceptions of medical tourism among Chinese, Japanese, and Korean tourists in Korea. *Tourism Management*, *33*(1), 80–88. doi:10.1016/j.tourman.2011.02.002

Zhang, L., Li, N., & Liu, M. (2012). Smart tourism basic concepts and theoretical system. *Journal of Tourism*, (5), 66–73.

Chapter 5
The Strategic Efficacy of Artificial Intelligence (AI) in Medical Tourism

Olabode Gbobaniyi
https://orcid.org/0000-0002-0722-0335
Global Banking School, UK

Daniela Tincani
https://orcid.org/0000-0002-5682-1372
Global Banking School, Malta

Peter Emelone
Global Banking School, UK

ABSTRACT

AI's integration and adoption in the sector have evolved to be a game-changer through operational revolutionisation regarding accessibility to advanced diagnosis and treatments, reduced waiting times, and cost savings. This chapter explores the strategic efficacy of AI in the context of medical tourism. Using the term "strategic efficacy," the authors encompass the concept of efficiency and effectiveness of AI in achieving a strategic outcome in medical tourism. The authors' purviews are that is important to ensure that an AI strategy in medical tourism not only looks good on paper but also continues to produce high success for the global practice. In this chapter, the authors discuss AI's emergence in the medical tourism industry, the strategic efficacy of AI in medical tourism, the categories of AI-system devices used in medical tourism, and the AI-system devices. Also discussed are AI systems applications to some major diseases in medical tourism.

DOI: 10.4018/979-8-3693-2248-2.ch005

1. INTRODUCTION

Medical tourism is a global phenomenon highly dependent on innovation and technology that has continued to develop (Subasinghe et al., 2020). Brazil, China, Costa Rica, India, Iran, Mexico, Panama, Singapore, South Africa, Taiwan, and Thailand are among the popular travel destinations for medical tourism (Lo, 2020). Travel firms and websites specialising in medical tourism sometimes promote package arrangements that portray recovery following surgery as a holiday. Offerings include psychiatric therapies and procedures like gender-reassignment surgeries that could be socially or culturally inappropriate and consequently inaccessible in other nations, as well as cosmetic, cardiac, eye, dental, and orthopaedic surgery (Rogers, 2023). Patients Beyond Borders indicated that the global medical tourism market was worth between \$65 and \$87.5bn, but by 2027, it is expected to grow to about \$207bn (Woodman, 2019). A wide range of medical procedures from cosmetic surgery and dentistry to cardiovascular procedures and experimental cancer treatments, have been identified to inspire medical tourism (Woodman, 2015). However, following the impact of COVID-19 in 2020, the United Nations World Tourism Organisation (UNWTO) estimated that the medical tourism sector had a 60 to 80 per cent decline, thus naming COVID-19 as the worst crisis that had ever confronted the industry (UNWTO, 2020a, 2020b). With the global economies crawling out from the devastating impact of COVID-19, the health and tourism industries in many affected countries have also increasingly continued to promote investments in technological advancements to the medical tourism sector in their attempt to increase patients' patronage of their country (Tileva, 2021). Simply as it is named, Ghassemi et al. (2020) described medical tourism as a process whereby patients travel outside of their country of residence to another country for medical care.

It is also important to emphasise that medical tourism is related to the broader notion of health tourism which, in some countries, has longstanding historical antecedents of spa towns, coastal localities, and other therapeutic landscapes (e.g., Greece, Turkey, Croatia) (Meghani, 2011). Some literature has considered both health and medical tourism as a combined phenomenon but with different emphases. For instance, Carrera and Bridges (2006) define health tourism as the organised travel outside one's local environment for the maintenance, enhancement, or restoration of an individual's well-being in mind and body. However, it can be said that Carrera and Bridges' definition encompasses medical tourism which is delimited to organised travel outside one's natural healthcare jurisdiction for the enhancement or restoration of the individual's health through medical intervention (Connel & Page, 2019; Lunt, et al., 2011).

Like, medical tourism, Artificial Intelligence (AI) has emerged as a potent and disruptive force, reshaping the methodologies and outcomes across diverse fields, particularly, in healthcare. AI techniques for intelligent mobile computing in healthcare have opened new opportunities not only for healthcare systems and the industry but also for the medical tourism industry. AI has many different definitions, and it is commonly defined as a scientific discipline in which machines can function appropriately and with foresight in their environment (Nilsson, 2009). AI approaches have recently had a significant impact on healthcare, and the debate on whether AI physicians will eventually take the role of human doctors in the future has continued to increase. It is believed that medical professionals will not be supplanted by robots in the future, but AI can undoubtedly help medical professionals make wiser clinical judgements, or even take over human discretion in specific functional domains in the medical area (e.g., radiology) (Jiang, et al., 2017). As a result, we have been able to see that the integration of AI into healthcare has ushered in a transformative era of advanced medical methodologies, discoveries, and innovation. Moreover, the recent effective uses of AI in healthcare have been made feasible by the expanding availability of healthcare data and the quick development of big data analysis techniques. Vast amounts of data including clinically important information that can help clinical decision-making can now be unlocked by sophisticated AI approaches guided by pertinent clinical queries are now incorporated into the medical tourism industry (Kolker, et al., 2016; Dilsizian & Siegel, 2014). In their study, Rajpukar et al. (2017) argued that AI has exceeded human capabilities and further indicated that with one month of system development, researchers at Stanford in 2016 were able to use AI to diagnose 14 types of medical conditions using frontal-view X-ray images, thus exceeding the human diagnostic accuracy for pneumonia. Additionally, researchers using AI-based approaches have been found to have used it for the early detection of dyslexia, diagnosis of autism spectrum disorder and attention deficit hyperactivity disorder (ADHD) (Boccanfuso et al., 2016; Drigas and Ioannidou, 2012).

Thus, the integration of AI into healthcare has ushered in a transformative era of advanced medical methodologies, discoveries, and innovation. For medical tourism, the integration of AI has evolved to be a game-changer through operational revolutionisation in terms of accessibility to advanced diagnosis and treatments, reduced waiting times, and cost savings. The purpose of this chapter therefore is to explore the strategic efficacy of artificial intelligence (AI) in the context of medical tourism. Using the term "strategic efficacy", the authors encompass the concept of efficiency and effectiveness of AI in achieving a strategic outcome in medical tourism. The authors' purviews are that is important to ensure that an AI strategy in medical tourism not only looks good on paper but also continues to produce high success for the global practice.

2. AI'S EMERGENCE IN THE MEDICAL TOURISM INDUSTRY

Advancements in AI technologies have shifted patient-practitioner "i.e., provider" diagnoses and treatments from being home-to-home country-based to home-to-foreign country-based (i.e., medical tourism). This is because AI can "learn" features from a vast amount of healthcare data collected from multiple countries using complex algorithms, and use the knowledge gained to improve clinical diagnoses and treatments (Jha & Topol, 2016). AI-powered systems can assist medical tourism physicians by providing up-to-date medical information from journals, textbooks, and clinical practices to proffer real-time inferences for health risk alerts and health outcome predictions before patients travel for any treatments (Darcy, et al., 2016).

The UNWTO data showed that in 2020 the medical tourism sector suffered a 60 to 80 per cent decline due to the COVID-19 pandemic, however, this was said to be a major factor that influenced the double increase in corporate funding for AI-powered healthcare in 2021 (PwC, 2021; UNWTO, 2020b). The increased funding has been identified to fund research and development into the training of AI systems on data generated from clinical activities such as screening, diagnosis, treatment assignment, and so forth before they can be used in healthcare applications. This is done to enable the systems to identify patterns among similar subject groups and relationships between subject characteristics and outcomes of interest. The clinical data inputted into the AI systems can take several forms, such as demographics, electronic records from medical equipment, physician notes, physical examinations, and clinical laboratory imaging that can help physicians or medical practitioners in the screening, diagnosis, and treatment planning of medical tourist from their home countries (Jiang, et al., 2017). For instance, radiologists in the medical tourism sector have continued to increasingly adopt AI technologies when analysing diagnostic images that contain vast data information (Gillies, et al., 2016). The funding and technological growth of AI in the sector has further spiralled with the increased transition from pre-consultation phone calls and emails between physicians and prospective medical tourists towards full video consultations, with both the patients and physicians in their respective home countries (Dryglas and Lubowiecki-Vikuk, 2019; Wang, et al., 2022). Thus, maintaining and ensuring patients' desire towards safety and convenience remain uncompromised. This has driven the adoption of virtual care and precision healthcare to a level that disrupts the traditional care delivery system (Nagar, et al., 2021). For instance, Sri Ramakrishna Hospital in Coimbatore is said to have positioned itself at the heart of medical tourism in southern India, incorporating AI into their established tele-diagnostic and teleradiology programme using tech from Tata's Global division, allowing international physicians and patients to request virtual diagnoses and second opinions, not only helping to reduce risk and costs for those considering travelling for care but also the access to specialised medical procedures and technologies (Lo, 2020). Additionally, advanced technologies have enabled remote physician-to-physician consultation for in-country diagnoses and treatments (Woodman, 2019). This virtual partnership builds in-country medical capacity and helps patients who do need to travel for a procedure better prepare for treatment and access greater support when they return to their home country. Recent literature in the medical tourism sector indicates that Turkey has emerged as a leading destination as the JCI has accredited more than 45 Turkish hospitals and ranks second worldwide for the number of hospitals receiving JCI (Joint Commission International) accreditation (Spencer Jolliffe, 2023). Additionally, the United Arab Emirates (UAE) is considered a leader in AI in medical tourism as the country has introduced its first AI policy and established its Artificial Intelligence Strategy (Spencer Jolliffe, 2023).

2.1 Strategic Efficacy of AI in Medical Tourism

1. **Improved treatment strategies and patient matching** - AI's ability to analyse large amounts of patient data, including their medical history, genetic information, diagnostic reports, unique needs and preferences, and treatment goals will not only ensure effective and efficient treatment strategies but also match the patient with the most suitable professionals for the required medical field. This provides an individualised approach that enhances treatment outcomes and patient satisfaction.
2. **AI-powered virtual assistants** – AI's sophisticated machine learning abilities will be able to assist and support medical tourists with their queries and concerns that may range from travel logistics to the foreign country to procedural information about their health issues (Nagar, et al., 2021). These AI-powered interfaces can assist with real-time treatment progress reports, appointment scheduling with human professionals either virtually or face-to-face, and travel arrangements (Yener & Köse, 2023). Medical tourism providers may increase operational efficiency and create a smooth experience for patients, irrespective of language or time zone difficulties, by automating regular processes.
3. **Risk assessment and predictive analytics** – The ability of AI's predictive analytical powers to examine enormous volumes of patient data to identify trends and anticipate potential dangers provides for the possibility of risk aversion through early treatment or placement of corrective measures. Medical tourism providers, using AI algorithms that leverage past data can effectively anticipate difficulties, patient outcomes, and post-treatment care requirements, thus allowing them to proactively address any prospective concerns. By taking a proactive approach, medical tourism providers and professionals will allocate

resources efficiently, reduce the likelihood of unfavourable outcomes and enhance patient safety.

4. **Virtual consultations and tele-healthcare** - Medical tourists can communicate with medical specialists remotely thanks to the significant advancements in artificial intelligence (AI) technology in tele-healthcare services. Without having to travel physically, patients may seek professional advice, go about treatment alternatives, and seek other opinions through virtual consultations. Medical pictures like X-rays and MRIs can be analysed by AI-powered diagnostic tools, which can provide insightful information to enable remote consultations. AI and tele-healthcare integration not only increase access to high-quality healthcare treatments but also provide efficient and effective diagnostic and treatment for those who live in rural places or have mobility issues (Subasinghe, et al., 2020).
5. **Post-treatment care and continuous monitoring** – As discussed above, medical tourism providers may now continually monitor patient health metrics thanks to AI-driven wearables and remote monitoring systems. Artificial intelligence (AI) systems can identify early warning signals of problems and initiate relevant therapies by gathering real-time data on vital signs, medication adherence, and recovery progress (Tileva, 2021). To guarantee the best long-term results for medical tourists, AI-powered post-treatment care systems may also offer individualised counsel, such as medication reminders, exercise regimens, and lifestyle suggestions.

2.2 Categories of AI-System Devices Used in Medical Tourism

AI-system devices in medical tourism are of two categories with the first being those based on the machine learning (ML) technique. Devices that use ML techniques can analyse structured data such as imaging, genetics, and electrophysiological (EP) data (i.e., data that contain unstructured narrative texts, "e.g., clinical notes" that are not directly analysable). In medical applications, ML techniques aim to group characteristics of patients together or estimate the likelihood of specific illness outcomes (Darcy, et al., 2016). There is a considerable increase in the deployment of devices based on ML techniques in the sector as they can eliminate medical tourists having to travel overseas for basic consultations, and they will only be required to visit medical facilities abroad, when necessary, thus enabling precision treatment plans and care for patients (Psiha & Vlamos, 2017).

The second category of AI-system devices in the sector are those that make use of natural language processing (NLP) methods. These devices use the NLP method to extract information from unstructured data such as clinical notes and medical journals to supplement and enrich structured medical data (Murff, et al., 2011). As a result, NLP methods are targeted at turning texts into machine-readable structured data that ML techniques can analyse. Examples of AI-powered/system devices based on the ML and NLP techniques in healthcare and medical tourism are the current 5G medical technology developments that enhance the patterns of production and consumption of health tourism services using smart wearable (e.g., clothes that measure heart rate, blood pressure, body temperature, skin moisture, etc.) and active device location tracking. This has been acknowledged as useful for health tourism, as it allows continuous communication and data updates through cloud computing between the patients in their home country and the healthcare professionals abroad where the patient's ongoing health data are continually shared, monitored, and reviewed by an AI and should any concerns arise, the human health practitioner is notified to engage with the patient overseas immediately (Hassan & Noaman, 2017).

2.2.1 AI-System Devices: Machine Learning "Classical and Deep Learning" (ML) and NLP (Nature Language Processing)

Here, we review the three categories of AI tools (or methods) that have proven helpful for practitioners and patients in the medical tourism industry.

2.2.1.1 Classical ML

ML builds algorithms for analysing data to extract features. Algorithms using ML can be fed patient "traits" and occasionally relevant medical outcomes. Patient traits include baseline data, such as age, gender, disease history and disease-specific data (i.e., diagnostic imaging, EP test, physical examination results, gene expressions, clinical symptoms, and medication), while medical outcomes include quantitative disease levels (i.e., tumour sizes) and patient's survival times. ML algorithms used in the healthcare and tourism sectors are of two major categories.

The first is unsupervised learning, which is suitable for feature extraction, while the second, supervised learning is suitable for predictive modelling through the building of relationships between the patient's traits "inputs" and the outcome of interest "outputs". This is the reason healthcare practitioners, in general, and those engaged in medical tourism, often use supervised learning because it provides clinically relevant results that high-level accurate answers for their patients.

2.2.1.2 Deep Learning ML

This is a modern extension of classical machine learning or the neural network technique. The deep learning ML networks are based on the rapid development of modern computing of neural networks with many layers that are infeasible for the classical ML networks (Amin, et al., 2021). As a result, deep learning ML is capable of exploring more complex non-linear patterns in data because they use more hidden layers of neural networks for their algorithm to handle complex data with various structures (Rathore & Mannepalli, 2021). Another fundamental reason for the popularity of deep learning is because of the analytic ability to identify patterns in complex and high volumes of global healthcare-related data (Mittal & Hasija, 2020).

The relevant literature in healthcare tourism indicates that the four commonly used AI deep learning tools are the convolution neural network (CNN), recurrent neural network, deep neural network, and the deep belief neural network (Kumar & Mahajan, 2020; Zhou, et al., 2021). CNN is said to be the most popular deep learning tool or algorithm used for diagnosis in healthcare tourism because compared to the classical ML algorithm that can only analyse a small number of data, it can analyse high dimensional data (i.e., data with many traits) (Mbunge & Batani, 2023). Currently, many software packages for healthcare such as Caffe from Berkeley AI Research (Liu, et al., 2018), CNTK from Microsoft (Seide & Agarwal, 2016), and TensorFlow from Google (Abadi, et al., 2016) are now able to implement the CNN. This has influenced elites from developing countries to travel to countries such as the US for medical tourism for certain disease diagnoses such as congenital cataracts through learning the ocular images (Lo, 2020; Long, et al., 2017). Also, due to the high proportion of correctly predicted malignant lesions (i.e., sensitivity) and benign lesions (i.e., specificity) of over 90%, the CNN has been introduced for many medical tourism destinations to identify skin cancer from clinical images (Esteva, et al., 2017). Lastly, another key diagnosis the CNN is popular for in the medical tourism industry is the detection of referable diabetic retinopathy through retinal fundus photographs (Gulshan, et al., 2016).

2.2.1.3 Natural Language Processing (NLP).

The NLP is a tool used for extracting useful information from the narrative text to assist in decision-making because a large percentage of clinical information is in the form of narrative text (i.e., clinical laboratory reports, physical examinations, operative notes, and discharge summaries) (Zhou, et al., 2022). The NLP pipeline, according to relevant literature, comprises two components. The first component is text processing, which enables the NLP to identify a series of disease-relevant keywords in the clinical notes based on historical databases (Lareyre, et al., 2023; Tariq, et al., 2022). Thereafter, a subset of the keywords is selected by examining their effects on the classification of the normal and abnormal cases. The validated keywords then enter and enrich the structured data to support clinical decision-making on diagnosis, treatment arrangements, and monitoring adverse effects (Omankwu, et al., 2023). For instance, Fiszman, et al. showed that using the NLP for chest X-ray reports would assist the antibiotic assistance system to alert physicians of the possible need for anti-infective therapy (2000). Likewise, Miller et al. indicated to have used the NLP to monitor laboratory-based side effects (2017) automatically. On disease diagnosis, through implementing the NLP on clinical notes, Castro et al. identified 14 cerebral aneurysms disease-associated variables (2017). On their part, Afzal *et al* achieved over 90% accuracy using the implementation of the NLP to extract the peripheral arterial disease-related keywords from narrative clinical notes that were then used to classify the normal and the patients with peripheral arterial disease (2017).

2.2.2 AI Systems Application to Diseases in Medical Tourism

2.2.2.1 Stroke

Stroke is a common and frequently occurring disease that affects more than 500 million people worldwide. It is classified as both a cardiovascular and neurological disease and is caused by the abrupt interruption of constant blood flow to the brain, causing a neurological function (Mouridsen, et al., 2020). As a result, it is indicated that stroke primarily manifests as a neurological condition due to its impact on the brain's blood supply and subsequent neurological consequences (Zelenak, et al., 2021). There is an increase in the research and literature from diverse fields, such as healthcare and medical tourism, on the focus of AI techniques on stroke. Particularly, based on research data shows that stroke is the leading cause of death in many countries including China and North America (Jiang et al., 2017). Additionally, the disease is identified to have cost over US$689 billion in medical expenses across the world, causing a heavy burden to countries and families (Elbagoury, et al., 2023; Saenger & Christenson, 2010).

Stroke have been identified to be caused by a thrombus in the cerebral infarction vessel. The disease's lack of judgement of its early symptoms results in only a few being diagnosed and receiving timely treatment. This according to Villar et al., (2015) inspired the development of a stroke-prediction movement-detecting device based on two ML algorithms. The detection process included a human activity recognition stage and a stroke-onset detection stage that alerts the medical provider or the health insurer once the movement of the patient is significantly different from the normal pattern. This process is a form of the predictive analytic capability of AI that has now influenced patients globally to visit medical facilities that have access to the AI tool. Additionally, Maninini et al. (2016) also developed a wearable device for collecting data on normal/pathological gaits for stroke prediction, where patients' data are extracted from the wearable device using Markov models and SVM. The algorithm is noted to identify correctly and classify 90.5% of the subjects to the right group. This device is an example of an AI-powered interface that assists with real-time diagnosis and treatment progress reports and appointment scheduling with a human professional either virtually or face-to-face, and travel arrangements, thus providing an individualised approach that enhances treatment outcomes and patient satisfaction (Yener & Köse, 2023). Another stroke diagnosis that has attracted many medical tourists to Singapore, the UK, and Canada is the use of neuroimaging techniques including Magnetic Resonance Imaging (MRI) and Computed Tomography (CT).

Like the use of AI-powered systems for early stroke detection, it is also being used for predicting and analysing treatment options. For instance, Jiang et al. indicated that as an emergency measure, intravenous thrombolysis (tPA) was strongly related to the prognosis and survival rate (2017). This method has been identified to perform better than the conventional radiology-based methods because the CT scan provided whole-brain images are inputted into the SVM to improve the clinal decision-making process of the tPA method (Bentley et al., 2014). This influences patients to travel from their home countries to facilities abroad for such treatment because it saves them time and resources and helps them to get to the post-treatment stage quicker.

2.2.2.2 Cancer

One of the main causes of death in every nation on Earth is cancer (Sung, et al., 2021). According to figures from the World Health Organisation (WHO) for the year 2022, cancer ranks second among noncommunicable illnesses in terms of mortality, accounting for 9.3 million deaths annually (WHO, 2022). Bi et al. indicated that cancer is a self-sustaining and adaptive process that interacts dynamically with its microenvironment and continues to thwart patients, researchers, and clinicians despite significant progress in understanding its biological underpinnings (2019). Moreover, while traditional diagnostic methods for cancer are indicated to be effective, they often have limitations in precisions and sometimes lead to overdiagnosis or overtreatment (Tewari, n.d). Based on the complexity of the disease, dilemmas arise at every stage of the disease management, including reliable early detection; accurate distinction of preneoplastic and neoplastic lesions; determination of infiltrative tumour margins during surgical treatment; tracking of tumour evolution and potential acquired resistance to treatments over time; and prediction of tumour aggressiveness, metastasis pattern, and recurrence.

Nonetheless, these complexities have continued to increase interest in the search for new diagnostics and treatments involving AI towards more targeted therapies (Bhinder, et al., 2021; Ho, 2020). Technological advances, and in particular, the integration of AI in cancer care in medical imaging and minimally invasive biomarkers have held promise in addressing challenges across the spectrum of cancer detection, treatment, and monitoring (Luchini, et al., 2022). AI-powered systems and technological advances have been found to transform cancer diagnosis, which has immensely contributed to the increase of patients engaging in medical tourism to other countries for more accurate and quicker diagnosis, cheaper diagnoses and treatment, or the need for specialised intensive care (Mital & Nguyen, 2022). For instance, AI's ML algorithms are being used to analyse data from a variety of sources such as electronic health records, genetic information, identify patterns, treatment outcomes, and environmental data, for predictive modelling and early detection of an individual's risk of developing cancer and to tailor prevention strategies (Huang, et al., 2019; Yang, et al., 2022). This data-driven approach enables the development of personalised treatment plans, fitted to the characteristics of the patient. The IBM Watson is an example of an innovative and reliable AI-powered system for oncology in assisting with the diagnosis of cancer, which has been identified to motivate millions of medical tourists to the UK and the US (Somashekhar, et al., 2017). Likewise, Esteva, et al. (2017) developed an AI system that analyses clinical images to identify skin cancer subtypes.

The United States remains at the forefront of AI-powered systems for cancer treatment, boasting in addition of highly trained oncology specialists and treatment centres. However, cancer treatment has surged globally with world-class centres established in many cities across the world that have influenced medical tourists. Destinations that have gained recognition in cancer diagnosis and treatment for medical tourism include Germany, Switzerland, South Korea, Singapore, Australia, and Japan (MTM, n.d). Additionally, there has been a considerable increase in the provision of more accurate and reliable diagnostics (Elemento, et al., 2021; García-Pola, et al., 2021), improved prognostics (Liang, et al., 2022), and aid in new drug discovery (Yang, et al., 2022). AI-powered systems predictive analytics are being used to anticipate cancer progression in patients and predict how the disease will develop, thus enabling the healthcare provider to make proactive decisions about the patient's care and treatment (Tewari, n.d). As a result, several areas of cancer care including cancer radiology and clinical oncology are currently benefiting from AI-related applications (Luchini, et al., 2022).

Although the application of AI in global cancer care is promising, as it helps in identifying patients at high risk and in implementing proactive treatment strategies, there are barriers to its widespread adoption that significantly impact medical tourism. These barriers include the underdeveloped ethical and regulatory issues associated with legal uncertainty about responsibility and accountability for AI-supported decisions in some countries, such as data privacy, algorithmic bias, and the need for substandard training for data analysis (Shreve, et al., 2022).

2.2.2.3 Dental Care

As in other areas of healthcare, AI in dental care has seen a considerable increase in recent years. Dental care is one of the biggest drivers of medical tourism. Likewise, the application of AI in dental care can also be classified into diagnosis, decision-making, treatment planning, and prediction of treatment outcomes (Ding, et al., 2023; Norgeot, et al., 2020).

Traditionally, dental professionals have used a precise set of criteria to identify caries using visual and tactile examinations or radiographic examinations. Deep fissures, strong interproximal contacts, and subsequent lesions make it difficult to identify early-stage lesions. Ultimately, a lot of lesions are only found when dental caries have progressed, necessitating more involved treatments like dental crowns, root canal therapy, or even implants. While dental radiography, including bitewing, periapical, and panoramic images, and explorers, sometimes known as dental probes, are frequently utilised and considered very dependable diagnostic instruments for identifying dental caries, a significant portion of the screening and ultimate diagnosis process often depends on the knowledge of dentists. Research on the identification of dental caries, vertical root fractures, apical lesions, pulp space volumetric assessment, and assessment of tooth wear has been conducted in the field of operational dentistry (Jaiswal & Bhirud, 2021). Additionally, Schwendicke et al., (2021) compared the cost-effectiveness of AI for proximal caries detection and found AI to be effective and less costly compared to traditional dentistry. Multiple studies have revealed that interdisciplinary cooperation between AI and clinicians results in promising, early detection, with high accuracy in identifying these defects compared to traditional dentistry. An example of the current use of AI in dental care in medical tourism is its utilisation to diagnose periodontitis and the classification of plausible periodontal disease types (Kim, et al., 2020).

As in the case of periodontitis, AI applications have been developed in the treatment planning and predictions of treatments for orthodontics such as simulating the changes in the appearance of pre- and post-treatment facial photographs. The experience and preferences of the orthodontists are typically taken into consideration while arranging orthodontic treatment (Tanikawa & Yamashiro, 2021). This is because every patient and orthodontist are different, thus the course of treatment is determined jointly by the two parties. However, the impact of the orthodontic treatment, the skeletal patterns, and anatomic landmarks in lateral cephalograms can be seen with the aid of AI-powered systems, thus assisting greatly in the communication between patients and dentists wherever they may be across the globe (Hwang, et al., 2020).

The biggest reasons for medical tourism in dental care is because of its high costs in most developed countries and the poor access to dental insurance in many countries. For instance, patients in the US and the UK can save as much as 80%

of the cost of root canal treatments, veneers, and dental implants if they opt to get the treatments in Mexico for US citizens, and Poland or Turkey for UK citizens (Spencer-Jolliffe, 2023). These destinations have become choice countries for patients seeking a range of dental procedures, with huge cost savings and access to quality treatments. Other countries that attract medical tourists for dental care and have integrated AI systems are Italy, which has the highest number of dentists per capita, Germany, and Sweden (Spencer-Jolliffe, 2023).

2.2.2.4 Cosmetic and Plastic Sugeries

Cosmetic and plastic surgeries have become increasingly popular worldwide, with more people seeking aesthetic procedures to enhance their appearance and boost self-confidence (Spencer-Jolliffe, 2023). Like other surgical specialities, cosmetic and plastic surgeries heal illnesses and enhance the appearance of an individual. AI technologies are increasingly applied in pre-operative consultations, diagnosis, therapeutic decisions, post-operative evaluation, and follow-up, with most of these processes being able to be done without the patient being physically present with the practitioner. These processes have been identified to be beneficial towards the reduction of medical errors, improving work efficiency, and decreasing resource waste on the part of both patients and medical facilities/practitioners (Qin & Gu, 2023). Most patients in the US are said to go through many of the processes with practitioners in Bogota, Columbia or Thailand before having to travel for the surgical procedure (Spencer-Jolliffe, 2023). According to Patel, et al. (2022), AI plays a significant role in practitioners' concerns about pre- and post-surgery, as a result, it is being used for "eligibility for surgery" and "options to revise surgery". At the consultation stage, patients interact with AI-powered virtual assistants (AIVA) using NLP technology to understand the intent of patients' speech and provide answers (Boczar, et al., 2020). Research shows that the use of the AIVA for cosmetic surgery indicates 92.3% accuracy and 83.3% of participants who have been engaged with it agreed that the answers were correct (Boczar, et al., 2020). This popularisation and application of AIVA have been identified to free cosmetic and plastic surgeons from primary medical consultation and provide patients with increased wellness and satisfaction as they can avoid the huge cost associated with such surgeries and long waiting times for appointments and surgeries in their home countries and receive prompt treatments in their destination countries (Khan, et al., 2016; Spencer-Jolliffe, 2023). Furthermore, most treatment plans for plastic and cosmetic surgery are extremely customised and take the patient's and the surgeon's aesthetic preferences into consideration. AI-powered systems are currently being utilised to create simulated visuals that serve as a surgical reference and enhance the treatment plan. For instance, BreastGAN is a portable AI-equipped tool that uses breast images of patients with bilateral augmentation to train an AI algorithm (Chartier, et al., 2022). Likewise, in rhinoplasty, AI models provide realistic simulations of rhinoplasty images that satisfy the surgeon's aesthetic standards (Chinski, et al., 2022). With the use of AI-driven predictions, these technologies enable patients to make better decisions that reduce the need for repeat procedures and increase patient post-surgery satisfaction. Studies have shown that the consistency of the simulated images and rhinoplasty outcomes is 92% (Qin & Gu, 2023).

Many tourists visit destinations such as South Korea, India, Brazil, Turkey, and Spain to enhance cosmetic surgery such as facial beauty (i.e., facelifts and liposuction). These destinations have now gained recognition with the use of AI to evaluate patients' facial beauty objectively. AI is now being used to prevent exaggerated or unnatural cosmetic or plastic surgery outcomes (Spencer-Jolliffe, 2023).

2.2.2.5 IVF and Fertility Treatment

Fertility tourists or intending parents now have a wide range of solutions to achieve conception thanks to the development of AI-assisted reproductive techniques and technologies. Destinations such as Spain, Israel, Australia, Greece, and Czechia have become global leaders in fertility treatment. Israel and Australia, particularly, are known for their pioneering research and advanced reproductive technologies with AI in the field (Spencer-Jolliffe, 2023). Spain is also a leading destination for a wide range of fertility services, including egg donation and surrogacy, and is becoming a choice destination for intending parents seeking surrogacy services. The rapid growth of AI technology in this field has led to a wide range of cutting-edge reproductive procedures in gamete selection, embryo selection and treatment regimens (Chow, et al., 2021). Other areas include preimplantation genetic testing (PGT), cytoplasmic transfer, cryopreservation, and embryoscopy (Chow, et al., 2021).

In gamete selection, the goal of clinical evaluation is to find early indicators of quality. This includes the visualisation of gametes via time-lapse videos, static photographs, or direct examination. The quality of female gametes is associated with the oocyte form, follicle size, and cytoplasmic properties. The morphology, concentration, and motility of the sperm are established characteristics that are directly connected to the effectiveness of the IVF or fertility treatment (Bouaziz, et al., 2018). AI has transformed the selection process as it has removed the subjectivity of human assessment from the decision-making process, and objectively rank gametes based on quality (Chow, et al., 2021). The greatest benefit of AI in the field comes from the selection of sperm for intracytoplasmic sperm injection (ICSI), a process currently performed by the embryologist. AI is also being used to compute the ideal sperm–egg combination to achieve the maximum success rate or figure perhaps determine whether IVF or ICSI is the best fertilisation approach (De Geyter, et al.,

2020). AI has now proven to be adept at forecasting fertility outcomes based on specific ultrastructural characteristics of mouse sperm (Kandel, et al. 2020).

In embryo selection, the most popular method for choosing embryos for transfers is morphological evaluation which occurs through direct visualisation with a light microscope or time-lapse imaging (Chen, et al., 2017). Both approaches grade embryos on their ability to reach particular stages of development on time. However, like in gamete selection, these evaluations are subjective, and there is a great deal of difference between operators and clinics. AI, using routinely generated images or time-lapse videos, objectively and accurately grade and rank embryos, thus, assisting in the decision-making process to transfer or freeze them. Further, AI now plays a significant role in analysing data from non-invasive metabolomic and secretory profiles from the embryo during culture which leads to improved culture media formulations and regimens (Fernandez, et al., 2020; VerMilyea, et al., 2020).

For treatment regimen, decision-making for an IVF cycle regimen is guided by patient age, gamete quality, medical history, and many more. The goal of this procedure is to increase the likelihood of getting pregnant and having a healthy baby (Chow, et al., 2021). From patient to patient, an IVF cycle might thus differ in stimulation protocol and mode of fertilisation (IVF vs intracytoplasmic sperm injection) as well as the potential for other procedures including assisted hatching and preimplantation genetic testing, amongst others. In planning for an IVF cycle, which is reliant upon input from the clinician, who may recommend an alternative course of treatment regimen depending on their clinical background, as well as variations in training and internal procedures between clinics (Chow, et al., 2021). AI integration with IVF and fertility treatment has aided fertility practitioners in their objective decision-making to optimise the treatment protocol for the best outcome. This has been found not to only provide an increased success rate for patients but also to reduce wastage and workload for clinicians and their facilities (Spencer-Jolliffe, 2023).

2.2.2.6 Orthopaedic Surgery

Orthopaedic surgery is the field of medicine dedicated to addressing, through invasive and non-invasive strategies, traumatic and pathological processes affecting the bones, joints, and adjacent connective tissues facilitating motor function. With wait times for elective orthopaedic procedures in many countries that could take multiple years, patients waiting for knee and hip replacement surgeries, or spinal fusion often wait in their pain to get a slot for the procedure (Spencer-Jolliffe, 2023). To bypass this barrier, patients opt to travel to destinations renowned for complex orthopaedic surgeries, including South Korea, Germany, India, and Turkey. Germany is renowned for its world-class healthcare system and cutting-edge medical technology, making it an ideal destination for orthopaedic surgery. South Africa also leverages its unique position as a technology powerhouse to advance orthopaedic surgery, offering even the most complex procedures, such as complex joint reconstruction (Spencer-Jolliffe, 2023).

Within the orthopaedic realm, orthopaedic surgeons have been able to improve their success rates by applying AI applications to their planning processes. The combination of technology and surgical skills has led to the improvement of patient recovery, as well as, saving time and medical resources (Slater & Sambo, 2023). The integration of various aspects of AI such as Machine Learning, Natural Language Processing, Visual Systems and Medical Robotics, all are working together to ensure that patients' treatment plan is optimised (Farhadi et al., 2022). For instance, in joint reconstruction, AI is being used to analyse images for automated diagnosis, implant evaluation, and predict clinical outcomes. A major example of joint disorder is osteoarthritis (OA) which affects around 10% of men and 13% of women over the age of 65 in the US (Swiecicki, et al., 2021). Based on the prevalence and long waiting time, many patients in the US engage in medical tourism to Canada, Germany, and South Korea for surgery. Practitioners in these countries now use AI's ML for automating the detection and staging of OA from imaging (Farhadi et al., 2022). Also, AI's convolutional neural network (CNN) is being used to automatically diagnose hip OA from radiographs, with a sensitivity of 95% and specificity of 90.7% compared to experienced practitioners. (Khan, et al., 2021).

AI's deep learning tools have been found to accurately predict life expectancy to enhance medical decision-making in orthopaedic oncology, such as helping to determine if surgery should be performed and, if so, which surgical treatment should be used and which country or distance the patient can travel to seek the medical treatment (Ogink, et al., 2021).

2.2.2.7 Organ Transplant Procedures

Long wait times for transplant procedures, huge costs of the procedure, and inaccessibility are driving thousands of patients beyond borders for transplant surgeries (Spencer-Jolliffe, 2023). In the UK, for example, the average wait time for a deceased donor kidney transplant is 2-3 years and could be as long as 5 years in other countries. Nonetheless, a few destinations have distinguished themselves as leaders in transplant surgery, pioneering cutting-edge procedures and transplant treatments. For example, The United States is a major player in the field with heavy research and advancements in organ transplant, including pancreas and heart transplants. Other top destinations for organ transplant procedures are India, Turkey, Spain, and South Korea (Spencer-Jolliffe, 2023).

Although AI in organ transplant procedures in medical tourism is in its infancy, its aim for its introduction into medical tourism was to facilitate the identification of hidden trends and complex relationships within large datasets to obtain logical results while optimising resources (Briceño, 2020). As a result, AI-based classifiers are being explored for the optimisation of four key areas: organ allocation and donor-recipient pairing, transplant oncology, real-time immunosuppression regimes, and precision transplant pathology (NetHope, 2022). This optimisation with AI applications has contributed to a lower incidence of organ rejection and the fine-tuning of the transplantation and organ preservation processes (Adedinsewo, et al., 2023). Key areas of organ transplant procedure in medical tourism where AI has become an important decision-making tool for practitioners include -

- Preventing the need for a transplant – AI is currently being used to detect the potential impacts of organ failure, the early interventions to delay or prevent it and the need for replacement therapies (Morales, et al., 2022). For instance, the Mayo Clinic in Florida, USA, is a major medical tourist centre for international patients (Mayoclinic.org, 2023). The centre indicates the use of AI to predict the ability of certain medications and devices to prevent kidney failure and pulmonary hypertension (Ruiz, et al., 2023).
- Improve the organ donor and recipient matching process – AI assists in identifying which given organ would result in a successful transplant, which helps to streamline the difficult process of connecting organ donors and recipients (NetHope, 2022). Using the AI-based classifier or scoring system to analyse kidney biopsies to determine inflammation levels of transplanted kidney health (Yasodhara, et al., 2021).
- Prevent organ rejection – An important problem in transplant medicine, and an even bigger one in medical tourism, is organ rejection. But the morbidity rate has decreased since AI can now identify individuals who are at risk of organ rejection (Ivanics, et al., 2020). For instance, for patients with heart transplants, specialised clinics that record a large number of visits from medical tourists such as the Mayo Clinic in the US, use AI-powered electrocardiography for the prediction of low-grade rejection risk, potentially without a biopsy (Mayoclinic.org, 2023). This procedure allows the identification of medical tourists that require closer follow-up potentially decreasing post-transplant complications (Tagde, et al., 2021).
- Increase organs usable for transplant – AI is now being used to determine which organs would benefit from perfusion systems and those that are unsuitable for donation (Ruiz, et al., 2023). AI-enhanced Perfusion systems, along with other new technologies, have enabled the transplant community to

expand the number of deceased donor organs available for transplant (Nagar, et al., 2021).

- Improve post-transplant care – AI-powered systems are now used to determine how a patient's body reacts to immunosuppressants and guide medication adjustments to administer only the minimum necessary dose (Nitski, et al., 2021). Additionally, AI's electrocardiography has been found to have a high success rate in the prediction of high- and at-risk patients post-heart surgery (Morales, et al., 2022).

2.2.2.8 Bariatric Surgery

Obesity has become a global epidemic, affecting millions of people worldwide. For many individuals struggling with weight loss, bariatric surgery is a life-changing solution. This has led to a surge in medical tourism, as patients seek out top-quality, affordable, and safe treatment options beyond their borders (Spencer-Jolliffe, 2023). More destinations around the world boast world-class bariatric centres that offer quality services provided by highly trained surgeons and clinicians. Travelling for bariatric surgery also comes with huge cost savings as well as other tourism perks. However, common destinations such as the United States, Turkey and South Korea provide a wide range of cutting-edge procedures, including sleeve gastrectomy, intra-gastric balloon, gastric bypass and bilopancreatic diversion with exceptional results (Spencer-Jolliffe, 2023).

The management of a patient for bariatric surgery is a very intricate topic for any medical establishment and in particular those that accommodate patients from other countries. This is because it requires evaluation by a multidisciplinary team of physicians consisting of internists, psychiatrists, general surgeons, and anesthesiologists, who will be involved in pre, intra, and postoperative evaluation that may be challenging due to the complexity of the patients suffering from obesity (Carron, et al., 2020). Additionally, careful preoperative screening is necessary for the management of obesity-related comorbidities such as obstructive sleep apnea (OSA), diabetes, heart disease, hypertension, and acid reflux disease (Pantelis, et al., 2021). Due to the intraoperative anesthesiologic control during bariatric surgery, which necessitates vigilance for the management of airways, breathing, and hemodynamics, the procedure can be more difficult than standard general surgery (Pantelis, et al., 2021). The use of AI's data mining tools to analyse impact at every stage of perioperative management, from presurgical evaluation to the postoperative period, has been suggested as a solution to these challenges in the treatment of obese patients undergoing bariatric procedures (Bellini, et al., 2022).

There has been an increase in the diffusion of AI-based technologies in bariatric surgery at each step of the clinical pathway from presurgical evaluation and intraoperative management to complications and outcome predictions (see Figure 6) (Bellini, et al., 2022). This is due to AI's capability to integrate a great amount of information for the assessment of patients suffering from different obesity-related comorbidities (Zhou, et al., 2021). Obese patients often suffer from chronic obstructive pulmonary disease (COPD), chronic lung disease, and asthma, which are usually detected by spirometry. However, AI-powered technologies have now proven to be equally effective in the investigation of these potential conditions during preoperative evaluation (Learning MUSM, 2020). Additionally, AI deep learning algorithms are now engaged for successful prediction of accurate surgery duration estimation and quality improvement (Twinanda, et al., 2018). Likewise, AI's neural networks have been developed for use in the clinical decision support system to predict early complications of one-anastomosis gastric bypass surgery based on patients' data using age, body mass index (BMI), smoking status, intra-operative complications, comorbidities, laboratory tests, sonography results, and endoscopy results (Cao, et al., 2020).

Lastly, a recent article posted by the University College Dublin indicates the development of a new AI-powered web tool that allows patients and their doctors to accurately predict weight loss over five years following different types of bariatric surgery (UCD, 2023). This is because predicting weight loss after bariatric surgery has been difficult. However, with this new tool, its prediction is the most accurate approach to predicting weight loss post-surgery. This advancement will be a milestone for both medical tourists and practitioners, as the tool will substantially reduce uncertainty for patients as they will be able to make more informed decisions about the surgery option before travelling abroad for surgery.

2.2.2.9 Eyecare

Eyecare is one of the fields that influence and attract many to medical tourism. This field has advanced in recent years, with state-of-the-art procedures and minimally invasive surgeries to treat complex eye conditions. Many eye centres around the world now offer complex refractive surgeries, such as laser-assisted in situ keratomileusis (LASIK) surgery, which is done to improve visual acuity. Glaucoma surgery as well as retinal and cataract surgeries have also become more advanced in recent years, with many destinations also boasting topline plastic and reconstructive eye surgeries. Top destinations that offer complex eye surgery include India, Cyprus, and the United Kingdom. Some of these destinations also offer affordable rates for such complex procedures, reaching 80% cost savings compared to rates in many much more developed countries (Spencer-Jolliffe, 2023).

The integration of AI technologies has brought about its use to analyse retinal images for automatic detection of diabetic eye disease, age-related macular degeneration, glaucoma, and systemic vascular and metabolic diseases (McNeil, 2019). AI has been able to achieve this revolutionalisation in eye care on several fronts and becoming increasingly important as populations age globally, leading to growing numbers of people who live with vision impairment, blindness, and major eye diseases. It is also helping to expand the physical abilities of eye clinics and hospitals and extend the expert availability of eye care providers such as optometrists and ophthalmologists using advanced technologies to reach nations that do not have such technologies and facilities.

Screening for diabetic retinopathy and macular oedema are the first areas to benefit from advances in AI (Bonner, 2021). AI and its ophthalmological applications are extending and expanding the use of digital ophthalmology and improving the accessibility, availability, and productivity of existing resources (Keng, et al., 2020). Paired with telemedicine, AI is decentralizing and scaling eye screening and primary care. AI screenings are seen as ways for residents in developing nations to easily receive care. The AI-assisted diabetic retinopathy screening is being used to address the acute shortage of specialists in developing countries and for people who cannot afford the high cost of eye care in some developed nations. AI has also worked its way into eye screening for disease severity in major eye diseases such as glaucoma (Mursch-Edlmayr, et al., 2020). The classification of ophthalmic imaging data such as colour fundus photography using deep learning algorithms has been used in the classification of age-related macular degeneration (AMD) in colour fundus photography to predict AMD progression using optical coherence tomography (OCT) and the classification of glaucoma in imaging (Dong, et al., 2021). As a result, eye care facilities in most countries that attract medical tourists use AI technologies that rely on fundus photos, and it has been found that a higher percentage of patients with diabetes tend to have diabetic retinopathy (Bonner, 2021).

The early detection of ocular diseases through AI screening tools allows for timely interventions, leading to improved patient outcomes and vision preservation. The capability of AI algorithms to integrate patient history, lifestyle factors, and genetic data facilitates tailored interventions and management strategies for patients' treatment plans (Al-Khaled, et al., 2020). This ability enhances patient satisfaction and improves treatment efficacy. In addition to enhancing patient care, AI optimises the workflow of eye care professionals in the medical tourism sector, as its intelligent systems can automate routine tasks such as analysing patient data, suggesting appropriate tests, and assisting in documentation (Scanzera, et al., 2022). By streamlining these processes, AI frees up valuable time for eyecare professionals, enabling them to focus on patient interaction and complex decision-making.

2.2.2.10 Cardiology

Cardiovascular disease (CVD) is the principal cause of mortality and morbidity globally. Clinical decision-making for cardiologists is difficult due to the demands for better treatment and the need to translate the most recent medical discoveries and knowledge into a workable strategy (Karatzia, et al., 2022). The potential of artificial intelligence (AI) in cardiology is demonstrated by the development of methods for accurate prediction of cardiovascular disease (CVD) outcomes, non-invasive diagnosis of coronary artery disease (CAD), detection of malignant arrhythmias through wearables, and diagnosis, treatment strategies, and outcome prediction for patients with heart failure (HF) (Itchhaporia, 2022). Pioneering advancements and innovations in cardiac surgery in the United States, with other destinations such as Germany, India, and South Korea also renowned for world-class and affordable heart surgery care. These destinations have recorded breakthroughs in minimally invasive cardiac procedures, including transcatheter cardiac valve replacements and laser angioplasty (Spencer-Jolliffe, 2023). Additionally, with long wait times for accessing cardiac care in many destinations and the huge cost of many cardiac procedures, medical tourists from around the world are leveraging the huge opportunities available abroad for their heart care.

AI algorithms are now increasingly being used for the diagnosis of an image, image segmentation and reconstruction, image quality control, patient prognostication, phenogrouping, and gaining scientific insight (Kagiyama, et al., 2019). The use of AI-based software devices and risk assessment tools are being used to assess patients' meta-data (demographics, co-morbidity). For instance, AI tools are now widely used to improve the accuracy of automated ECG interpretation and aid patient stratification and prognostication. Additionally, atrial fibrillation detection is considered a difficult task because its current diagnostic methods (i.e., pulse palpitation, ECG, ambulatory Holter monitoring) have some limitations (Karatzia, et al., 2022). However, the use of various mobile devices such as smartphones, smart bands, smartwatches, earlobe sensors, and handheld electrocardiogram devices are now used to monitor atrial fibrillation. These devices are user-friendly and allow uninterrupted monitoring and instantaneous individual analysis of ECG signals. The Apple Watch and AliveCor are distinguished examples of these wearables.

Echocardiogram is another AI-powered tool that has impacted medical tourism to the US, as elites from developing nations travel for diagnosis and treatment of heart-related diseases. The echocardiogram is an imaging modality that provides real-time imaging of the heart and instant identification of any structural defects (Ouyang, et al., 2020). The AI-powered devise enables the enhancement of imaging measure measurements' accuracy, via the decrease in inter- and intra-operator inconsistency and by offering information, subtle to be detected merely by the human eye (Kusunose,

2021). Furthermore, the advancement of AI has led to the development of ML-based risk prediction models for improved predictions of Coronary artery disease (CAD), another echocardiogram-like AI tool, "cardiac computed tomography angiography (CTA)". The CTA is a non-invasive imaging investigation that permits a direct evaluation of the patency of coronary arteries that has been vital in ascertaining the incidence of CAD and consequent prognostication (Al'Aref, et al., 2020). These AI-based devices allow for accurate prediction of cardiovascular outcomes, non-invasive diagnosis of coronary artery disease, detection of malignant arrhythmias and diagnosis, treatment, and prediction of outcomes for heart failure patients. This has made cardiac surgery a critical component of medical tourism, as patients around the world search for high-quality, affordable care for their heart health needs (Spencer-Jolliffe, 2023).

2.2.2.11 Neurology

Neurosurgical procedures include those for brain and spinal cord conditions, such as tumours, congenital malformations, and aneurysms. These medical treatments are some of the most expensive medical procedures and come with great risks to life and functionality. Therefore, accessing the best centres with the most skilled experts for this is a major challenge for health tourists. Destinations such as the United States, India, and Turkey rank high as some of the best destinations for receiving neurosurgical care. These destinations have centres with strong collaborations with neurosurgery centres of excellence around the world and with access to the best training and resources available (Spencer-Jolliffe, 2023). The field of neurology is currently facing various challenges, with the shortage of neurologists being the major (Klebanoff & Safdieh, 2023). It has been identified that the field is currently witnessing an 11 per cent supply shortage of neurologists which is estimated to rise to 19 per cent by 2025 (Majersik, et al., 2021). However, it has been identified that AI is changing the landscape of the neurology sector and that of medical tourism. Foremost, is in neuro-oncology, where brain tumours are one of the most misdiagnosed diseases (Haq, et al., 2022). Misdiagnoses of brain tumours occur for multiple reasons including inaccurate analysis of medical reports and wrong interpretation of symptoms. Currently, AI-powered optical imaging and deep convolutional neural networks (CNNs) are being used to accurately predict and diagnose brain tumours (Hollon, et al., 2020). Israel is said to be a top destination for medical tourists concerning the diagnosis and treatment of brain tumours due to the country having specialised neurologists and the cost of diagnosis is not as high as that of the US.

The second disease AI is becoming a strong support for is in the use of its applications for neuro-vascular disorders/diseases that can affect the blood supply in the brain or spinal cord (Klebanoff & Safdieh, 2023). The most dangerous and common neurovascular condition is a stroke, and it is ranked as the second leading cause of death (Donkor, 2018). AI-enabled CT scanning for haemorrhagic patients allows automation of lesion segmentation and detection of haemorrhagic expansion. In the US, AI-enabled tools such as BrainQ and Google are also being used to treat patients who are affected by stroke by providing them with personalised treatments (Klebanoff & Safdieh, 2023). While the affordability of this treatment option is out of reach for most medical tourists who look for cheaper healthcare treatment options, the elites and affluent do go for this medical journey to the US for treatment.

Third is AI for traumatic brain injuries (TBI), which are injuries that happen to the brain due to an accident. Globally, TBI is said to affect an estimated 60 million people yearly with short- and long-term health consequences in terms of their severity. Patients suffering from different types of brain injuries are taken to countries to scam medical images and detect the type of brain injuries the patient has successfully and accurately (Monteiro, et al., 2020). Additionally, AI is indicated to help reduce the overuse of medical imaging which is identified as a problem in paediatric TBI. This is because in most cases after a mild injury, precautionary CT scans are performed on the child to understand the severity of the injury. However, only about ten per cent of children are said to end up having TBI, while the other ninety per cent that are exposed to radiation caused by medical imaging produce nothing significant (Ellethy, et al., 2022). AI's deep neural networks are now being used to predict the need for CT scans in mild paediatric TBI (Ellethy, et al., 2022).

Lastly, AI involvement in neurosurgery or brain surgery. Neurosurgery or brain surgery is one of the riskiest and most complicated medical procedures in the field of medicine, where the most experienced surgeon can also make errors and mistakes that can have serious consequences, thus leading to further health complications for the patient. AI tools have now been developed to help improve, pre-, intra-, and post-operative phases of brain surgery (Chandra, et al., 2022). At the preoperative surgery stage, AI helps to improve the accuracy of diagnoses and create a better surgical plan based on historical data. At the intra-operative stage, AI enables fast and accurate analysis of brain tissues during the procedure. At the postoperative stage, AI helps predict postoperative complications to improve recovery and aftercare.

3. CONCLUSION

From a medical tourism viewpoint, several factors influence a patient to engage in the practice because it provides an array of benefits and options for individuals seeking medical care. One of the primary motivations for medical tourism by patients is its cost-saving propensity compared to medical treatments in their home countries. Consumers are increasingly becoming conscious of saving money. Healthcare costs have risen steadily, and it is expected to grow at a 2.7% compound annual growth rate (CAGR) (Spencer-Jolliffe, 2023). The cost of diagnoses and treatments can be more affordable in foreign countries compared to patients' home countries, making it an attractive option for those seeking quality care at a lower cost (Rogers, 2023). Another key motivation for medical tourism is that some places act as countries of origin and destination in the medical tourism marketplace. As a result, while high-income countries may service overseas elites, other citizens may choose to travel as medical tourists to lower and middle-income countries for treatments (Seow, et al., 2017). For example, while the likes of the Harley Street Hospital in the UK and the Mayo and Cleveland Clinics in the United States have longstanding reputations for medical tourism for elites of other countries, the emergence of lower-cost treatments in Thailand, India or parts of Eastern Europe attract individuals from higher incomes countries who pursue treatments based on lower cost (Lunt, et al., 2011).

Other motivations include medical tourists seeking high-quality healthcare services that are provided by well-trained medical professionals, cutting-edge medical technologies and treatment, and state-of-the-art facilities, including having to avoid long waiting times for appointments and surgeries and receiving prompt treatments in their destination countries. Additionally, some medical treatments and procedures may not be available or are expensive in their home countries, prompting individuals to travel abroad for the specialised care they require (Seow, et al., 2017). While medical tourism allows patients to combine necessary medical care with leisure, to create a unique and enjoyable experience, medical tourists are also attracted to some of these destinations because they offer personalised and holistic medical care that emphasises wellness and patient satisfaction (Khan, et al., 2016).

Finally, this chapter explored the capacity of AI's strategic efficacy in the global medical tourism industry. It highlights the goals of AI in the medical tourism sector as it enhances human capabilities from the practitioner-patient perspective, streamlines medical processes, and acceleration of both scientific and medical discoveries.

REFERENCES

Abadi, M., Agarwal, A., Barham, P., Brevdo, E., Chen, Z., Citro, C., & Zheng, X. (2016). Tensorflow: Large-scale machine learning on heterogeneous distributed systems. *arXiv preprint arXiv:1603.04467.*

Adedinsewo, D., Hardway, H. D., Morales-Lara, A. C., Wieczorek, M. A., Johnson, P. W., Douglass, E. J., Dangott, B. J., Nakhleh, R. E., Narula, T., Patel, P. C., Goswami, R. M., Lyle, M. A., Heckman, A. J., Leoni-Moreno, J. C., Steidley, D. E., Arsanjani, R., Hardaway, B., Abbas, M., Behfar, A., & Yamani, M. (2023). Non-invasive detection of cardiac allograft rejection among heart transplant recipients using an electrocardiogram based deep learning model. *European Heart Journal. Digital Health*, *4*(2), 71–80. doi:10.1093/ehjdh/ztad001 PMID:36974261

Afzal, N., Sohn, S., Abram, S., Scott, C. G., Chaudhry, R., Liu, H., Kullo, I. J., & Arruda-Olson, A. M. (2017). Mining peripheral arterial disease cases from narrative clinical notes using natural language processing. *Journal of Vascular Surgery*, *65*(6), 1753–1761. doi:10.1016/j.jvs.2016.11.031 PMID:28189359

Al-Khaled, T., Valikodath, N., Cole, E., Hallak, J., Campbell, J. P., Chiang, M. F., & Chan, R. V. P. (2020). Evaluation of physician perspectives of artificial intelligence in ophthalmology: A pilot study. *Investigative Ophthalmology & Visual Science*, *61*(7), 2023–2023.

Al'Aref, S. J., Maliakal, G., Singh, G., van Rosendael, A. R., Ma, X., Xu, Z., Alawamlh, O. A. H., Lee, B., Pandey, M., Achenbach, S., Al-Mallah, M. H., Andreini, D., Bax, J. J., Berman, D. S., Budoff, M. J., Cademartiri, F., Callister, T. Q., Chang, H.-J., Chinnaiyan, K., & Shaw, L. J. (2020). Machine learning of clinical variables and coronary artery calcium scoring for the prediction of obstructive coronary artery disease on coronary computed tomography angiography: Analysis from the CONFIRM registry. *European Heart Journal*, *41*(3), 359–367. doi:10.1093/eurheartj/ehz565 PMID:31513271

Amin, R., Al Ghamdi, M. A., Almotiri, S. H., & Alruily, M. (2021). Healthcare techniques through deep learning: Issues, challenges, and opportunities. *IEEE Access : Practical Innovations, Open Solutions*, *9*, 98523–98541. doi:10.1109/ACCESS.2021.3095312

Bellini, V., Valente, M., Turetti, M., Del Rio, P., Saturno, F., Maffezzoni, M., & Bignami, E. (2022). Current applications of artificial intelligence in bariatric surgery. *Obesity Surgery*, *32*(8), 2717–2733. doi:10.1007/s11695-022-06100-1 PMID:35616768

Bentley, P., Ganesalingam, J., Jones, A. L. C., Mahady, K., Epton, S., Rinne, P., & Rueckert, D. (2014). Prediction of stroke thrombolysis outcome using CT brain machine learning. *NeuroImage. Clinical*, *4*, 635–640. doi:10.1016/j.nicl.2014.02.003 PMID:24936414

Bhinder, B., Gilvary, C., Madhukar, N. S., & Elemento, O. (2021). Artificial Intelligence in Cancer Research and Precision Medicine. *Cancer Discovery*, *11*(4), 900–915. doi:10.1158/2159-8290.CD-21-0090 PMID:33811123

Bi, W. L., Hosny, A., Schabath, M. B., Giger, M. L., Birkbak, N. J., Mehrtash, A., Allison, T., Arnaout, O., Abbosh, C., Dunn, I. F., Mak, R. H., Tamimi, R. M., Tempany, C. M., Swanton, C., Hoffmann, U., Schwartz, L. H., Gillies, R. J., Huang, R. Y., & Aerts, L. (2019). Artificial intelligence in cancer imaging: Clinical challenges and applications. *CA: a Cancer Journal for Clinicians*, *69*(2), 127–157. doi:10.3322/caac.21552 PMID:30720861

Boccanfuso, L., Barney, E., Foster, C., & Ahn, Y. A., Chawarska, K., Scassellati, B., & Shic, F. (2016). Emotional Robot to Examine Differences in Play Patterns and Affective Response of Children with and Without ASD. In *The Eleventh ACM/IEEE International Conference on Human Robot Interaction*, (pp. 19–26). IEEE Press. https://dl.acm.org/citation.cfm?id=2906831.2906837

Boczar, D., Sisti, A., Oliver, J. D., Helmi, H., Restrepo, D. J., Huayllani, M. T., Spaulding, A. C., Carter, R., Rinker, B. D., & Forte, A. J. (2020). Artificial intelligent virtual assistant for plastic surgery patient's frequently asked questions: A pilot study. *Annals of Plastic Surgery*, *84*(4), 16–21. doi:10.1097/SAP.0000000000002252 PMID:32032120

Bonner, A. (2021). Four ways artificial intelligence will affect your eye care practice. Evecare Leaders. https://eyecareleaders.com/artificial-intelligence-ophthalmology/.

Bouaziz, J., Mashiach, R., Cohen, S., Kedem, A., Baron, A., Zajicek, M., Feldman, I., Seidman, D., & Soriano, D. (2018). How artificial intelligence can improve our understanding of the genes associated with endometriosis: Natural language processing of the PubMed Database. *BioMed Research International*, *2018*, 20186217812. doi:10.1155/2018/6217812 PMID:29750165

Briceño, J. (2020). Artificial Intelligence and Organ Transplantation: Challenges and Expectations. *Current Opinion in Organ Transplantation*, *25*(4), 393–398. doi:10.1097/MOT.0000000000000775 PMID:32487888

Cao, Y., Montgomery, S., Ottosson, J., Näslund, E., & Stenberg, E. (2020). Deep learning neural networks to predict serious complications after bariatric surgery: Analysis of scandinavian obesity surgery registry data. *JMIR Medical Informatics*, *8*(5), e15992. doi:10.2196/15992 PMID:32383681

Carrera, P., & Bridges, J. (2006). Health and medical tourism: What they mean and imply for health care systems. *Geneva Association Information Newsletter*, *6*(4), 11–15.

Carron, M., Leppariello, G., Safaee, F. B., & Foletto, M. (2020). Perioperative care of the obese patient: A successful approach. *British Journal of Surgery, 107*(6), 772–773. doi:10.1002/bjs.11541 PMID:32339281

Castro, V. M., Dligach, D., Finan, S., Yu, S., Can, A., Abd-El-Barr, M., Gainer, V., Shadick, N. A., Murphy, S., Cai, T., Savova, G., Weiss, S. T., & Du, R. (2017). Large-scale identification of patients with cerebral aneurysms using natural language processing. *Neurology, 88*(2), 164–168. doi:10.1212/WNL.0000000000003490 PMID:27927935

Chandra, J., Rangaswamy, M., Banerjee, B., Prajapati, A., Akhtar, Z., Sakauye, K., & Joseph, A. (2022). Applications of artificial intelligence to neurological disorders: current technologies and open problems. In *Augmenting Neurological Disorder Prediction and Rehabilitation Using Artificial Intelligence* (pp. 243–272). Academic Press. doi:10.1016/B978-0-323-90037-9.00005-9

Chartier, C., Watt, A., Lin, O., Chandawarkar, A., Lee, J., & Hall-Findlay, E. (2022). BreastGAN: Artificial intelligence-enabled breast augmentation simulation. *Aesthetic Surgery Journal. Open Forum, 4*, 52. doi:10.1093/asjof/ojab052 PMID:35072073

Chen, M., Wei, S., Hu, J., Yuan, J., & Liu, F. (2017). Does time-lapse imaging have favourable results for embryo incubation and selection compared with conventional methods in clinical in vitro fertilization? A meta-analysis and systematic review of randomized controlled trials. *PLoS One, 12*(6), e0178720. doi:10.1371/journal.pone.0178720 PMID:28570713

Chinski, H., Lerch, R., Tournour, D., Chinski, L., & Caruso, D. (2022). An artificial intelligence tool for image simulation in rhinoplasty. *Facial Plastic Surgery, 38*(2), 201–206. doi:10.1055/s-0041-1729911 PMID:34051703

Chow, D. J. X., Wijesinghe, P., Dholakia, K., & Dunning, K. R. (2021). Does Artificial intelligence have a role in IVF clinics? *Reproduction & Fertility, 2*(3), 29–34. doi:10.1530/RAF-21-0043 PMID:35118395

Connell, J., & Page, S. (2019). Tourism, ageing and the demographic time bomb-the implications of dementia for the visitor economy: A perspective paper. *Tourism Review, 75*(1), 81–85. Advance online publication. doi:10.1108/TR-02-2019-0070

Darcy, A.M., Louie, A.K., & Roberts, L.W. (2016). Machine Learning and the Profession of Medicine. *JAMA, 2*(1), 315-551.

De Geyter, C., Wyns, C., Calhaz-Jorge, C., de Mouzon, J., Ferraretti, A. P., Kupka, M., Nyboe Andersen, A., Nygren, K. G., & Goossens, V. (2020). 20 years of the European IVF-monitoring consortium registry: What have we learned? A comparison with registries from two other regions. *Human Reproduction (Oxford, England)*, *35*(12), 352832–352849. doi:10.1093/humrep/deaa250 PMID:33188410

Dilsizian, S. E., & Siegel, E. L. (2014). Artificial intelligence in medicine and cardiac imaging: Harnessing big data and advanced computing to provide personalized medical diagnosis and treatment. *Current Cardiology Reports*, *16*(1), 441. doi:10.1007/s11886-013-0441-8 PMID:24338557

Ding, H., Wu, J., Zhao, W., Matinlinna, J. P., Burrow, M. F., & Tsoi, J. K. H. (2023). Artificial Intelligence in dentistry – A review. *Frontiers in Dental Medicine*, *4*, 1085251. doi:10.3389/fdmed.2023.1085251

Dong, L., Yang, Q., Zhang, R. H., & Wei, W. B. (2021). Artificial intelligence for the detection of age-related macular degeneration in colour fundus photographs: A systematic review and meta-analysis. *EClinicalMedicine*, *35*, 100875. doi:10.1016/j.eclinm.2021.100875 PMID:34027334

Donkor, E. S. (2018). Stroke in the century: A snapshot of the burden, epidemiology, and quality of life. *Stroke Research and Treatment*, 2018. PMID:30598741

Drigas, A., & Ioannidou, R. (2012). Artificial Intelligence in Special Education: A Decade Review. *International Journal of Engineering Education*, *28*(6), 1366–1372.

Dryglas, D., & Lubowiecki-Vikuk, A. (2019). Image of Poland as perceived by German and British medical tourists. *Tourism Review*, *74*(4), 861–871. doi:10.1108/TR-07-2018-0105

Elbagoury, B. M., Vladareanu, L., Vlădăreanu, V., Salem, A. B., Travediu, A. M., & Roushdy, M. I. (2023). A Hybrid Stacked CNN and Residual Feedback GMDH-LSTM Deep Learning Model for Stroke Prediction Applied on Mobile AI Smart Hospital Platform. *Sensors (Basel)*, *23*(7), 3500. doi:10.3390/s23073500 PMID:37050561

Elemento, O., Leslie, C., Lundin, J., & Tourassi, G. (2021). Artificial intelligence in cancer research, diagnosis, and therapy. *Nature Reviews. Cancer*, *21*(12), 747–752. doi:10.1038/s41568-021-00399-1 PMID:34535775

Ellethy, H., Chandra, S. S., & Nasrallah, F. A. (2022). Deep neural networks predict the need for CT in pediatric mild traumatic brain injury: A corroboration of the PECARN rule. *Journal of the American College of Radiology*, *19*(6), 769–778. doi:10.1016/j.jacr.2022.02.024 PMID:35381190

Esteva, A., Kuprel, B., Novoa, R. A., Ko, J., Swetter, S. M., Blau, H. M., & Thrun, S. (2017). Dermatologist-level classification of skin cancer with deep neural networks. *Nature, 542*(7639), 115–118. doi:10.1038/nature21056 PMID:28117445

Farhadi, S., Afzal, P., Boveiri Konari, M., Daneshvar Saein, L., & Sadeghi, B. (2022). Combination of Machine Learning Algorithms with Concentration-Area Fractal Method for Soil Geochemical Anomaly Detection in Sediment-Hosted Irankuh Pb-Zn Deposit, Central Iran. *Minerals (Basel), 12*(6), 689. doi:10.3390/min12060689

Fernandez, E. I., Ferreira, A. S., Cecilio, M. H. M., Cheles, D. S., de Souza, R. C. M., Nogueira, M. F. G., & Rocha, J. C. (2020). Artificial intelligence in the IVF laboratory: Overview through the application of different types of algorithms for the classification of reproductive data. *Journal of Assisted Reproduction and Genetics, 37*(10), 372359–372376. doi:10.1007/s10815-020-01881-9 PMID:32654105

Fiszman, M., Chapman, W. W., Aronsky, D., Evans, R. S., & Haug, P. J. (2000). Automatic detection of acute bacterial pneumonia from chest X-ray reports. *Journal of the American Medical Informatics Association : JAMIA, 7*(6), 593–604. doi:10.1136/jamia.2000.0070593 PMID:11062233

García-Pola, M., Pons-Fuster, E., Suárez-Fernández, C., Seoane-Romero, J., Romero-Méndez, A., & López-Jornet, P. (2021). Role of Artificial Intelligence in the Early Diagnosis of Oral Cancer. A Scoping Review. *Cancers (Basel), 13*(18), 4600. doi:10.3390/cancers13184600 PMID:34572831

Ghassemi, M., Nejad, M. G., & Aghaei, I. (2020). Knowledge management orientation and operational performance relationship in medical tourism (overview of the model performance in the COVID-19 pandemic and post-pandemic era). *Health Services Management Research*, 1–15. PMID:33183088

Gillies, R. J., Kinahan, P. E., & Hricak, H. (2016). Radiomics: Images are more than pictures, they are data. *Radiology, 278*(1), 563–577. doi:10.1148/radiol.2015151169 PMID:26579733

Gulshan, V., Peng, L., Coram, M., Stumpe, M. C., Wu, D., Narayanaswamy, A., Venugopalan, S., Widner, K., Madams, T., Cuadros, J., Kim, R., Raman, R., Nelson, P. C., Mega, J. L., & Webster, D. R. (2016). Development and validation of a deep learning algorithm for detection of diabetic retinopathy in retinal fundus photographs. *Journal of the American Medical Association, 316*(22), 2402–2410. doi:10.1001/jama.2016.17216 PMID:27898976

Haq, A. U., Li, J. P., Khan, S., Alshara, M. A., Alotaibi, R. M., & Mawuli, C. (2022). DACBT: Deep learning approach for classification of brain tumors using MRI data in IoT healthcare environment. *Scientific Reports*, *12*(1), 15331. doi:10.1038/s41598-022-19465-1 PMID:36097024

Hassan, V., & Noaman, S. (2017). *Relation between Tourism and Health: Case Study AIDS in Lebanon*. Atiner Conference Paper Series, No: TOU2017-2262. https://www.atiner.gr/paper-series.

Ho, D. (2020). Artificial intelligence in cancer therapy. *Science*, *367*(6481), 982–983. doi:10.1126/science.aaz3023 PMID:32108102

Hollon, T. C., Pandian, B., Adapa, A. R., Urias, E., Save, A. V., Khalsa, S. S. S., Eichberg, D. G., D'Amico, R. S., Farooq, Z. U., Lewis, S., Petridis, P. D., Marie, T., Shah, A. H., Garton, H. J. L., Maher, C. O., Heth, J. A., McKean, E. L., Sullivan, S. E., Hervey-Jumper, S. L., & Orringer, D. A. (2020). Near real-time intraoperative brain tumor diagnosis using stimulated Raman histology and deep neural networks. *Nature Medicine*, *26*(1), 52–58. doi:10.1038/s41591-019-0715-9 PMID:31907460

Huang, S., Yang, J., Fong, S., & Zhao, Q. (2019). Artificial intelligence in cancer diagnosis and prognosis: Opportunities and challenges. *Cancer Letters*, *471*, 61–71. doi:10.1016/j.canlet.2019.12.007 PMID:31830558

Hwang, H. W., Park, J. H., Moon, J. H., Yu, Y., Kim, H., Her, S. B., Srinivasan, G., Aljanabi, M. N. A., Donatelli, R. E., & Lee, S.-J. (2020). Automated identification of cephalometric landmarks: Part 2-might it be better than human? *The Angle Orthodontist*, *90*(1), 69–76. doi:10.2319/022019-129.1 PMID:31335162

Itchhaporia, D. (2022). Artificial intelligence in cardiology. *Trends in Cardiovascular Medicine*, *32*(1), 34–41. doi:10.1016/j.tcm.2020.11.007 PMID:33242635

Jaiswal, P., & Bhirud, S. (2021). Study and analysis of an approach towards the classification of tooth wear in dentistry using machine learning technique. *IEEE International conference on technology, research, and innovation for the betterment of society (TRIBES)*. IEEE. 10.1109/TRIBES52498.2021.9751650

Jha, S., & Topol, E. J. (2016). Adapting to Artificial Intelligence: Radiologists and pathologists as information specialists. *Journal of the American Medical Association*, *316*(22), 2353–2354. doi:10.1001/jama.2016.17438 PMID:27898975

Jiang, F., Jiang, Y., Zhi, H., Dong, Y., Li, H., Ma, S., Wang, Y., Dong, Q., Shen, H., & Wang, Y. (2017). Artificial intelligence in healthcare: Past, present, and future. *Stroke and Vascular Neurology*, *2*(4), 230–243. doi:10.1136/svn-2017-000101 PMID:29507784

Kagiyama, N., Shrestha, S., Farjo, P. D., & Sengupta, P. P. (2019). Artificial intelligence: Practical primer for clinical research in cardiovascular disease. *Journal of the American Heart Association*, *8*(17), e012788. doi:10.1161/JAHA.119.012788 PMID:31450991

Kandel, M. E., Rubessa, M., He, Y. R., Schreiber, S., Meyers, S., Matter Naves, L., Sermersheim, M. K., Sell, G. S., Szewczyk, M. J., Sobh, N., Wheeler, M. B., & Popescu, G. (2020). Reproductive outcomes predicted by phase imaging with computational specificity of spermatozoon ultrastructure. *Proceedings of the National Academy of Sciences of the United States of America*, *117*(31), 11718302–11718309. doi:10.1073/pnas.2001754117 PMID:32690677

Karatzia, L., Aung, N., & Aksentijevic, D. (2022). Artificial intelligence in cardiology: Hope for the future and power for the present. *Frontiers in Cardiovascular Medicine*, *9*, 945726. doi:10.3389/fcvm.2022.945726 PMID:36312266

Keng, J. (2020). *Autonomous Diabetic Retinopathy Screening System Gains Approval.* American Academy of Ophthalmology.

Khan, A. A., Laghari, A. A., & Awan, S. A. (2021). Machine learning in computer vision: A review. *EAI Endorsed Scal Inf Syst*, *8*(32), e4. doi:10.4108/eai.21-4-2021.169418

Khan, M. J., Chelliah, S., & Haron, M. S. (2016). Medical Tourism Destination Image Formation Process: A Conceptual Model. *International Journal of Healthcare Management*, *9*(2), 134–143. doi:10.1080/20479700.2016.1142046

Kim, E. H., Kim, S., Kim, H. J., Jeong, H. O., Lee, J., Jang, J., Joo, J.-Y., Shin, Y., Kang, J., Park, A. K., Lee, J.-Y., & Lee, S. (2020). Prediction of chronic periodontitis severity using machine learning models based on salivary bacterial copy number. *Frontiers in Cellular and Infection Microbiology*, *10*, 698. doi:10.3389/fcimb.2020.571515 PMID:33304856

Klebanoff, L. M., & Safdieh, J. E. (2023). Modern Neurology Training Is Failing Outpatients. *JAMA Neurology*, *80*(4), 333–334. doi:10.1001/jamaneurol.2022.5490 PMID:36848066

Kolker, E., Özdemir, V., & Kolker, E. (2016). How Healthcare can refocus on its Super-Customers (Patients, $n = 1$) and Customers (Doctors and Nurses) by Leveraging Lessons from Amazon, Uber, and Watson. *OMICS: A Journal of Integrative Biology*, *20*(1), 329–333. doi:10.1089/omi.2016.0077 PMID:27310474

Kumar, Y., & Mahajan, M. (2020). Recent advancement of machine learning and deep learning in the field of healthcare system. *Computational intelligence for machine learning and healthcare informatics, 1*, 77.

Kusunose, K. (2021). Steps to use artificial intelligence in echocardiography. *Journal of Echocardiography, 19*(1), 21–27. doi:10.1007/s12574-020-00496-4 PMID:33044715

Lareyre, F., Behrendt, C. A., Chaudhuri, A., Lee, R., Carrier, M., Adam, C., Lê, C. D., & Raffort, J. (2023). Applications of artificial intelligence for patients with peripheral artery disease. *Journal of Vascular Surgery, 77*(2), 650–658. doi:10.1016/j.jvs.2022.07.160 PMID:35921995

Learning, M. U. S. M. (2020). Obstructive sleep apnea: a prediction model using supervised machine learning method. *The Importance of Health Informatics in Public Health during a Pandemic, 272*, 387.

Liang, F., Wang, S., Zhang, K., Liu, T. J., & Li, J. N. (2022). Development of artificial intelligence technology in diagnosis, treatment, and prognosis of colorectal cancer. *World Journal of Gastrointestinal Oncology, 14*(1), 124–152. doi:10.4251/wjgo.v14.i1.124 PMID:35116107

Liu, L., Wu, Y., Wei, W., Cao, W., Sahin, S., & Zhang, Q. (2018, July). Benchmarking deep learning frameworks: Design considerations, metrics and beyond. In *2018 IEEE 38th International Conference on Distributed Computing Systems (ICDCS)* (pp. 1258-1269). IEEE.

Lo, C. (2020). Medical tourism: How is digital tech reshaping the industry? Retrieved from: https://medical-technology.nridigital.com/medical_ technology_jan20/medical_tourism_how_is_digital_tech_reshaping_the_industry

Long, E., Lin, H., Liu, Z., Wu, X., Wang, L., Jiang, J., & Liu, Y. (2017). An artificial intelligence platform for the multihospital collaborative management of congenital cataracts. *Nature biomedical engineering, 1*(2), 0024.

Luchini, C., Pea, A., & Scarpa, A. (2022). Artificial intelligence in oncology: Current applications and future perspectives. *British Journal of Cancer, 126*(1), 4–9. doi:10.1038/s41416-021-01633-1 PMID:34837074

Lunt, N., Smith, R. D., Exworthy, M., Green, S. T., Horsfall, D. G., & Mannion, R. (2011). *Medical tourism: treatments, markets, and health system implications: a scoping review.*

Majersik, J. J., Ahmed, A., Chen, I. H. A., Shill, H., Hanes, G. P., Pelak, V. S., Hopp, J. L., Omuro, A., Kluger, B., & Leslie-Mazwi, T. (2021). A shortage of neurologists–we must act now: A report from the AAN 2019 Transforming Leaders Program. *Neurology*, *96*(24), 1122–1134. doi:10.1212/WNL.0000000000012111 PMID:33931527

Mannini, A., Trojaniello, D., Cereatti, A., & Sabatini, A. M. (2016). A machine learning framework for gait classification using inertial sensors: Application to elderly, post-stroke and huntington's disease patients. *Sensors (Basel)*, *16*(1), 134. doi:10.3390/s16010134 PMID:26805847

MayoClinic.org. (2023). *Five ways artificial intelligence promises to transform organ transplant*. Medical Professionals – Transplant Medicine. https://www.mayoclinic.org/medical-professionals/transplant-medicine/news/five-ways-artificial-intelligence-promises-to-transform-organ-transplant/mac-20548221.

Mbunge, E., & Batani, J. (2023). Application of deep learning and machine learning models to improve healthcare in sub-Saharan Africa: Emerging opportunities, trends and implications. *Telematics and Informatics Reports*, 100097.

McNeil, R. (2019). Coming to terms with AI. *Eye News*, *26*(2), 6–14.

Meghani, Z. (2011). A Robust, Particularist Ethical Assessment of Medical Tourism. *Developing World Bioethics*, *11*(1), 16–29. doi:10.1111/j.1471-8847.2010.00282.x PMID:20491867

Miller, T. P., Li, Y., Getz, K. D., Dudley, J., Burrows, E., Pennington, J., Ibrahimova, A., Fisher, B. T., Bagatell, R., Seif, A. E., Grundmeier, R., & Aplenc, R. (2017). Using electronic medical record data to report laboratory adverse events. *British Journal of Haematology*, *177*(2), 283–286. doi:10.1111/bjh.14538 PMID:28146330

Mital, S., & Nguyen, H. V. (2022). Cost-effectiveness of using artificial intelligence versus polygenic risk score to guide breast cancer screening. *BMC Cancer*, *22*(1), 501. doi:10.1186/s12885-022-09613-1 PMID:35524200

Mittal, S., & Hasija, Y. (2020). Applications of deep learning in healthcare and biomedicine. *Deep learning techniques for biomedical and health informatics*, 57-77.

Monteiro, M., Newcombe, V. F., Mathieu, F., Adatia, K., Kamnitsas, K., Ferrante, E., Das, T., Whitehouse, D., Rueckert, D., Menon, D. K., & Glocker, B. (2020). Multiclass semantic segmentation and quantification of traumatic brain injury lesions on head CT using deep learning: An algorithm development and multicentre validation study. *The Lancet. Digital Health*, *2*(6), e314–e322. doi:10.1016/S2589-7500(20)30085-6 PMID:33328125

Morales, J. R., Nativi-Nicolau, J., Jang, J., Patel, P., Yip, D., Leoni-Moreno, J., & Goswami, R. (2022). Artificial intelligence 12 lead ECG based heart age estimation and 1-year outcomes after heart transplantation. *The Journal of Heart and Lung Transplantation*, *41*(4), S213. doi:10.1016/j.healun.2022.01.1671

Mouridsen, K., Thurner, P., & Zaharchuk, G. (2020). Artificial intelligence applications in stroke. *Stroke*, *51*(8), 2573–2579. doi:10.1161/STROKEAHA.119.027479 PMID:32693750

Murff, H. J., FitzHenry, F., Matheny, M. E., Gentry, N., Kotter, K. L., Crimin, K., Dittus, R. S., Rosen, A. K., Elkin, P. L., Brown, S. H., & Speroff, T. (2011). Automated identification of postoperative complications within an electronic medical record using natural language processing. *Journal of the American Medical Association*, *306*(8), 848–855. doi:10.1001/jama.2011.1204 PMID:21862746

Mursch-Edlmayr, A. S., Ng, W. S., Diniz-Filho, A., Sousa, D. C., Arnould, L., Schlenker, M. B., Duenas-Angeles, K., Keane, P. A., Crowston, J. G., & Jayaram, H. (2020). Artificial intelligence algorithms to diagnose glaucoma and detect glaucoma progression: Translation to clinical practice. *Translational Vision Science & Technology*, *9*(2), 55–55. doi:10.1167/tvst.9.2.55 PMID:33117612

Nagar, R., Singh, Y., & Jaglan, V., & Meenakshu (2021). A review on machine learning applications in medical tourism. In *2021 Fourth International Conference on Computational Intelligence and Communication Technologies (CCICT)* (pp. 208-215). IEEE. 10.1109/CCICT53244.2021.00048

NetHope. (2022). *How IoT Helps Transplant Surgeons Track Organ Shipments*. NetHope. https://www.networkworld.com/article/3098879/how-iot-helps-transplant-surgeons-track-organ-shipments.html.

Nilsson, N. (2009). *The Quest for Artificial Intelligence: A History of Ideas and Achievement*. Cambridge University Press. doi:10.1017/CBO9780511819346

Nitski, O., Azhie, A., Qazi-Arisar, F. A., Wang, X., Ma, S., Lilly, L., Watt, K. D., Levitsky, J., Asrani, S. K., Lee, D. S., Rubin, B. B., Bhat, M., & Wang, B. (2021). Long-term Mortality Risk Stratification of Liver Transplant Recipients: Real-Time Application of Deep Learning Algorithms on Longitudinal Data. *The Lancet. Digital Health*, *3*(5), e295–e305. doi:10.1016/S2589-7500(21)00040-6 PMID:33858815

Norgeot, B., Quer, G., Beaulieu-Jones, B. K., Torkamani, A., Dias, R., Gianfrancesco, M., Arnaout, R., Kohane, I. S., Saria, S., Topol, E., Obermeyer, Z., Yu, B., & Butte, A. J. (2020). Minimum information about clinical artificial intelligence modeling: The MI-CLAIM checklist. *Nature Medicine*, *26*(9), 1320–1324. doi:10.1038/s41591-020-1041-y PMID:32908275

Ogink, P. T., Groot, O. Q., Karhade, A. V., Bongers, M. E. R., Oner, F. C., Verlaan, J. J., & Schwab, J. H. (2021). Wide range of applications for machine-learning prediction models in orthopedic surgical outcome: A systematic review. *Acta Orthopaedica*, *92*(5), 526–531. doi:10.1080/17453674.2021.1932928 PMID:34109892

Omankwu, O. C., Osodeke, E. C., & Ubah, V. I. (2023). Artificial Intelligence a Paradigm Shift in Healthcare Sector Past, Present and Future Prospects. *Journal of Science and Technology Research*, *5*(3).

Ouyang, D., He, B., Ghorbani, A., Yuan, N., Ebinger, J., Langlotz, C. P., Heidenreich, P. A., Harrington, R. A., Liang, D. H., Ashley, E. A., & Zou, J. Y. (2020). Video-based AI for beat-to-beat assessment of cardiac function. *Nature*, *580*(7802), 252–256. doi:10.1038/s41586-020-2145-8 PMID:32269341

Pantelis, A. G., Stravodimos, G. K., & Lapatsanis, D. P. (2021). A scoping review of artificial intelligence and machine learning in bariatric and metabolic surgery: Current status and future perspectives. *Obesity Surgery*, *31*(10), 4555–4563. doi:10.1007/s11695-021-05548-x PMID:34264433

Patel, R., Tseng, C. C., Choudhry, H. S., Lemdani, M. S., Talmor, G., & Paskhover, B. (2022). Applying machine learning to determine popular patient questions about mentoplasty on social media. *Aesthetic Plastic Surgery*, *46*(5), 2273–2279. doi:10.1007/s00266-022-02808-8 PMID:35201377

Psiha, M. M., & Vlamos, P. (2017). IoT applications with 5G connectivity in medical tourism sector management: Third-party service scenarios. *Advances in Experimental Medicine and Biology*, *989*, 141–154. doi:10.1007/978-3-319-57348-9_12 PMID:28971423

PwC. (2021). *Can AI help us understand AI-powered healthcare?* PwC. https://www.pwc.com/m1/en/publications/strategic-foresight/sector-strategies/healthcare-intellectual-capital/can-ai-help-us-understand-ai-powered-healthcare.html. Accessed: 01/11/2023.

Qin, F., & Gu, J. (2023). Artificial intelligence in plastic surgery: Current developments and future perspectives. *Plastic and Aesthetic Research*, *10*(3), 3. doi:10.20517/2347-9264.2022.72

Rajpurkar, P., Irvin, J., Zhu, K., Yang, B., Mehta, H., Duan, T., & Ng, A. Y. (2017). Chexnet: Radiologist-level pneumonia detection on chest x-rays with deep learning. Doi: *arXiv preprint arXiv:1711.05225*.

Rathore, D. K., & Mannepalli, P. K. (2021). Recent Trends in Machine Learning for Health Care Sector. *International Journal of Innovative Research in Technology and Management*, *5*(2).

Rogers, K. (2023). Medical tourism. *Encyclopedia Britannica*. https://www.britannica.com/science/medical-tourism.

Ruiz, J., Desai, S., Paghdar, S., Malkani, S., Nativi, J., Yip, D., Patel, P., Leoni, J., Lyle, M., & Goswami, R. (2023). The Impact of Axillary Mechanical Circulatory Support in Patients Awaiting Heart Transplantation with Pulmonary Hypertension. *The Journal of Heart and Lung Transplantation*, *42*(4), S354. doi:10.1016/j.healun.2023.02.820

Saenger, A. K., & Christenson, R. H. (2010). Stroke biomarkers: Progress and challenges for diagnosis, prognosis, differentiation, and treatment. *Clinical Chemistry*, *56*(1), 21–33. doi:10.1373/clinchem.2009.133801 PMID:19926776

Scanzera, A. C., Shorter, E., Kinnaird, C., Valikodath, N., Al-Khaled, T., Cole, E., Kravets, S., Hallak, J. A., McMahon, T., & Chan, R. P. (2022). Optometrist's perspectives of Artificial Intelligence in eye care. *Journal of Optometry*, *15*, S91–S97. doi:10.1016/j.optom.2022.06.006 PMID:36137899

Schwendicke, F., Rossi, J., Göstemeyer, G., Elhennawy, K., Cantu, A., Gaudin, R., Chaurasia, A., Gehrung, S., & Krois, J. (2021). Cost-effectiveness of artificial intelligence for proximal caries detection. *Journal of Dental Research*, *100*(4), 369–376. doi:10.1177/0022034520972335 PMID:33198554

Seide, F., & Agarwal, A. (2016, August). CNTK: Microsoft's open-source deep-learning toolkit. In *Proceedings of the 22nd ACM SIGKDD international conference on knowledge discovery and data mining* (pp. 2135-2135). ACM. 10.1145/2939672.2945397

Seow, A. N., Choong, Y. O., Moorthy, K., & Chan, L. M. (2017). Intention to visit Malaysia for medical tourism using the antecedents of Theory of Planned Behaviour: A predictive model. *International Journal of Tourism Research*, *19*(3), 383–393. doi:10.1002/jtr.2120

Shreve, J. T., Khanani, S. A., & Haddad, T. C. (2022). Artificial Intelligence in Oncology: Current Capabilities, Future Opportunities, and Ethical Considerations. *American Society of Clinical Oncology Educational Book*, *42*(42), 1–10. doi:10.1200/EDBK_350652 PMID:35687826

Slater, G., & Sambo, T. (2023). *Applications of AI in orthopaedic treatments.* Orthopedic Surgeon. https://orthopaedic-surgeon.com.au/applications-of-ai-in-orthopaedic-treatments/#:~:text=AI%20can%20identify%20any%20anomalies,diagnose%20and%20plan%20patient%20treatment.&text=It%20is%20always%20important%20to,procedure%20prior%20to%20its%20implementation. Accessed - 19/11/2023.

Somashekhar, S. P., Kumarc, R., Rauthan, A., Arun, K. R., Patil, P., & Ramya, Y. E. (2017). Abstract S6-07: Double-blinded validation study to assess the performance of IBM artificial intelligence platform, Watson for oncology in comparison with manipal multidisciplinary tumour board? The first study of 638 breast cancer cases. *Cancer Research, 77*(4), S6–S07. doi:10.1158/1538-7445.SABCS16-S6-07

Spencer-Jolliffe, N. (2023). *AI prompts wave of medical tourism.* Medical Advice Network.https://www.medicaldevice-network.com/features/ai-prompts-wave-of-medical-tourism/.

Subasinghe, M., Magalage, D., Amadoru, N., Amarathunga, L., Bhanupriya, N., & Wijekoon, J. L. (2020, November). Effectiveness of artificial intelligence, decentralized, and distributed systems for prediction and secure channelling for Medical Tourism. In *2020 11th IEEE Annual Information Technology, Electronics and Mobile Communication Conference (IEMCON)* (0314-0319). IEEE. 10.1109/IEMCON51383.2020.9284898

Sung, H., Ferlay, J., Siegel, R. L., Laversanne, M., Soerjomataram, I., Jemal, A., & Bray, F. (2021). Global Cancer Statistics 2020: GLOBOCAN Estimates of Incidence and Mortality Worldwide for 36 Cancers in 185 Countries. *CA: a Cancer Journal for Clinicians, 71*(3), 209–249. doi:10.3322/caac.21660 PMID:33538338

Swiecicki, A., Li, N., O'Donnell, J., Said, N., Yang, J., Mather, R. C., Jiranek, W. A., & Mazurowski, M. A. (2021). Deep learning-based algorithm for assessment of knee osteoarthritis severity in radiographs matches performance of radiologists. *Computers in Biology and Medicine, 133*, 104334. doi:10.1016/j.compbiomed.2021.104334 PMID:33823398

Tagde, P., Tagde, S., Bhattacharya, T., Tagde, P., Chopra, H., Akter, R., Kaushik, D., & Rahman, M. H. (2021). Blockchain and Artificial Intelligence Technology in E-Health. *Environmental Science and Pollution Research International, 28*(38), 52810–52831. doi:10.1007/s11356-021-16223-0 PMID:34476701

Tanikawa, C., & Yamashiro, T. (2021). Development of novel artificial intelligence systems to predict facial morphology after orthognathic surgery and orthodontic treatment in Japanese patients. *Scientific Reports*, *11*(1), 1–11. doi:10.1038/s41598-021-95002-w PMID:34349151

Tariq, A., Santos, T., & Banerjee, I. (2022). Natural Language Processing for Cardiovascular Applications. In *Artificial Intelligence in Cardiothoracic Imaging* (pp. 231–243). Springer International Publishing. doi:10.1007/978-3-030-92087-6_23

Tewari, A. (n.d.). The future of prostate cancer care: AI and Machine Learning. *Medical Tourism Magazine.* https://www.magazine.medicaltourism.com/article/the-future-of-prostate-cancer-care-ai-and-machine-learning.

Tileva, I. G. (2021). How technologies will affect the future of international medical tourism. *Вестник Пермского национального исследовательского политехнического университета. Социально-экономические науки, 3*(1), 308-318.

Twinanda, A. P., Yengera, G., Mutter, D., Marescaux, J., & Padoy, N. (2018). RSDNet: Learning to predict remaining surgery duration from laparoscopic videos without manual annotations. *IEEE Transactions on Medical Imaging*, *38*(4), 1069–1078. doi:10.1109/TMI.2018.2878055 PMID:30371356

UCD, University College Dublin. (30 August 2023). *New AI tool makes it possible to predict weight loss after bariatric surgery.* UCD. https://www.ucd.ie/newsandopinion/news/2023/august/30/newai toolmakes itpossible topredict weightloss afterbariatricsurgery/ #:~:text= New%20AI %20tool %20makes %20 it %20possible %20to %20predict %20weight %20loss %20after %20bariatric %20 surgery, -Posted %2030 %20August &text=First %2Dtime %20patients %20in %20 need, bypass %20or %20sleeve %20gastrectomy %20operation

UNWTO. (2020a). *Impact assessment of the COVID-19 outbreak on international tourism*. UN. https://www.unwto.org/impact-assessment-of-the-covid-19-outbreak-on-international-tourism.

UNWTO. (2020b). *International tourism growth continues to outpace the global economy*. UN. https://www.unwto.org/international-tourism-growth-continues-to-outpace-the-economy.

VerMilyea, M., Hall, J. M. M., Diakiw, S. M., Johnston, A., Nguyen, T., Perugini, D., Miller, A., Picou, A., Murphy, A. P., & Perugini, M. (2020). Development of an artificial intelligence-based assessment model for prediction of embryo viability using static images captured by optical light microscopy during IVF. *Human Reproduction (Oxford, England)*, *35*(4), 35770–35784. doi:10.1093/humrep/deaa013 PMID:32240301

Villar, J. R., González, S., Sedano, J., Chira, C., & Trejo-Gabriel-Galan, J. M. (2015). Improving human activity recognition and its application in early stroke diagnosis. *International Journal of Neural Systems*, *25*(04), 1450036. doi:10.1142/S0129065714500361 PMID:25684369

Wang, K., Kong, H., Bu, N., Xiao, H., Qiu, X., & Li, J. (2022). AI in health tourism: Developing a measurement scale. *Asia Pacific Journal of Tourism Research*, *27*(9), 954–966. doi:10.1080/10941665.2022.2142620

Woodman, J. (2015). *Patients Beyond Borders: Everybody's Guide to Affordable, World-Class Medical Travel*. Healthy Travel Media.

Woodman, J. (2019). Eight drivers for medical tourism. Insights. -https://insights.omnia-health.com/hospital-management/eight-drivers-medical-tourism-2019.

World Health Organization (WHO). (2022). *Noncommunicable Diseases*. WHO. https://www.who.int/news-room/fact-sheets/detail/noncommunicable-diseases..

Yang, F., Darsey, J. A., Ghosh, A., Li, H. Y., Yang, M. Q., & Wang, S. (2022). Artificial Intelligence and Cancer Drug Development. *Recent Patents on Anti-Cancer Drug Discovery*, *17*(1), 2–8. doi:10.2174/1574892816666210728123758 PMID:34323201

Yasodhara, A., Dong, V., Azhie, A., Goldenberg, A., & Bhat, M. (2021). Identifying Modifiable Predictors of Long-Term Survival in Liver Transplant Recipients with Diabetes Mellitus Using Machine Learning. *Liver Transplantation*, *27*(4), 536–547. doi:10.1002/lt.25930 PMID:33113221

Yener, S., & Köse, İ. (2023). Treatment Prediction Using Artificial Intelligence-Integrated Chatbots in Call Centers in Medical Tourism. *Turkish Studies-Economics, Finance. Politics*, *18*(2).

Zeleňák, K., Krajina, A., Meyer, L., & Fiehler, J., & Vukasinovic, I. (2021). How to improve the management of acute ischemic stroke by modern technologies, artificial intelligence, and new treatment methods. *Life (Chicago, Ill.)*, *11*(6), 488. PMID:34072071

Zhou, B., Yang, G., Shi, Z., & Ma, S. (2022). Natural language processing for smart healthcare. *IEEE Reviews in Biomedical Engineering*. PMID:36170385

Chapter 6
Artificial Intelligence-Driven Enhancements in Medical Tourism:
Opportunities, Challenges, and Future Prospects

Gaurav Bathla
CT University, India

Ashish Raina
https://orcid.org/0000-0001-5812-5920
CT University, India

Varinder Singh Rana
https://orcid.org/0000-0001-6158-9195
City University, Ajman, UAE

ABSTRACT

This chapter explores the opportunities, challenges, and future prospects associated with the integration of AI-driven enhancements in the realm of medical tourism. As globalisation continues to advance, individuals in search of medical remedies frequently consider alternatives situated outside their immediate vicinity. Medical tourism is also known as wellness tourism, medical travel, or health tourism was first used by travel firms and the media to refer to the quickly expanding trend of people traveling abroad for medical treatment. The health care industry is becoming more globalized, and many developing nations like India which have the means

DOI: 10.4018/979-8-3693-2248-2.ch006

and potential to meet this demand. By integrating AI technologies in this particular context, uncharted territories are revealed, and the overall patient experience, medical outcomes, and operational efficiency are all enhanced. Telemedicine and AI-powered virtual consultations facilitate uninterrupted communication between patients and medical practitioners, surpassing the limitations imposed by physical distance.

1. INTRODUCTION

Medical Tourism refers to the practice of clients traveling globally to acquire medical treatment due to the rapid expansion of the sector. Tourists receive medical attention and also have the opportunity to see unique cultural, historical, and geographical attractions in the host nations (Mueller & Kaufmann, 2001). Medical tourism should be considered a subset within health tourism. Medical tourism involves individuals traveling to another nation to receive medical, dental, or surgical treatment, as defined by the Medical Tourism Association (2011). The quality of care provided would be on par with or superior than the care available in their home country. Medical tourism and its associated activities are considered very lucrative ventures within the hospitality industry of numerous countries, especially developing ones (Han & Hyun, 2015).

The growing popularity of medical tourism has attracted the attention of policy makers, researchers, and the media. The term originally referred to patients who travel from less developed countries to developed countries to receive treatment not available in their home country. This industry is growing rapidly, becoming a global trend and contributing billions of dollars to the economy. Although it represents a small niche market in the broader tourism sector, its impact is significant (Ghasemi et al., 2021). Health tourism can be broadly interpreted as any type of tourism that emphasizes fun and incorporates stress-reducing activities. This suggests that health tourism provides fun and energizing encounters. Advances in artificial intelligence (AI) and methods are playing an important role in improving patient well-being by providing them with advanced self-care resources. Artificial intelligence empowers individuals to take responsibility for their own health and well-being and improves healthcare professionals' understanding of patients' daily patterns and needs. This increased understanding allows us to provide more effective feedback, guidance, and support to promote overall well-being (Horne et al., 2020). Incorporating artificial intelligence improvements into the medical tourism sector presents a variety of potential benefits, barriers and opportunities for growth. The potential of artificial intelligence to revolutionize global healthcare is demonstrated by its profound impact on personalized medicine, telemedicine and research. However ethical, privacy and regulatory issues must be carefully considered if the benefits of AI in medical tourism are to be leveraged in a responsible and comprehensive manner thereby contributing significantly to medical advances around the world.

1.1 Medical tourism in India

Artificial Intelligence plays a critical function in the field of medical tourism but its complete utilisation has not yet occurred in India. However, the healthcare industry in India is progressively acknowledging the immense potential of artificial intelligence and is incrementally but consistently imbuing medical tourism with it maintaining the momentum is essential. As the globe continues to expand, the tourism industry is gradually emerging from the challenges it faced during the pandemic. There is currently an observed resurgence in the hospitality, healthcare, and tourism industries. Numerous sectors were constrained to transition from manual to digital operations in response to the pandemic. An increasing number of enterprises adopted diverse online services that were crucial for the uninterrupted operation of their operations amidst the COVID-19 pandemic, catalysing technological advancements (Banerjee, 2022).

However, the attention was diverted to the medical and business sectors, which capitalised on the opportunity by devising numerous methods to streamline a significant portion of their operations. From dated appointments for RT-PCR tests to blood reports, everything was organised and scheduled with the assistance of a number of applications that were operational at the time. Existing actors in this industry perceived the pandemic as a chance to advance their standing and attract an entirely new user base. Among numerous others Pharm Easy, Healthify, and LivLong were examples of online application services that offered medical facilities. Historically, India has attracted a significant number of tourists interested in its medical facilities. India is selected due to its comparatively lower-cost medical services in contrast to the exorbitant costs observed in Western and European nations. Recent endeavours such as the 'Heal in India' initiative appear to indicate a positive trajectory for the situation (Banerjee, 2022).

India has become a prominent centre for medical tourism in recent years and is currently ranked among the six most valuable travel destinations for medical purposes worldwide. Medical tourism-related international visitor arrivals increased from 1,39,447 in 2014 to 4,95,056 in 2017. India is an appealing medical travel destination due to several factors. These include reputable healthcare facilities and professionals, health care workers who are fluent in English, state-of-the-art global technology, personalised service, affordable rates, minimal waiting periods, and an abundance of cultural and recreational opportunities (CGI, 2023).

By the year 2025, the value of medical tourism to India is anticipated to reach nine billion U.S. dollars. In 2016 and 2017, this Consulate approved a total of 523 and 545 medical visas, respectively. The following are some of the initiatives that our Ministry of Tourism has effectively executed:

- E Visa Supplement to the Existing Medical and Medical Attendant Visas for Medical Tourists
- Patients not required to personally register with the FRRO.
- Facilitation counters at 3 international airports and 5 more are on the anvil.
- NABH accreditation was granted to Medical Value Travel Facilitators, a new category whose coordination function would be comparable to that of tour operators (CGI, 2023).

Figure 1. System overview diagram of the proposed system

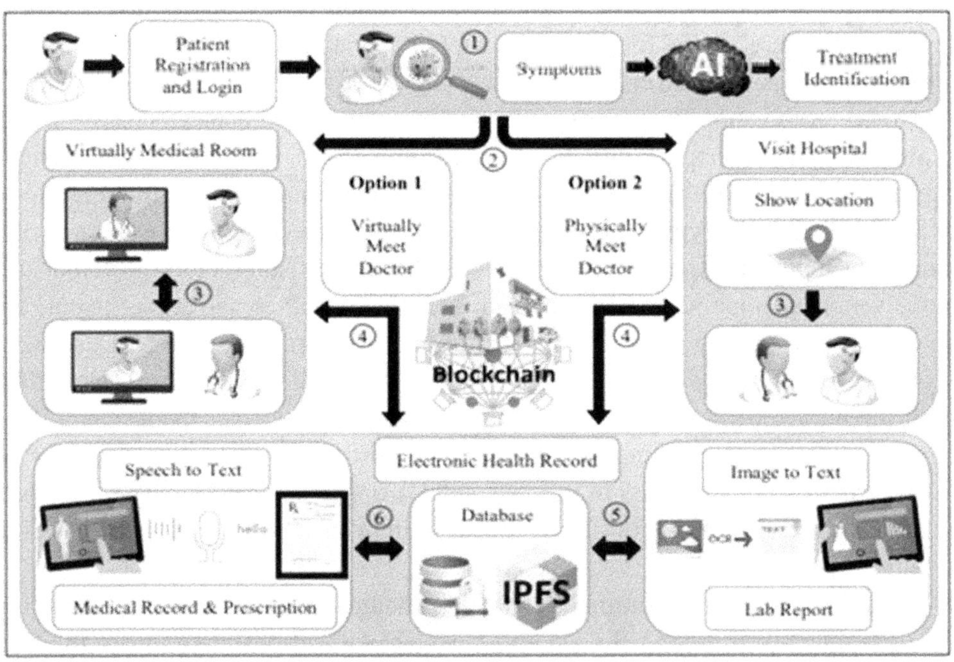

As illustrated in Fig. 1, the operational principle of the proposed system facilitates the tasks performed by its users in a novel fashion. Each of the subcomponents of the system contributes significantly to the achievement of the objective. Utilising cutting-edge technologies to optimise the medical tourism and healthcare industry is the overarching goal. As previously stated, the initial procedure of the proposed system involves providing symptoms as inputs (refer to Step 1) in order to generate an AI-based treatment identification. (Subasinghe et al., 2020). Upon accessing the system, patients are presented with a questionnaire in which they are requested to provide information regarding any symptoms they are aware of. Using the information provided, the patient is informed of the optimal treatment recommendation. Subsequently, the patient is presented with the choice between an in-person or virtual consultation with the physician. The system determines which option is most viable in light of the predicted treatment's level of complexity. The patient's selection (see Step 2) will inform the system's recommendation of a list of physicians most qualified to specialise in the treatment that was previously identified. The patient will be presented with a list of the nearest hospitals and institutions if they choose to undergo a physical visit (see Step 3) (Ahuja, 2019). For any operation to be executed within the system, patients are allocated two distinct identifiers. Users are unable to access any system feature in the absence of the ids. Upon initial patient registration, the system generates and assigns the 'Patient Id', the first identifier that is subsequently stored in the system. The 'EHR Id', which is another identifier, is produced each time a patient logs into the 'Virtual Clinic Room'. As an illustration, the 'Patient Id' will serve as the password to access their logon credentials, treatment particulars, and transactions pertaining to the uploading of medical documents. Conversely, the 'EHR Id' will be required to view and authenticate transactions associated with their Electronic Health Record (EHR) system. Subsequently, users will be redirected to the 'Virtual Clinic Room' (see Step 3), which contains all the feasible technologies for conducting a physical or video consultation with the doctor. Prior to commencing a treatment session, the user will utilise the Patient Portal, which includes an audio call, messaging, and document prediction system, as well as the file upload and download capabilities of the EHR system. Nonetheless, video conferencing would be available exclusively to users participating in a virtual treatment session; messaging and file uploading capabilities would be integrated into the platform as well. During the virtual consultation, the physician may request that the patient's coronary risk level be assessed (Subasinghe et al., 2020). The physician obtains the patient's Systolic Blood Pressure (SBP), Serum Total Cholesterol to HDL cholesterol ratio (TC: HDL), and plots these values on the coronary risk prediction chart in order to perform this procedure. The SBP value may be obtained during a physical examination, whereas the TC and HDL values may be extracted via image-to-text processing from the blood report submitted by the patient. The corresponding chart for each individual patient is determined by parameters including age, smoking status, and diabetes status. After the values are supplied, the system will select a graph to display them on, and the resulting risk level will be computed. All the data gathered throughout the procedure will be archived in the database (see Step 4) within the electronic health record (EHR) of the patient. The physician generates a fresh medical record and prescription by incorporating pertinent information obtained from the patient and lab reports (see Step 5, if available). The health records and prescriptions are generated utilising the speech-to-text service (see Step 6). Once the physician has finished inputting and submitted the form with all the requisite information, the data is transmitted and entered into the database. The newly entered information is accessible to the user on the electronic health record page that is queried for database data. In addition to the aforementioned functionalities, the blockchain and IPFS systems (see Step 4) would operate concurrently. Blockchain, despite its scalability limitations, is a more secure method for storing transactions and is compatible with the world state (remote database). As such, it will function as the transaction recorder for user registrations, user treatment sessions, file uploads and downloads, and electronic health records (EHR systems). As previously explicated, IPFS is a viable alternative for utilisation as a decentralised data storage and file system. Users will be granted the ability to upload documents, lab reports, and electronic health records (EHRs), while maintaining complete control over all access. Therefore, the user could only peruse and delete the EHRs, and only physicians could contribute information to the EHRs. Consequently, physicians are also intended to utilise this proposed system; system administrators will manually register them in accordance with their respective hospitals and medical centres. Furthermore, physicians are provided with advanced features, including voice-to-text and voice-commanding capabilities to facilitate the generation of new prescriptions and the uploading of electronic health records (EHRs), in addition to 'Virtual Clinic Room' features that enable them to monitor and complete the medication process. Physicians are occupied with the administration of medications; therefore, these functionalities would enable them to proceed with their duties while also optimising their time. Both users and physicians will have access to patient and doctor portals, respectively, which will incorporate user-friendly functionalities (Subasinghe et al., 2020).

1.2 Artificial Intelligence (AI)

AI is a technology that is highly regarded for its potential to enhance various industries. In the medical field, AI is seen as a tool that can improve access to high-quality healthcare by advancing diagnostics, personalized treatment, disease prevention, and the development of new therapies. The utilization of AI in medicine is projected to grow 10 times within the next five years (Perry, 2016).

AI is the utilization of coded computer software procedures, known as algorithms, to carry out tasks typically requiring human cognitive abilities. This program can assist individuals in comprehending and analyzing language, distinguishing noises, recognizing objects, and utilizing learning algorithms to address issues. Machine learning (ML) is a method of iteratively improving an algorithm. The refinement process utilizes extensive data and is automated, enabling the algorithm to adapt in order to enhance the accuracy of the artificial intelligence (Zandi, 2019). Artificial Intelligence allows computers to mimic intelligent behavior with little human involvement and has demonstrated superior performance to humans in some tasks. In 2017, for instance, it was reported that deep neural networks (a branch of AI) had been used successfully to analyse skin cancer images with greater accuracy than a dermatologist, and to diagnose diabetic retinopathy (DR) from retinal images (The Lancet, 2017).

The definition of AI is changing. In addition to the technical definition provided before, AI is also seen as an entity that mimics human intelligence and aims to surpass the capability of any one technology. It is a technology interface that enables a machine to perform a task in a way that resembles human behavior. Artificial General Intelligence (AGI) refers to a machine's capacity to accomplish tasks that are within the capabilities of a human. AGI systems are modeled after the human brain. AGI has not been achieved yet; scientists predict it will emerge by 2060 (Joshi, 2019).

1.3 History of AI in Medical Field

Considerable progress has been achieved in the application of artificial intelligence systems to the diagnosis of patients. As an illustration, clinical imaging data has been utilised to develop classification models that assist physicians in the diagnosis of skin cancer, skin lesions, and psoriasis in the field of visually oriented specialties such as dermatology. A deep convolutional neural network (DCNN) model was trained on 129,450 images to classify them as malignant melanoma or benign nevus, or keratinocyte carcinoma or seborrhoeic keratosis (also known as a binary classification problem in machine learning). Additionally, it was demonstrated that the DCNN demonstrated efficacy equivalent to that of twenty-one board-certified dermatologists. The findings of their study illustrated that artificial intelligence (AI) systems could accurately classify skin cancers with an equivalent level of proficiency as dermatologists. Furthermore, training the model for AI systems took a significantly shorter time compared to the years that physicians spent in medical school and the decades of experience they gained through patient diagnosis (Kumar et al., 2023).

Additionally, considerable effort has been devoted to AI and patient prognosis. As an illustration, a DCNN was constructed and trained by Google researchers using 128,175 retinal fundus images. Its purpose was to classify images depicting diabetic retinopathy and macular edoema in adults with diabetes. Numerous benefits accrue from the presence of such an artificially intelligent model, including the following:

- As a result of automated grading of diabetic retinopathy, a greater number of patients can be diagnosed in a reduced period of time;
- Serving as a second opinion opthalmologists;
- Early detection of diabetic retinopathy owing to the model's capacity to analyse images at the particulate level, which is unattainable for a human opthalmologist;
- Vast coverage of screening programs reducing barriers to access (Kaul et al., 2020).

In the application of AI systems to drug discovery and the provision of personalised treatment options, tremendous progress has been made. Verge Genomics and similar organisations employ machine-learning algorithms to analyse human genomic data with the objective of identifying cost-effective pharmaceuticals to treat neurological disorders, including Parkinson's, Alzheimer's, and amyotrophic lateral sclerosis (ALS) (Basu et al., 2020).

The healthcare industry is also implementing artificially intelligent systems to improve patient care and experience, as well as to assist physicians via AI assistants. BotMD and similar businesses have developed systems that can assist around-the-clock with clinical issues pertaining to:

- Efficiently determining the availability of attending physicians and arranging subsequent appointments, the AI system is capable of querying various scheduling systems across multiple institutions.
- Answering prescription related questions, like drug availability and cost-effective alternative drugs
- Facilitating physicians' access to hospital protocols, a catalogue of available clinical instruments, and available medications via a mobile application, thereby enhancing hospital workflow (Basu et al., 2020).

1.4 Transforming Medical Tourism

Over the past few years, medical tourism has witnessed substantial expansion, as individuals seek healthcare services overseas for a variety of reasons, including the ability to save money, access cutting-edge treatments, endure reduced wait times, and more. With the ongoing growth of this sector, the incorporation of Artificial Intelligence (AI) has emerged as a transformative development, fundamentally altering the operational procedures of medical tourism. PwC reports that India experienced the greatest surge in the implementation of AI during the pandemic. In fact, the adoption of AI by the pharmaceutical and healthcare industries in India increased by 73%. In addition, 71% of physicians in New Zealand and Australia believed that AI had a positive impact on the discipline of medicine. Consequently, medical tourism providers can optimise treatment outcomes, expedite operations, and enhance patient experiences through the implementation of AI technologies. The present article delves into the profound implications that AI may have on the domain of medical tourism (Lamba, 2023).

1. Enhanced Patient Matching and Customized Treatment Plans:

Algorithms powered by artificial intelligence are capable of analysing extensive volumes of patient data, such as genetic information, medical history, and diagnostic reports, in order to facilitate the efficient matching of patients with the most appropriate healthcare providers. By employing advanced machine learning methodologies, artificial intelligence (AI) has the capability to evaluate the distinct requirements, inclinations, and objectives of each patient, thereby facilitating the formulation of individualised treatment strategies. The implementation of this individualised strategy not only enhances patient contentment but also optimises the likelihood of favourable treatment results (Bohr & Memarzadeh, 2020).

2. AI-Powered Chatbots and Virtual Assistants:

Medical travellers commonly possess a multitude of inquiries and apprehensions, encompassing travel logistics as well as procedural details. Chatbots and virtual assistants enabled by AI provide round-the-clock assistance by delivering precise and punctual replies to inquiries from patients. Real-time updates on treatment progress and assistance with appointment scheduling and travel arrangements are all functionalities of these AI-powered interfaces. Medical tourism providers have the potential to optimise operational efficiency and ensure patients have a smooth experience, irrespective of language barriers or time zone disparities, through the automation of commonplace tasks (Guzman et al., 2020).

3. Predictive Analytics and Risk Assessment:

Medical tourism providers can utilise the predictive analytical capabilities of AI to analyse extensive volumes of healthcare data in order to detect patterns and forecast potential risks. Through the utilisation of past data, artificial intelligence algorithms are capable of predicting complications, patient outcomes, and post-treatment care needs. This empowers medical tourism providers to take preventative measures against potential issues. Adopting a proactive approach not only enhances patient safety but also allows healthcare providers to effectively allocate resources and reduce the frequency of adverse events (Amjad et al., 2023).

4. Telemedicine and Remote Consultations:

The implementation of AI technologies has significantly enhanced the efficiency of telemedicine services, allowing medical tourism to establish remote connections with healthcare professionals. Patients have the ability to obtain second opinions, seek expert guidance, and discuss treatment alternatives via virtual consultations—all without the necessity of making an in-person journey. Medical image analysis, including X-rays and MRIs, can be performed by diagnostic tools powered by AI, yielding significant insights that facilitate remote consultations. This integration of artificial intelligence and telemedicine increases the availability of high-quality healthcare services, especially for those who reside in remote areas or have mobility issues (Haleem et al., 2021).

5. Continuous Monitoring and Post-Treatment Care:

The utilisation of AI-powered peripheral devices and remote monitoring systems empowers medical tourism providers to consistently observe and assess patient health parameters. Through the acquisition of up-to-date information regarding vital signs, adherence to medication, and progress in recuperation, AI algorithms have the capability to identify potential complications in advance and initiate suitable interventions. In addition, after-treatment care platforms powered by artificial intelligence can offer customised guidance to medical tourism, encompassing exercise regimens, medication reminders, and lifestyle suggestions, in order to guarantee the most favourable long-term results.

Artificial intelligence possesses the capacity to fundamentally transform the domain of medical tourism, presenting an array of advantages for healthcare providers and patients alike. The incorporation of AI technologies offers numerous benefits to healthcare organisations, including optimised treatment outcomes, enhanced patient experiences, and operational efficiency (e.g., telemedicine solutions and personalised treatment plans). In light of the ongoing evolution of the medical tourism sector, the utilisation of artificial intelligence (AI) will assume a progressively critical role in providing patient-centric, high-quality healthcare to individuals seeking services overseas (Lamba, 2023).

1.4 AI Maturity in the Hospitality and Healthcare Sectors

The rate of AI adoption in India has increased by 45 percent, which is the highest among key economies such as the United States, the United Kingdom, and Japan. In the travel and hospitality industries, approximately 89% of businesses have implemented AI in some capacity.

Figure 2. Emerging technologies in medical tourism

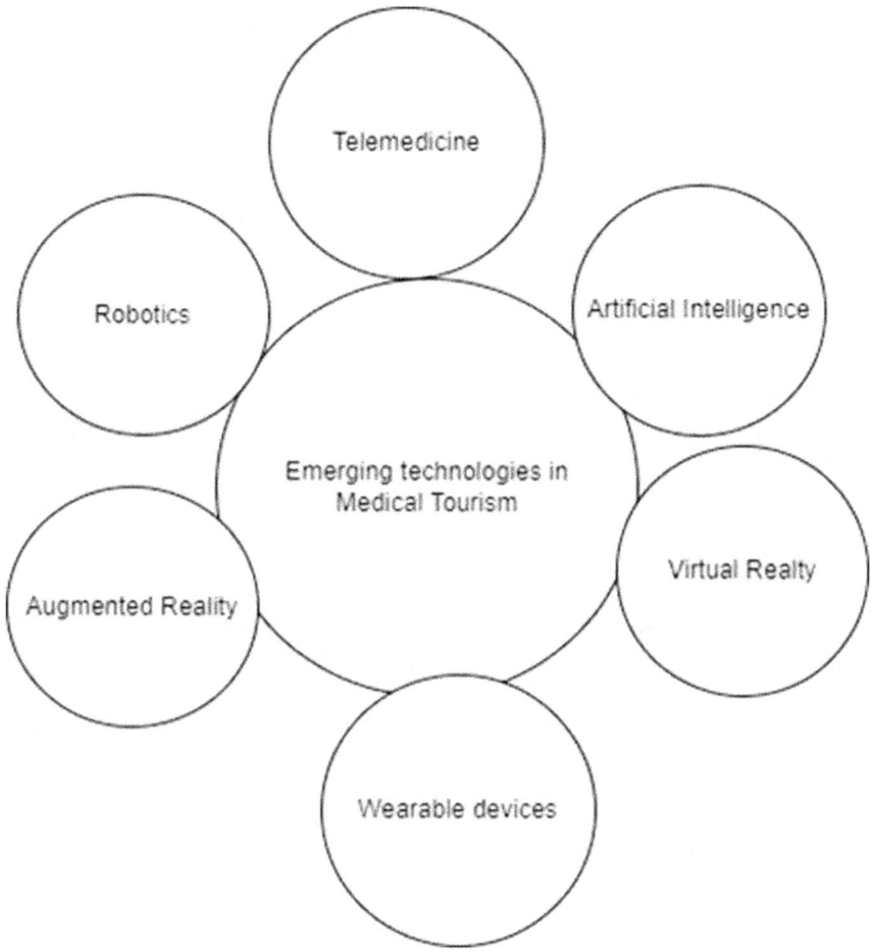

Table 1. Comparative analysis of different technologies for medical tourism

Technology	AI	Application
Intrgrating data and predictive Analysis	✓	AI based Information System
Technology in Mental Health	✓	AI based Rapid Recognition
Remote patient Monitoring & Virtual care	✓	IOT Based intelligent medication system
Digital Tharapeutics	✓	Personalized electronic treatments System.
Augmented & Virtual Reality in Healthcare	✓	Virtual reality in orthopaedic therapy.
Artificial Intelligence	✓	Medical Tourism

Figure 3. AI maturity

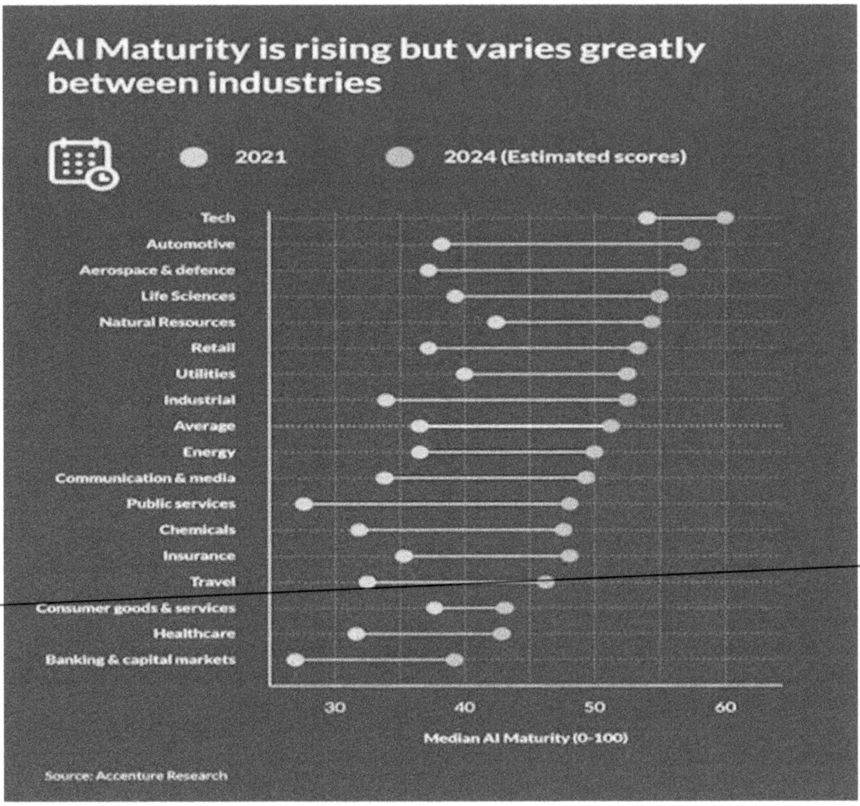

The utilisation of technology by the Indian Government and medical sectors to combat COVID-19 was highly effective. Healthcare organisations nationwide have increased their reliance on technology to facilitate prompt responses and the rapid distribution of information (Banerjee, 2022).

India, according to a global study by PwC India, has experienced the greatest increase in the use of artificial intelligence since the outbreak. According to the study, 73% of pharmaceutical and healthcare organisations in India had implemented AI. Health emergencies that occurred during COVID-19, according to experts, are the impetus for placing AI at the forefront. The government undertook numerous initiatives, including the implementation of AI-powered chatbots utilised by MyGov to combat the pandemic by ensuring rapid communications. The Indian Council of Medical Research (ICMR), in an additional effort to combat COVID, implemented the 'Watson Assistant' on their website to aid frontline personnel and data input operators. It accomplished this by addressing their inquiries from multiple diagnostic and testing facilities located throughout the nation. Nevertheless, much will transpire in its wake due to the severe dearth of awareness and resources. It is anticipated that organisations undergoing AI-driven transformations will utilise intelligence systems to enhance human intelligence. However, each of these is currently undergoing active development (Neil Banerjee, 2022).

2. ADVANTAGES AND CHALLENGES OF ARTIFICIAL INTELLIGENCE IN HEALTH CARE

The utilisation of AI, like everything else, ought to be balanced. Although this technology offers numerous benefits in the healthcare industry, it also possesses substantial disadvantages. In addition to the benefits, you will now consider the challenges that AI poses in healthcare.

2.1 Advantages

- Enhanced diagnosis
- Better serves remote areas
- Improved clinical procedures
- Helps in streamlining a number of processes
- Early detection of disease
- AI can analyse tests, X-rays, CT scans, data entry, and other monotonous jobs faster and more precisely. The amount of data to evaluate in cardiology and radiology can be intimidating and time-consuming.
- AIs have been developed to assess data, including notes and reports from patient files, outside research, and clinical experience, to help choose the best and most individualised treatment course.
- A medical consultation based on a user's personal medical history and general medical knowledge can be provided through an app.
- People can monitor the patient's status and continue with therapies in between doctor appointments with the aid of a digital nurse.
- Wearable fitness trackers keep an eye on exercise and heart rate. They can notify users to move more and share this information with medical professionals (and AI systems) to gather more information on patient behaviours and requirements(Amisha et al., 2019).

2.2 Challenges

Although artificial intelligence has made significant strides, its utilisation by illiterates remains limited. These limitations must and will be promptly surmounted through the implementation of innovative approaches, latest technologies, and revised policies. Regardless of the domains in which AI supplants human labor—giving rise to the debate regarding its potential to supplant human intelligence—many unresolved issues persist across numerous commercial sectors. Although "Artificial Intelligence" is delivering a distinctive customer experience and enabling businesses to reduce expenses through employee replacements, it is not yet capable of rivalling human intelligence due to the nascent stage of development in this domain (Tater, Umaid; Jain, 2022).

- AI poses a threat to data security and privacy.
- Artificial Intelligence technologies are very expensive so it's difficult for small service providers to purchase these technologies.
- A key issue with these software-controlled services is that a minor malware threat can interrupt the software programmes and operations of the service providers, resulting in trouble.
- Chatbots are limited to answer the simple questions. When it comes to complex issues customer still rely on human help (Tater, Umaid; Jain, 2022).

2.3 Overcoming Challenges in Implementing AI in Medical Tourism

Notwithstanding the myriad benefits that AI imparts to the medical tourism sector, obstacles persist that must be surmounted during its execution. Integration of artificial intelligence (AI) systems with pre-existing healthcare IT infrastructure stands as a principal obstacle. Extensive coordination and collaboration between healthcare providers and technology specialists are necessary to ensure uninterrupted data exchange and connectivity (Dania, 2023).

Regulatory and ethical considerations pertaining to the implementation of AI in healthcare constitute an additional barrier. The autonomous decision-making of AI systems gives rise to inquiries regarding accountability and liability in the event of errors or unfavourable consequences. Concerns pertaining to patient privacy and data security must also be attended to in order to safeguard the confidence and secrecy of patients' data.

Moreover, the implementation of AI within the medical tourism sector necessitates comprehensive education and training for healthcare practitioners. Healthcare providers must possess a comprehensive understanding of interpreting AI-generated insights and integrating them into clinical practice in order to optimise the utilisation of AI technology.

By tackling these obstacles, the medical tourism sector can efficiently exploit the capabilities of artificial intelligence to augment the calibre of treatment and satisfaction of patients, thereby solidifying its appeal as a more desirable choice for individuals in search of healthcare services on a global scale (Dania, 2023).

2.4 The Benefits of AI in Medical Tourism Industry

As the medical tourism sector expands, travel agencies and healthcare providers equally are increasingly compelled to integrate artificial intelligence (AI). By leveraging the power of AI, the medical tourism industry can provide patients and providers with a variety of benefits.

Efficiency gains are a significant benefit of AI in the medical tourism industry. Administrative tasks, including medical record administration and appointment scheduling, can be automated by AI-powered systems, enabling healthcare providers to streamline operations and reduce the risk of human error.

Moreover, through the identification of patterns and trends in patient health through the analysis of enormous quantities of data, AI can improve personalised patient care. This capability empowers healthcare providers to administer individualised treatment plans and suggestions, ultimately leading to enhanced patient outcomes (García-Madurga & Grilló-Méndez, 2023).

In addition, artificial intelligence can be of great assistance in overcoming the frequent language barriers that plague the medical tourism sector. The utilisation of AI-driven translation tools facilitates efficient communication between medical professionals and patients, thereby guaranteeing the precise transmission and comprehension of vital medical information.

In summary, the integration of artificial intelligence (AI) into the medical tourism sector yields numerous advantages, such as heightened operational effectiveness, tailored treatment, and improved correspondence. In light of AI's ongoing evolution, healthcare providers and travel agencies must wholeheartedly adopt this technological progression in order to maintain a competitive edge in this burgeoning sector. In the following section, we will examine the numerous AI technologies utilised in the medical tourism industry in greater detail (García-Madurga & Grilló-Méndez, 2023).

2.5 Engaging and Investing in AI

Medical tourism necessitates the online upload of substantial volumes of user data to cloud management systems. Medical travellers also utilise AI technology to plan their journeys, obviating the necessity to visit travel agencies in order to reserve accommodations or make flight reservations. Health consultants have been supplanted by AI assistants and intelligent algorithms, enabling travellers to make reservations for accommodations, flights, and health packages. AI serves as the foundation for technological advancements and expedited service access.(Bulchand-Gidumal, 2020).

A wide range of AI capabilities encompass fundamental domains like machine learning, as well as more sophisticated areas like computer vision, decision making, and motion. Numerous processes and functions of healthcare organisations are aided by these AI capabilities, including patient administration and scheduling, medical diagnosis, surgery, and other operating room operations (Natasha Spencer-Jolliffe, 2023). By integrating and developing artificial intelligence (AI) into cloud systems, social media platforms can enhance their ability to monitor patterns and comprehend user behaviours. This empowers them to personalise the content that is displayed on each user's interface, eliminating the need for proactive user engagement.

Through the analysis of user engagement, healthcare organisations, institutions, and agencies can leverage artificial intelligence to tailor content concerning medical procedures and tourism to the preferences of prospective patients worldwide. By enabling targeted advertisements on social media platforms and web browsers, AI increases the number of medical travellers and the likelihood of reaching more individuals.

Sectors across the travel and tourism industry utilise AI to perform a variety of administrative and customer-facing duties, according to a 2022 research study. AI development is of utmost importance in medical tourism as it assists medical agencies in establishing connections between prospective medical tourists and physicians. Turkey has established itself as a prominent hub for medical tourism. By number of hospitals accredited by the JCI, Turkey is ranked second globally, with over 45 hospitals having received accreditation from the JCI.

Figure 4. Global medical tourism market size (in billion USD)

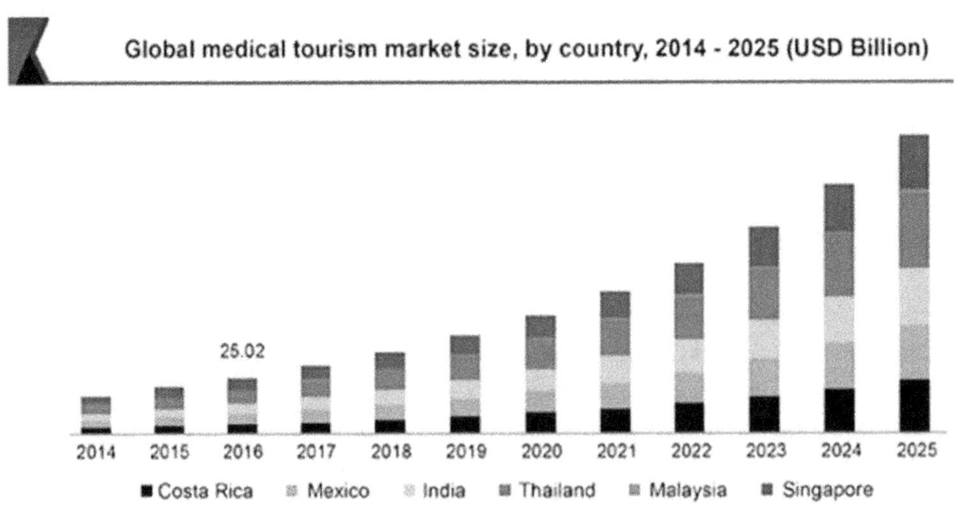

In medical tourism, the United Arab Emirates (UAE) is regarded as a champion in artificial intelligence (AI), having implemented the first AI policy to regulate the development of its healthcare sector. Furthermore, the UAE has named the inaugural Minister of State for AI and established its own Artificial Intelligence Strategy. Artificial intelligence (AI), a technology of the era of the Fourth Industrial Revolution, will advance global healthcare by facilitating secure and real-time access to readily available health records and allowing patients to receive medical treatments abroad that are more affordable and convenient (Natasha Spencer-Jolliffe, 2023).

2.6 Industry Revolution 4.0 Technologies

The implementation of IR 4.0 technologies across the healthcare sector as a whole will have a positive impact on the growth of health tourism by transforming the travel preferences and routines of patients, especially during the initial consultation and subsequent follow-up phases. This benefit assumes greater significance for the elderly and individuals with mobility limitations when it comes to travel. Investments in digitalized and connected health care have increased in recent years, and the pervasive adoption of technology-enabled health care has contributed to the realisation of the "Smart Hospital" concept. The demand for the implementation of blockchain technology in the health tourism sector to improve the security and efficiency of health data utilisation has garnered considerable interest in recent times. By eliminating the need for health tourists to attend follow-up sessions in person after returning to their home country, technology may enable the provision of more convenient and cost-effective in-person consultations. Healthcare technologies and expertise are accessible in numerous regions across the globe. Additionally, the accessibility of IR 4.0 technologies improves the virtual mobility of experts (e.g., physicians, nurses, professors, and carers) and strengthens the preparedness of developing countries to warmly welcome health tourism. Potentially, the implementation of IR 4.0 technologies within health tourism establishments could eliminate the need for travellers to physically travel to other locations in order to obtain health services, including basic health exams and advice. The current worldwide travel disruption caused by the coronavirus pandemic serves as a compelling illustration of how the future of health tourism ought to progress in tandem with technological advancements, particularly the IR 4.0 technologies that are presently accessible. Health care has expanded beyond municipal boundaries to encompass cross-national contexts, both virtually and physically (Hassan & Bellos, 2022).

Figure 5. The importance of medical tourism

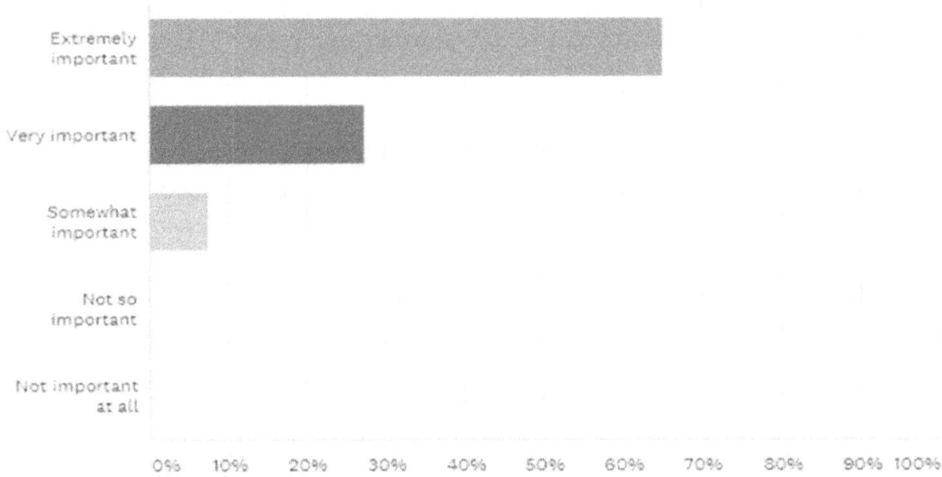

2.7 Advent of Technology

The incorporation of IR 4.0 technologies throughout the healthcare sector will have a positive impact on the growth of health tourism by transforming the travel preferences and routines of patients, especially during the initial consultation and subsequent follow-up phases. This benefit is especially significant for the elderly and those with potential mobility issues when it comes to travelling.

Investments in connected and digitalized health care have increased in recent years, and the pervasive adoption of technology-enabled health care has contributed to the realisation of the "Smart Hospital" concept. The demand for the health tourism sector to benefit from blockchain technology in order to improve the security and efficacy of health data utilisation has garnered increased interest in recent times. Additionally, the provision of IR 4.0 technologies improves the virtual mobility of experts and strengthens the preparedness of developing countries to welcome healthy tourism.

India, according to PwC, experienced the greatest increase in the utilisation of AI during the pandemic. In India, the implementation of AI by the pharmaceutical and healthcare industries increased by 73%. In addition, 71% of physicians in New Zealand and Australia believed that AI had a positive impact on the discipline of medicine. Consequently, medical tourism providers can optimise treatment outcomes, expedite operations, and enhance patient experiences through the implementation of AI technologies.

2.8 Changes AI Brings

By analysing vast quantities of patient data, such as genetic information, medical history, and diagnostic reports, AI algorithms can facilitate the efficient matching of patients with the most qualified healthcare providers. By employing advanced machine learning methodologies, artificial intelligence (AI) has the capability to evaluate the distinct requirements, inclinations, and objectives of each patient, thereby facilitating the formulation of personalised treatment strategies (Raja, 2023).

Similarly, chatbots enabled by AI offer round-the-clock assistance by delivering precise and punctual replies to inquiries from patients. Predictive analytics are another capability of AI. By analysing healthcare data, medical tourism providers are able to recognise patterns and anticipate potential dangers.

Telemedicine services were significantly facilitated by AI, which enabled remote access to the services. Prompt interventions can be prompted in response to early warning indicators of complications identified by AI algorithms.

AI possesses the capacity to fundamentally transform the domain of medical tourism, presenting an array of advantages for both patients and healthcare practitioners. The integration of AI technologies maximises treatment outcomes, optimises operational efficiency, and enhances patient experiences through the implementation of personalised treatment plans, AI-powered chatbots, and telemedicine solutions (Raja, 2023).

2.9 Future of AI in Medical Tourism

According to researchers AI appears to be the industry's future as the technology prioritises putting patients at the centre of health and wellbeing and ensuring that healthcare professionals are aware of their daily routines and requirements in order to offer enhanced guidance, support and feedback. A research study conducted in 2023 examined the significance of data within the medical tourism industry and discovered that there is a scarcity of information pertaining to patients who travel internationally for premeditated medical treatment. Medical tourism healthcare providers could potentially benefit from AI if they consent to the transfer of data to a centralised registry. Online patient outreach for hospitals and clinics can be expanded with the aid of AI. Utilising AI can increase the visibility and promotion of medical tourism providers, thereby accelerating sector expansion (Natasha Spencer-Jolliffe, 2023).

3.0 CONCLUSION

The medical tourism business is increasingly reliant on artificial intelligence (AI) for healthcare providers and travel agencies. The medical tourism business can provide several advantages to patients and providers by utilizing AI technology. AI in the medical tourism industry provides enhanced efficiency as a key advantage. AI technologies can automate administrative chores in healthcare, like appointment scheduling and medical record administration, to help physicians simplify operations and minimize human error. AI can improve tailored patient care by analyzing extensive data to detect patterns and trends in patient health. This allows healthcare providers to create customized treatment plans and suggestions, thereby enhancing patient results. AI can be essential in overcoming language difficulties in the medical tourism sector. AI-driven translation technologies facilitate effective communication between patients and medical providers, enabling accurate transmission and comprehension of crucial medical information.

The present chapter offers an all-encompassing examination of the convergence of artificial intelligence (AI) and the swiftly progressing domain of medical tourism. Chapter also examines the prospective advantages, intrinsic difficulties, and future developments that may emerge as a result of incorporating AI technologies within the realm of international healthcare travel.

The chapter commences by underscoring the expanding importance of medical tourism, with a particular focus on the rising pattern of individuals pursuing healthcare services overseas. The statement establishes the context by recognising the critical impact that artificial intelligence (AI) has on numerous sectors, including healthcare, and emphasises how it has the capacity to completely transform the medical tourism industry.

Integrating artificial intelligence into the medical tourism industry has a substantial advantage in improving decision-making capabilities. AI-powered systems can help healthcare providers make informed decisions about diagnosis, treatment strategies, and patient care by utilizing the large amount of available healthcare data.AI can utilize sophisticated algorithms to examine medical records, test outcomes, and research findings to uncover patterns and connections that may elude human healthcare providers. This enables the development of precise and tailored treatment strategies, resulting in enhanced patient results. Moreover, AI has the ability to consistently acquire knowledge and adjust based on fresh data, staying current with the most recent medical progress and study. This guarantees that healthcare providers can obtain the latest information while making crucial judgments.

Additionally, the introduction explores the potential benefits that artificial intelligence (AI) can bring to the field of medical tourism. The text examines the impact of artificial intelligence (AI) on healthcare outcomes, streamlines administrative procedures for healthcare providers and patients, and facilitates personalised treatment plans. The chapter highlights the potential of AI-driven innovations to enhance the efficiency, cost-effectiveness, and patient-centricity of the medical tourism ecosystem.

Although AI offers several benefits to the medical tourism sector, there are still obstacles that must be addressed for its successful integration. One of the main obstacles is the incorporation of AI technologies into current healthcare IT infrastructure. Extensive coordination and collaboration between healthcare providers and technology experts are necessary to ensure smooth connectivity and data exchange. Regulatory and ethical considerations present another challenge in the implementation of AI in healthcare. As AI systems autonomously make judgments, concerns emerge around accountability and culpability in the event of errors or negative consequences. Concerns around patient privacy and data security must be resolved to maintain the confidence and confidentiality of patients' information.

Moreover, implementing AI in the medical tourism sector necessitates thorough training and education for healthcare workers. Healthcare professionals must possess a strong understanding of interpreting AI-generated insights and integrating them into clinical practice in order to properly deploy AI technology. The medical tourism business may improve the quality of care and patient experience by overcoming these difficulties and utilizing AI technology, thereby making it a more attractive choice for those seeking healthcare services in India.

REFERENCES

Ahuja, A. S. (2019). The impact of artificial intelligence in medicine on the future role of the physician. *PeerJ*, *7*(10), e7702. doi:10.7717/peerj.7702 PMID:31592346

Amisha, M., Malik, P., Pathania, M., & Rathaur, V. K. (2019). Overview of artificial intelligence in medicine. *Journal of Family Medicine and Primary Care*, *8*(7), 2328. doi:10.4103/jfmpc.jfmpc_440_19 PMID:31463251

Amjad, A., Kordel, P., & Fernandes, G. (2023). A Review on Innovation in Healthcare Sector (Telehealth) through Artificial Intelligence. *Sustainability, 15*(8), 6655. doi:10.3390/su15086655

Banerjee, N. (2022). *India's medical tourism is gearing up for AI*. SME. https://smefutures.com/indias-medical-tourism-is-gearing-up-for-ai/

Basu, K., Sinha, R., Ong, A., & Basu, T. (2020). Artificial Intelligence: How is It Changing Medical Sciences and Its Future? *Indian Journal of Dermatology*, *65*(5), 365. doi:10.4103/ijd.IJD_421_20 PMID:33165420

Bohr, A., & Memarzadeh, K. (2020). The rise of artificial intelligence in healthcare applications. *Artificial Intelligence in Healthcare*, *25*, 25–60. doi:10.1016/B978-0-12-818438-7.00002-2

Bulchand- Gidumal. J. (2020). Impact of Artificial Intelligence in Travel, Tourism, and Hospitality. Handbook of E-Tourism, 1–20. doi:10.1007/978-3-030-05324-6_110-1

CGI. (2023). *Consulate General of India - Dubai : Medical Tourism in India*. https://www.cgidubai.gov.in/page/medical-tourism-in-india/

Dania. (2023). *The Role of Artificial Intelligence in Medical Tourism: Enhancing Patient Experience.* Dania Medicare Dania Medicare. https://www.daniamedicare.com/blogs/the-role-of-artificial-intelligence-in-medical-tourism-enhancing-patient-experience/

García-Madurga, M. Á., & Grilló-Méndez, A. J. (2023). Artificial Intelligence in the Tourism Industry: An Overview of Reviews. *Administrative Sciences, 13*(8), 172. doi:10.3390/admsci13080172

Ghasemi, M., Nejad, M. G., & Aghaei, I. (2021). Knowledge management orientation and operational performance relationship in medical tourism (overview of the model performance in the COVID-19 pandemic and post-pandemic era). *Health Services Management Research*, *34*(4), 208–222. doi:10.1177/0951484820971438 PMID:33183088

Guzman, B. MS, Metzger, I., MS, Aphinyanaphongs, Y., D., M., D., P., Grover, H., & D, P. (2020). *Assessment of Amazon Comprehend Medical: Medication Information Extraction*. http://arxiv.org/abs/2002.00481

Haleem, A., Javaid, M., Singh, R. P., & Suman, R. (2021). Telemedicine for healthcare: Capabilities, features, barriers, and applications. *Sensors International*, *2*, 100117. doi:10.1016/j.sintl.2021.100117 PMID:34806053

Han, H., & Hyun, S. S. (2015). Customer retention in the medical tourism industry: Impact of quality, satisfaction, trust, and price reasonableness. *Tourism Management*, *46*, 20–29. doi:10.1016/j.tourman.2014.06.003

Hassan, V., & Bellos, G. (2022). COVID-19: Reshaping Medical Tourism through Artificial Intelligence (AI) and Robotics. *Athens Journal of Tourism*, *9*(2), 77–98. doi:10.30958/ajt.9-2-2

Horne, B. D., Nevo, D., Adali, S., Manikonda, L., & Arrington, C. (2020). Tailoring heuristics and timing AI interventions for supporting news veracity assessments. *Computers in Human Behavior Reports*, *2*, 100043. doi:10.1016/j.chbr.2020.100043

Joshi, N. (2019, June). How Far Are We From Achieving Artificial General Intelligence? *Forbes, 10*. https://www.forbes.com/sites/cognitiveworld/2019/06/10/how-far-are-wefrom-achieving-artificial-general-intelligence/#8f678606dc4d

Kaul, V., Enslin, S., & Gross, S. A. (2020). History of artificial intelligence in medicine. *Gastrointestinal Endoscopy*, *92*(4), 807–812. doi:10.1016/j.gie.2020.06.040 PMID:32565184

Kumar, Y., Koul, A., Singla, R., & Ijaz, M. F. (2023). Artificial intelligence in disease diagnosis: A systematic literature review, synthesizing framework and future research agenda. *Journal of Ambient Intelligence and Humanized Computing*, *14*(7), 8459–8486. doi:10.1007/s12652-021-03612-z PMID:35039756

Lamba, D. (2023). *Revolutionizing Medical Tourism with Artificial Intelligence*. Invest India. https://www.investindia.gov.in/team-india-blogs/transforming-medical-tourism-harnessing-power-artificial-intelligence

Mueller, H., & Kaufmann, E. L. (2001). Wellness tourism: Market analysis of a special health tourism segment and implications for the hotel industry. *Journal of Vacation Marketing*, *7*(1), 5–17. doi:10.1177/135676670100700101

Natasha Spencer-Jolliffe. (2023). *AI prompts wave of medical tourism*. Medical Device Network. https://www.medicaldevice-network.com/features/ai-prompts-wave-of-medical-tourism/

Perry, P. (2016). *How Artificial Intelligence will Revolutionize Healthcare*. Big Think. https://bigthink.com/philip-perry/how-artificial-intelligence-will-revolutionize-healthcare

Raja, A. (2023). *Reshaping medical tourism with AI*. India AI. https://indiaai.gov.in/article/reshaping-medical-tourism-with-ai

Subasinghe, M., Magalage, D., Amadoru, N., Amarathunga, L., Bhanupriya, N., & Wijekoon, J. L. (2020). Effectiveness of artificial intelligence, decentralized and distributed systems for prediction and secure channelling for Medical Tourism. *11th Annual IEEE Information Technology, Electronics and Mobile Communication Conference, IEMCON 2020, November*, (pp. 314–319). IEEE. 10.1109/IEMCON51383.2020.9284898

Tater, Umaid; Jain, A. T. (2022). Artificial Intelligence must be a Revolutionary Technique of Marketing in. [IJBA]. *Indian Journal of Business Administration*, *15*(1), 39–45.

The Lancet. (2017). Artificial intelligence in health care: within touching distance. *The Lancet, 390*(10114), (accessed 30 Mar. 2020). doi:10.1016/S0140-6736(17)31540-4

Zandi, D., Reis, A., Vayena, E., & Goodman, K. (2019). New ethical challenges of digital technologies, machine learning and artificial intelligence in public health: A call for papers. *Bulletin of the World Health Organization*, *97*(1), 1–72. doi:10.2471/BLT.18.227686

Chapter 7
Impact of AI and Robotics on the Medical Tourism Industry

Simon Abou Fayad
https://orcid.org/0000-0002-8777-9023
Lebanese University, Lebanon

ABSTRACT

The internet and its proliferation have had a profound impact on various fields such as science, industry, and medical computer networks. This has resulted in improved medical treatments. Telemedicine serves as a connection between the fields of medicine and engineering, allowing the medical community to utilize engineering resources in order to enhance the well-being of the population. Smart surgery is a form of remote surgery where a surgeon operates on a patient from a distance. The surgeon operates remotely via communication tools, highlighting the intriguing potential of this technology. Telemedicine can be employed during surgery to establish communication with another specialist regarding the feasibility of the procedure. Televisions can serve as a means for postoperative monitoring, while robotic surgeons enable proficient medical professionals to do minimally invasive surgery (MIS) with remote-controlled surgical robots, delivering expert treatment from distant sites.

1. INTRODUCTION

Through the utilization of contemporary technology and software simulation, it is feasible to assist the medical tourism sector in providing medical care to patients worldwide, particularly those who are unable to relocate, in order to treat their patients. Thus, in this chapter, we examine all the crucial factors necessary to achieve

DOI: 10.4018/979-8-3693-2248-2.ch007

optimal conditions for medical tourism utilizing robotic and artificial intelligence (AI) systems.

In recent times, the field of information technology has permeated many sectors and significantly altered various phenomena. The utilization of advanced robots is increasingly imperative in the domains of medicine, engineering, and tourism. Telemedicine is a broad term that encompasses several aspects of remote medical care. The primary concept of telemedicine involves the transmission of information through electrical signals and the automation of clinical services and counseling worldwide using advanced and intelligent technology. The utilization of robots and artificial intelligence (AI) in the development of medical robots designed to aid physicians. The successful design and construction of medical robots implies that the implementation of these ideas should not come as a surprise (Shahmirzadi et al., 2022)

Today, robots are being utilized in diverse medical domains, enabling tasks such as creating small cavities without requiring extensive incisions and performing intricate surgery. Telemedicine aims to enhance patient care, provide medical services to distant regions, and provide improved access to physicians for counseling (Lanfranco, 2004).

Telemedicine encompasses several applications such as distant counseling, e-learning, remote monitoring, remote surgery, treatment of skin illnesses, telemedicine ultrasound imaging, pathology, and remote therapy of cognitive disorders. It is widely acknowledged by researchers that the use of advanced telemedicine and remote surgery has made it feasible for a proficient surgeon in one country, equipped with reliable internet connections and accurate technical equipment, to do surgeries expeditiously (Shahmirzadi et al., 2022).

1.1 Artificial Intelligence and Robotics

The field of study that focuses on the development of intelligent machines and systems, known as Artificial Intelligence (AI), as well as the branch of technology that deals with the design and construction of robots, known as robotics.

Artificial intelligence and robotics is a field of study that focuses on the development of intelligent machines and systems, known as artificial intelligence (AI), as well as the branch of technology that deals with the design and construction of robots, known as robotics. Artificial intelligence and robotics share a common origin and have a lengthy history of interconnection and scholarly discourse. The emergence of artificial intelligence and robotics occurred concurrently in the 1950s, and initially, there was a lack of obvious differentiation between the two fields. The concept of an "intelligent machine" inherently gives rise to the fields of robots and robotics. It might be contended that not all machines qualify as robots because artificial intelligence encompasses virtual agents as well (i.e., agents that do not possess a physical form) (Chella et al., 2006)

However, artificial intelligence research does not address many of the technological difficulties and answers required for robot creation. In the 1970s, a distinct division between the sciences of robotics and artificial intelligence emerged. Robotics began to prioritize industrial automation, while artificial intelligence utilized robots to showcase their ability to function in everyday settings (Chella et al., 2006).

Subsequently, the challenges faced in developing robotic systems that can operate in unrestricted situations caused AI researchers to reject robotics as a favored platform for testing artificial intelligence. In contrast, the research conducted in the field of robotics has resulted in the advancement of increasingly intricate industrial robots.

The influence of technology on health tourism has evolved through advancements in technology and the proliferation of new technologies. AI assists in processing the data provided by an outpatient to a software or application regarding their medical condition. The utilization of artificial intelligence (AI) in monitoring health tourists has the potential to enhance the precision and scientific nature of the process (Ghadirinejad et al., 2021).

By utilizing artificial intelligence, patients can obtain enhanced and precise guidance by inputting their age, weight, and existing medical issues. Conversely, compact medical devices and gadgets utilizing IoT technology offer medical facilities the capability to remotely check the health status of their patients from any location worldwide (Saeedi et al., 2020).

Despite being a relatively recent field, robotic surgery has made significant advancements over time. Therefore, a significant portion of the scientific investigations conducted in the medical engineering departments of prestigious institutions worldwide have focused on the development of medical robots and the advancement of robotic surgery.

Mahjoory (2022) asserted that including robots in surgical procedures will enhance the precision of surgeons. Surgical robots can also significantly contribute to time-saving and lowering the duration of intricate surgical procedures.

A team of scientists has successfully created a preliminary version of a robotic capsule that can securely adhere to the gut without causing any harm to the surrounding tissue. The adhesive properties of the robotic capsule allow it to securely adhere to the gastrointestinal tract without causing any harm to the tissue. This adhesive robot, capable of being ingested like a typical pill, subsequently navigates the digestive system to reach the bowels. This type of robot has the capability to offer precise and painless internal examination of the body, leading to a significant transformation in the effectiveness of procedures such as tissue biopsies, drug administration, heart therapy, and many other medical applications (Fada et al., 2021).

Put simply, the robotic arms are designed to move without any vibration, allowing for precise incisions to be made at their intended locations. Nevertheless, the treating physician may face constraints when the surgeon, when performing the surgery, experiences hand tremors while creating the incision at the surgical site.

Furthermore, robotic surgeons possess the capability to reach deep within the human body. Hence, by making a minor surgical cut, the procedure can be executed with precision on the patient, eliminating the necessity for extensive incisions.

Under such circumstances, the patient will experience a more expedited recovery following the surgical procedure. The console, designated as the physician's workstation, empowers the clinician to exercise control over the surgical operation (Malekzadeh et al., 2021). The console offers a simulated environment of surgical circumstances to allow physicians to closely see patients during surgery at a magnification level of 20x. The surgeon's robotic system transmits data on factors such as flexibility, pressure, and tissue strength at the surgical site to the computer console. The physician then utilizes this information to calculate the precise movements of the robotic arm (Barbaz-I, 2021).

In 2016, Shahgholi asserted that surgical robots have the benefit of minimizing the duration of postoperative recuperation. Surgical pain is experienced after a medical procedure. Duration of hospitalization and associated expenses. Utilizing robots in surgical procedures enhances the precision of the surgeon's actions. Robots provide the capability to generate precise incisions as needed by the surgeon, devoid of any vibration, and with exceptional accuracy. Additionally, the utilization of these robots results in time savings. The advantages of this approach for the patient include diminished postoperative discomfort, minimized scarring, reduced bleeding, decreased risk of infection, shorter hospitalization, expedited healing, and quicker resumption of normal activities. An advantage of this technology is that it allows the surgeon to assume a seated position, facilitating easier monitoring of the surgery. These artificially intelligent robots can be utilized for the medical treatment of astronauts (Ghasemi et al., 2022).

Artificial intelligence and robots enable the execution of remote surgeries, which will be further elaborated in the next paragraph

1.2 Remote Surgery

A surgeon from New York City performed the first tele robotic surgery on September 7, 2001, in Strasbourg, France. The surgery was a laparoscopic cholecystectomy. While the "Lindbergh operation" achieved success, the practicality of regularly employing this technology was doubted by many.

Figure 1. Telerobotic room at St. Joseph's Hospital in Hamilton
Source: https://www.ncbi.nlm.nih.gov/pmc/articles/PMC1356984/

Figure 2. Setup of the Zeus TS Robotic Arms in the operating room of North Bay General Hospital
Source: https://www.ncbi.nlm.nih.gov/pmc/articles/PMC1356984/

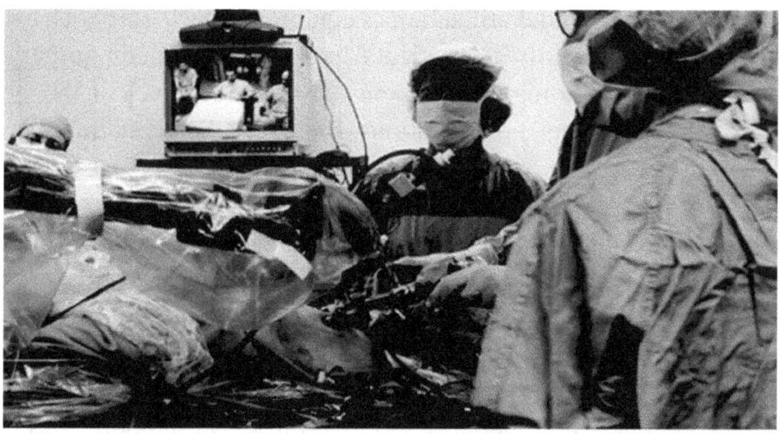

For instance, The Zeus TS microjoint system, developed by Computer Motion Inc in Santa Barbara, CA, is utilized to offer telepresence to the telerobotic surgeon in Hamilton, Ontario. This surgeon is highly skilled in performing advanced laparoscopic procedures (see Figure 1). In North Bay, Ontario, the local laparoscopic surgeon (Figure 2) sets up the 3 arms of the robot for each case. The surgeon has received training in laparoscopic surgery through brief training courses and has also participated in 5 mentoring and telementoring sessions. Both surgeons utilize a Stryker 988 digital camera, manufactured by Stryker Canada in Waterdown, Ontario, to obtain the laparoscopic view (Marescaux et al, 2002).

Inquiries were made regarding the implementation of telerobotic remote surgery for complex laparoscopic procedures, its feasibility in underserved rural areas where a specific ATM (asynchronous transfer mode) bandwidth (as utilized in the "Lindbergh operation") is unavailable, and the simplicity of assembling the robotic arms by non-experts in robotics who are present at the side of a remote patient (Marescaux et al., 2002).

Talebi et al., (2021) asserted that a crucial concern in remote sensing systems is the diminished value of surgeon's robots compared to the surgeon's natural hand. While these robots enhance the surgeon's dexterity and provide a superior visual perspective of the operating field, they compromise the surgeon's tactile perception of tissue properties.

During surgical procedures, surgeons rely on their tactile perception to assess the condition of the tissue being operated on and make informed decisions (Morovvati, 2010). The absence of sensory input, the costly nature of the procedure, the physician's limited comprehension of the patient's surroundings, delays in controlling power-feedback systems, as well as delays in the projection of the operating field video, all have an impact on performance throughout the procedure.

Provide instructions for the utilization of equipment, such as robots, to carry out surgical procedures universally. The patients' inclination to undergo surgery is driven by the advent of new technology. Implement a 25-second latency on the monitors in accordance with the doctors' specifications. Reducing the delay time between the surgeon's visual perception of the knife movement and the actual incision to less than one-fifth of a second enhances the risk of making an incorrect cut. Currently, the robot has a maximum operational range of around 300 km when connected via a wire and 35 km when connected wirelessly (Monfared, 2018).

Robots have a significant role in assisting patients with cardiac problems, reducing the requirement for direct nursing intervention, and minimizing the risk of respiratory diseases such as COVID-19. Hence, both patients and tourists were provided with exceedingly excellent accommodations (Ghasemi & Hasanpoor, 2021). Various forms of telecommunications are accessible for the purpose of conducting remote surgeries, which will be elaborated on in the next paragraph.

1.3 Type of Telecommunications

Telemedicine programmers have the initial and challenging task of selecting a telecommunications method, which encompasses a range of options in terms of bandwidth and telecommunication technology, including microwave, satellite, wireless, cable, and the Internet (Kamarian et al., 2022). The selection is contingent upon the program's requirements, the accessibility of expenses, and the future technological strategy (Lucchini et al., 2012).

Figure 3. The IP/VPN network used to provide the telecommunication connection between the St. Joseph's telerobotic room and North Bay General Hospital
Source: https://www.ncbi.nlm.nih.gov/pmc/articles/PMC1356984/

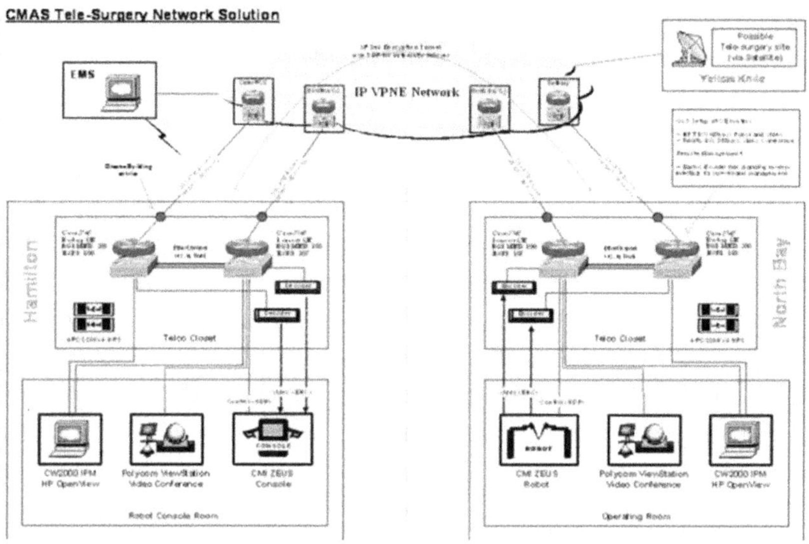

In order to demonstrate fully the example of St. Joseph's Hospital in Hamilton a commercially available IP/VPN (Internet Protocol-Virtual Private Network) network with QOS (Quality of Service) is used to link the 2 hospitals at a bandwidth of 15 Mbps (Figure 3).

This system operates the local Internet and facilitates secure communication among banks, government offices, and other comparable entities. This service provides both an operational connection and a completely redundant line, allowing the tele robotic surgeon to seamlessly switch to the backup line in the event of a malfunction in the primary line. Tele robotic surgeries are conducted with the utmost priority on quality of service (QoS), which is determined by the network. This ensures that the signal is transmitted between the two sites as quickly as possible. At the time, the surgical signals were given more priority than any other network traffic (Marescaux et al., 2002).

The successful clinical use of telemedicine relies on the two-way transmission of audio and video information. Therefore, the development of this technology, together with AI and robots, as previously indicated, is crucial.

1.4 Tourism Industries and Network

medical tourism refers to the practice of seeking healthcare services outside of one's usual region of residence. The concept of "outside of the area" is subject to varying perspectives. It can refer to being outside of the patient's own nation, or it can encompass cases when the patient receives treatment in a different region or province. In such instances, the term "medical tourism" may be applicable.

The global yearly revenue generated by medical tourism is projected to reach around 850 trillion Rupiah, while in the ASEAN region, it is expected to reach approximately 150 trillion rupiah. Based on Patients Beyond Borders (2019), India, Costa Rica, Israel, Mexico, Singapore, South Korea, Turkey, Taiwan, Thailand, and the United States are recognized as the leading destinations for medical tourism (Ayuningtyas and Ariwibowo, 2020).

Patients opt for medical tourism over local medical facilities due to a multitude of factors. Patients in industrialized countries often find that receiving medical treatment abroad is more cost-effective than receiving treatment in their home region or country, especially when factoring in transportation and accommodation expenses. Additionally, the quality of care is typically comparable.

Countries that have emerged as medical tourism destinations often offer more extensive medical services and advanced equipment compared to the healthcare professionals available in the patient's home area. In addition to medical concerns, people may also desire to travel abroad for vacation while receiving medical treatment (Ayuningtyas and Ariwibowo, 2020).

In the present era, due to the advancements in Internet and communication technology, individuals may conveniently purchase things from any part of the globe using internet-enabled devices and their credit cards. The medical industry's current access to such chances stands as one of the most significant contributions of science and technology to human existence. Individuals can access the services of a highly experienced physician from any location worldwide, without the need to physically travel, and have the ability to communicate with them and receive medical care remotely (Carrera and Bridges, 2006).

The World Health Organization, when asked about the concept of telemedicine, provides the following explanation: The World Health Organization (2010) defined 'telemedicine' as a word created in the 1970s to specifically refer to remote recuperation.

Telemedicine facilitates the acquisition of medical information about patients by health professionals through the utilization of communication technologies.

Garshnek and Hassell (2000) provided the following definitions to differentiate between telehealth and telemedicine: telehealth encompasses a wide array of technology and services aimed at delivering patient care and enhancing the overall healthcare system. Telehealth and telemedicine are distinct from each other; telehealth encompasses a wider range of information. In the current era of rapid technical advancements, the utilization of technology services has become a prevalent occurrence, especially within the healthcare sector.

Telehealth and telemedicine are widely utilized across various healthcare domains, employing information and communication technologies.

Efficient and reliable connections between remote and medical facilities are crucial for the effective use of telemedicine, therefore necessitating the provision of communication networks with suitable technical considerations. The network must possess the capability to monitor data transit and establish connections with other networks. The subset mentioned in Talebi et al. (2021) encompasses technical attributes such as bandwidth, information processing, information security, and information flow.

The tourism industries in various countries, such as North Cyprus, India, and Iran, have witnessed the emergence of new medical technologies, particularly the development of surgical robots. This advancement has garnered the attention of knowledge-based companies and experts in the field. However, it is important to note that this technology is relatively new in Iran. While robots have been utilized in the medical field for over a decade, the majority of the equipment used has been imported from other countries (Fazlollahi, 2021).

The system comprises two primary components: the remote surgical console and the surgeon robots positioned on the patient's bed. The surgical console's handles capture the surgeon's hand motions, which are then replicated by the surgeon's robots on the patient's bed. The surgical console's guiding robots are connected to the patient's bedside robots through the Internet, enabling surgical control and performance at remote locations, such as the farthest regions of the country or even on a remote ocean-going ship. Additionally, this technology possesses the capability to analyze and contrast the motions of the surgeon's hands, hence enhancing the precision and dexterity required for executing highly intricate maneuvers. Furthermore, the utilization of 3D printing technology has progressively established a strong position in the field of medical education. In order to offer a substitute for a real human body in the training of surgeons and medical students, a technique has been devised that utilizes 3D printing to fabricate synthetic organs. The organs appeared, had the tactile sensation of genuine organs, and even exhibited bleeding akin to that of real organs. Subsequently, the utilization of 3D printing in educational establishments has broadened due to its enhancement of future physicians' comprehension of anatomy and pathology, facilitation of a more comprehensive understanding of procedures, and augmentation of the trainee's proficiency and self-assurance (Khandan and Esmaeili, 2019).

Utilizing simulation software for precise mechanical forecasting can enhance the efficacy of metallic devices and equipment manufactured in the medical engineering and mechanical engineering sectors. Utilizing materials that possess appropriate mechanical properties and vibration responses can be greatly facilitated by the application of finite element analysis (FEA) and molecular dynamics (MDs) simulation (Torkan and Pirmoradian, 2019).

The field of tourism, particularly medical tourism, holds significant potential from a fresh economic standpoint. Countries may support their populations by leveraging tourism science. By promoting tourism, it is feasible to incentivize the local population and community to deter their migration through the utilization of technology. These communities exhibit contentment with the promotion of tourism and possess a positive image and forward-thinking perspective (Shahmirzadi 2017).

Tourism has the potential to deter indigenous people from migrating to urban centers, since the favorable climate in rural areas might serve as an incentive for them to remain there. This form of ecotourism contributes to the expansion of the economy, particularly in sectors that rely on knowledge and traditional practices (Shahmirzadi 2012).

2. CONCLUSION

Given the rapid advancements in artificial intelligence (AI) and robotics, it is now feasible to assist the medical tourism sector by providing remote medical treatment to patients worldwide, particularly those who are unable to travel.

Today, the utilization of robots in diverse medical domains, ranging from minimally invasive procedures that don't require extensive cuts to intricate surgeries, has become feasible. Telemedicine aims to enhance patient care, expand medical services in rural regions, and provide improved access to clinicians for counseling.

The influence of technology on health tourism has evolved through the advancement of technological frontiers and the proliferation of novel technologies.

Artificial intelligence and robotics enable the execution of remote procedures, which will significantly impact the tourism business, particularly medical tourism.

REFERENCES

Ayuningtyas, D., Ariwibowo, D. (2020). The strategic role of information communication technology in succeeding medical tourism. *Enfermería Clínica, Volume 30*(6).

Barbaz-I, R. (2014). *Experimental determining of the elastic modulus and strength of composites reinforced with two nanoparticles* [Doctoral dissertation, Iran University of Science and Technology, Tehran, Iran].

Carrera, P. M., & Bridges, J. F. (2006). Globalization and healthcare: Understanding health and medical tourism. [CrossRef] [PubMed]. *Expert Review of Pharmacoeconomics & Outcomes Research*, *6*(4), 447–454. doi:10.1586/14737167.6.4.447 PMID:20528514

Chella, A., Iocchi, L., Macaluso, I., & Nardi, D. (2006). Artificial Intelligence and Robotics. *Intelligenza Artificiale*, *3*(1-2), 87–93.

Fada, R., Farhadi Babadi, N., Azimi, R., Karimian, M., & Shahgholi, M. (2021). Mechanical properties improvement and bone regeneration of calcium phosphate bone cement, Polymethyl methacrylate and glass ionomer. *Journal of Nanoanalysis*, *8*(1), 60–79.

Fazlollahi, M., Morovvati, M. R., & Mollaei Dariani, B. (2019). Theoretical, numerical and experimental investigation of hydro- mechanical deep drawing of steel/polymer/steel sandwich sheets. *Proceedings of the Institution of Mechanical Engineers. Part B, Journal of Engineering Manufacture*, *233*(5), 1529–1546. doi:10.1177/0954405418780173

Ghadirinejad, N., Nejad, M. G., & Alsaadi, N. (2021). A fuzzy logic model and a neuro-fuzzy system development on supercritical CO2 regeneration of Ni/Al2O3 catalysts. *Journal of CO2 Utilization, 54,* 101706.

Ghasemi, M., & Hasanpoor, E. (2021). *COVID-19 attack on the world's tourism industry*. Evidence Based Health Policy, Management and Economics.

Ghasemi, M., Nejad, M. G., Alsaadi, N., Abdel-Jaber, M. T., Yajid, A., Shukri, M., & Habib, M. (2022). Performance measurment and lead- time reduction in epc project-based organizations: A mathematical modeling approach. *Mathematical Problems in Engineering*, *2022*, 2022. doi:10.1155/2022/5767356

Kamarian, S., Bodaghi, M., Isfahani, R. B., & Song, J. I. (2022). *A comparison between the effects of shape memory alloys and carbon.*

Khandan, A., & Esmaeili, S. (2019). Fabrication of polycaprolactone and polylactic acid shapeless scaffolds via fused deposition modelling technology. *Journal of Advanced Materials and Processing*, *7*(4), 16–29.

Lanfranco, A. R., Castellanos, A. E., Desai, J. P., & Meyers, W. C. (2004). Robotic surgery: A current perspective. *Annals of Surgery*, *239*(1), 14–21. doi:10.1097/01.sla.0000103020.19595.7d PMID:14685095

Lucchini, R., Carnelli, D., Gastaldi, D., Shahgholi, M., Contro, R., & Vena, P. (2012). A damage model to simulate nanoindentation tests of lamellar bone at multiple penetration depth. In *6th European Congress on Computational Methods in Applied Sciences and Engineering, ECCOMAS 2012* (pp. 5919-5924). IEEE.

Mahjoory, M., Shahgholi, M., & Karimipour, A. (2022). The effects of initial temperature and pressure on the mechanical properties of reinforced calcium phosphate cement with magnesium nanoparticles: A molecular dynamics approach. *International Communications in Heat and Mass Transfer*, *135*, 106067. doi:10.1016/j.icheatmasstransfer.2022.106067

Malekzadeh, R., Abedi, G., Abedini, E., Haghgoshayie, E., Hasanpoor, E., & Ghasemi, M. (2021). Ethical predictability of patient safety in Iranian hospitals. *International Journal of Risk & Safety in Medicine*, *32*(1), 51–60. doi:10.3233/JRS-200022 PMID:32597821

Marescaux, J., Leroy, J., Rubino, F., Smith, M., Vix, M., Simone, M., & Mutter, D. (2002). Mutter Transcontinental robot-assisted remote telesurgery: Feasibility and potential applications. *Annals of Surgery*, *235*(4), 300–301. doi:10.1097/00000658-200204000-00005 PMID:11923603

Monfared, R. M., Ayatollahi, M. R., & Isfahani, R. B. (2018). Synergistic effects of hybrid MWCNT/nanosilica on the tensile and tribological properties of woven carbon fabric epoxy composites. *Theoretical and Applied Fracture Mechanics*, *96*, 272–284. doi:10.1016/j.tafmec.2018.05.007

Morovvati, M. R., Mollaei-Dariani, B., & Haddadzadeh, M. (2010). Initial blank optimization in multilayer deep drawing process using GONNS. *Journal of Manufacturing Science and Engineering*, *132*(6), 061014. doi:10.1115/1.4003121

Saeedi, M. R., Morovvati, M. R., & Mollaei-Dariani, B. (2020). Experimental and numerical investigation of impact resistance of aluminum–copper cladded sheets using an energy-based damage model. *Journal of the Brazilian Society of Mechanical Sciences and Engineering*, *42*(6), 1–24. doi:10.1007/s40430-020-02397-0

Shahgholi, M., Oliviero, S., Baino, F., Vitale-Brovarone, C., Gastaldi, D., & Vena, P. (2016). Mechanical characterization of glass-ceramic scaffolds at multiple characteristic lengths through nanoindentation. *Journal of the European Ceramic Society*, *36*(9), 2403–2409. doi:10.1016/j.jeurceramsoc.2016.01.042

Shahmirzadi, E. K. (2012). *Community Based Tourism (CBT) Planning and Possibilities: The Case of Shahmirzad, Iran* [Doctoral dissertation, Eastern Mediterranean University (EMU)].

Shahmirzadi, E. K. (2017). *Pro- Tourism and Anti-Tourism Community Groups at a World Heritage Site in Turkey*. Pamukkale.

TalebiM.Abbasi-RadS.MalekzadehM.ShahgholiM.ArdakaniA. A.FoudehK., & Rad,

Talebi, M., Abbasi-Rad, S., Malekzadeh, M., Shahgholi, M., Ardakani, A. A., Foudeh, K., & Rad, H. S.H. S. (2021). Cortical bone mechanical assessment via free water relaxometry at 3 T. *Journal of Magnetic Resonance Imaging*, *54*(6), 1744–1751. doi:10.1002/jmri.27765 PMID:34142413

Torkan, E., & Pirmoradian, M. (2019). Efficient higher-order shear deformation theories for instability analysis of plates carrying a mass moving on an elliptical path. *Journal of Solid Mechanics*, *11*(4), 790–808.

World Health Organization. (2010). *Telemedicine: Opportunities and Developments in Member States. Report on the Second Global Survey on EHealth*. World Health Organization.

Chapter 8
Redefining Industry Dynamics:
The Impact of AI and Robotics on Competitive Advantage

Shakeel Basheer
https://orcid.org/0000-0001-7854-8285
Lovely Professional University, India

Ishfaq Tramboo
https://orcid.org/0000-0002-8303-4323
Lovely Professional University, India

Sheezan Farooq
Islamic University of Science and Technology, Indonesia

Zahra Behboodi
Kharazmi University, Iran

ABSTRACT

This study examines how AI and robots shift competitive advantage across industries. Through a literature analysis, case studies, and a robust methodological framework, this study shows how new technologies change operational paradigms, redefine innovation, and put organizations at the forefront of industry dynamics. Strategic AI and robot integration is essential for competitive advantage in manufacturing, healthcare, banking, logistics, and tourism. This theoretical framework helps scholars, practitioners, and policymakers navigate the intricate relationships between AI, robots, and industry dynamics. Finally, the chapter suggests further research on ethical issues, interdisciplinary studies, long-term effects, global perspectives, and human-AI collaboration. This research provides a solid foundation for understanding how AI and robotics are changing the commercial landscape at the technological frontier.

DOI: 10.4018/979-8-3693-2248-2.ch008

1. INTRODUCTION

In the ever-changing environment of modern industries, the combination of artificial intelligence (AI) and robots is a revolutionary force, overturning established paradigms and redefining competitive landscapes (Doborjeh et al., 2022). As we traverse the complicated interplay between technology breakthroughs and market factors, it becomes clear that the convergence of AI and robots is more than a trend; it is a fundamental change with far-reaching repercussions (Subasinghe et al., 2020). This chapter digs into the dynamic junction of these technologies and their far-reaching effects on competitive advantage across industries (Javed et al., 2023). The acceleration of technological progress, particularly in AI and robots, has become a feature of the twenty-first century (Vinuesa et al., 2020). From automated manufacturing processes to intelligent decision-making systems, the range of applications is extensive, affecting many aspects of industry. This technological explosion can be traced back to the convergence of processing power, big data analytics, and machine learning techniques (Yigitcanlar et al., 2021). As sectors face unprecedented challenges and opportunities, understanding the implications of these breakthroughs for competitive advantage becomes critical. The modern corporate environment is characterized by fierce competition, with attaining and maintaining a competitive edge being the crux of organizational success. Traditional competitive advantages, like as cost leadership or product differentiation, are changing as AI and robotics change the rules of engagement (Boguszewicz-Kreft et al., 2020). The essential question arises: How do these technical developments alter the nature of competition, and what strategic imperatives do they impose on enterprises wishing to prosper in this new era (Filieri et al., 2021).

This investigation has far-reaching implications beyond academic discourse; it pervades corporate boardrooms and entrepreneurial initiatives (Camaréna, 2021). Understanding the revolutionary influence of AI and robotics is more than just an academic exercise; it is required for strategic decision-making. Businesses who recognize the consequences early on will have a significant advantage over their competitors, while those that fail to adapt risk obsolescence in an era marked by technological dynamism (Buhalis & Karatay, 2022). This chapter will methodically investigate and analyze the multidimensional relationship between AI, robotics, and competitive advantage. It aims to contribute to the growing corpus of literature on technology-driven industry transitions by synthesizing current research, investigating real-world examples, and developing conceptual frameworks (S. Park & Kim, 2022). Through this exploration, we hope to provide scholars, practitioners, and policymakers with insights that might help them make strategic decisions, stimulate innovation, and negotiate the obstacles posed by the inexorable march of technological advancement.

As we embark on this intellectual adventure, the chapters that follow will chronicle the story of technological disruption, strategic adaptation, and the redefining of industry dynamics in the age of AI and robotics (Farooq et al., 2023). In doing so, we ask readers to join us in deciphering the complexities of a new industrial era defined by the convergence of artificial intelligence and robots.

2. LITERATURE REVIEW

The interaction of artificial intelligence (AI), robotics, and competitive advantage has become a focus of scholarly research in recent years. A variety of literature has evolved, reflecting the rising understanding that these technologies are transformative forces altering sectors around the world, rather than simply novelties (Rokach, 2010). To understand the current discussion, it is necessary to consider the historical trajectory of technology disruptions in industries (Prawiyogi et al., 2022). The industrial revolution was the first wave of mechanization, clearing the way for mass production and ushering in an era of economic transition (Hartwell, 2014). Today's age, known as the Fourth Industrial Revolution, builds on this historical foundation (Hassan, n.d.). The incorporation of AI and robotics into corporate processes represents a paradigm change similar to past technology revolutions, albeit of unparalleled magnitude and complexity. Despite the profusion of research, fundamental gaps remain in our knowledge of how AI and robots redefine competitive advantage (Ayuningtyas & Ariwibowo, 2020). While many studies focus on the benefits of single technologies, few comprehensive analyses explore the systemic consequences of their convergence. Furthermore, research frequently focuses on success stories, leaving out the hurdles, setbacks, and ethical implications involved in the use of these technologies.

Theoretical frameworks help scholars evaluate and analyze the complex links between AI, robotics, and competitive advantage (Basheer, Walia, Mehraj, et al., 2023). The resource-based view (RBV), dynamic capacities theory, and technological acceptance models are often used to explain how firms might use these technologies to obtain a strategic advantage (Heydari Fard et al., 2021). However, more sophisticated frameworks are required that take into consideration the interaction of technology capabilities, organizational structures, and external environmental influences (Gursoy et al., 2022). Traditional sources of competitive advantage, such as cost leadership and differentiation, are evolving in the wake of AI and robotics. Early research suggests that cost gains stem not only from labour savings through automation but also from the increased precision and efficiency of AI-driven operations (Basheer, Walia, Farooq, et al., 2023). Similarly, product distinctiveness is no longer limited to physical features; it also includes the sophisticated algorithms that power products and services.

AI and robotics applications exist in a wide range of industries, each with its own set of difficulties and potential (Wu et al., 2016). Manufacturing, for example, has undergone a paradigm shift with the rise of collaborative robots (cobots) that work alongside humans to increase efficiency. AI-powered diagnostics and robotic surgeries are transforming medical practices (Basheer, Farooq, et al., 2023). The financial sector uses AI algorithms for predictive analytics and risk management. Despite the multiplicity of applications, some themes emerge: the possibility for increased efficiency, cost savings, and the development of new value propositions (Noh, 2022) . While the literature frequently stresses the benefits of AI and robotics adoption, a growing corpus of research provides light on the obstacles and ethical implications that accompany this shift. Issues such as employment displacement owing to automation, algorithmic prejudice, and the ethical implications of AI decision-making are fundamental to arguments about the technologies' societal influence (Kye et al., 2021). Researchers emphasize the importance of responsible AI development and proactive strategies to mitigate the unforeseen repercussions of widespread deployment.

A recurring topic in the literature is the need for organizations to adjust strategically to maximize on the potential provided by AI and robotics (Subasinghe et al., 2020). This includes cultivating an innovative culture, investing in personnel reskilling, and implementing strong cybersecurity measures. Organizations that see these technologies as fundamental parts of their strategic vision, rather than just operational tools, are more likely to realize their full competitive advantage (Mehraj, Qureshi, et al., 2023). Finally, the literature assessment indicates a diverse body of studies on the transformative influence of AI and robotics on competitive advantage . As we progress through this chapter, it becomes clear that, while existing studies provide valuable insights, a complete understanding necessitates a holistic examination of the interconnections, challenges, and opportunities that arise from the integration of these technologies into various industries (Mehraj, Ul Islam, et al., 2023). The next parts will help to fill these gaps by giving a detailed analysis based on both empirical evidence and theoretical frameworks.

At the heart of our investigation into the influence of AI and robotics on competitive advantage is the requirement for precise definitions and concepts (S. M. Park & Kim, 2022). Artificial intelligence comprises a wide range of technologies, including machine learning algorithms capable of predictive analytics and natural language processing systems. Robotics, on the other hand, is the integration of physical machines capable of automating and making autonomous decisions. In this chapter, AI and robotics are viewed as complementary forces driving technological innovation in industries.

3. CONCEPTUAL FRAMEWORK

Our conceptual foundation is based on known theoretical frameworks, including the Resource-Based View (RBV), Dynamic Capabilities Theory, and Technology Acceptance Models (Bolger, 2021). The RBV aids in understanding how companies use their unique resources, such as AI and robotics skills, to obtain a competitive advantage. Dynamic Capabilities Theory helps by emphasizing the importance of organizational adaptability and innovation in the face of technological change. Technology Acceptance Models lead our investigation into how users, including businesses and employees, perceive and accept these technologies (Um et al., 2022). Our conceptual framework suggests that the relationship between AI, robotics, and competitive advantage is dynamic and multidimensional. AI and robotics allow competitive advantage through three main mechanisms: efficiency benefits, innovation, and strategic adaptability. The automation of typical tasks results in increased efficiency, lower costs, and improved operational effectiveness (Monaco & Sacchi, 2023). The ability of these tools to analyse large datasets, provide insights, and promote product/service differentiation fosters innovation. The quickness with which these technologies may respond to market changes promotes strategic adaptability.

Within this framework, we highlight strategic imperatives for firms looking to capitalize on the potential of AI and robots for competitive advantage (Buhalis et al., 2023). This includes cultivating an environment of constant learning and flexibility, investing in staff training and reskilling initiatives, and encouraging cross-functional collaboration. The framework recognizes that the successful integration of AI and robotics into business processes necessitates not only technological expertise, but also cultural and organizational readiness. Recognizing the dynamic nature of industries, our conceptual framework takes into account external environmental elements that determine the impact of AI and robots on competitive advantage, such as regulatory contexts, market dynamics, and public expectations. Organizations must navigate and adapt to these external factors effectively to maximize the benefits and reduce the risks associated with AI and robotics adoption

4. METHODOLOGY

The research design for this chapter is extensive, incorporating parts of literature evaluation, case study analysis, and empirical evidence synthesis. This combination of methodologies enables a comprehensive assessment of the influence of AI and robots on competitive advantage across industries. By combining qualitative and quantitative data, we hope to provide a nuanced exploration that goes beyond theoretical frameworks.

Given the study's broad reach, the population under examination spans multiple industries, including manufacturing, healthcare, finance, and others. Within each industry, a purposive sample technique will be used to identify cases that represent a variety of AI and robotics deployment situations. This ensures a balanced and representative examination of the difficulties, possibilities, and competitive dynamics presented by these technologies. Quantitative survey data will be analysed statistically to detect trends, correlations, and patterns. Thematic analysis will be applied to qualitative data from interviews in order to uncover nuanced perspectives and real-world tales.

5. AI AND ROBOTICS IN INDUSTRY

In the world of medical tourism, robotic surgery has emerged as a breakthrough technique, providing precise and minimally invasive treatments (Casillo et al., 2019). The use of surgical robots allows for greater accuracy, shorter recovery times, and better post-operative outcomes. This draws foreign patients looking for cutting-edge medical treatments with the added benefit of modern robotic surgical procedures. Furthermore, AI plays an important role in personalizing medical therapies through precision medicine. AI algorithms assist generate individualized treatment plans by analysing large datasets and genetic information, ensuring that medical tourists receive care that is tailored to their specific health profiles.

Figure 1. Companies mentioning AI transformation
Source: https://www.business2community.com/statistics-pages/ai-in-tourism

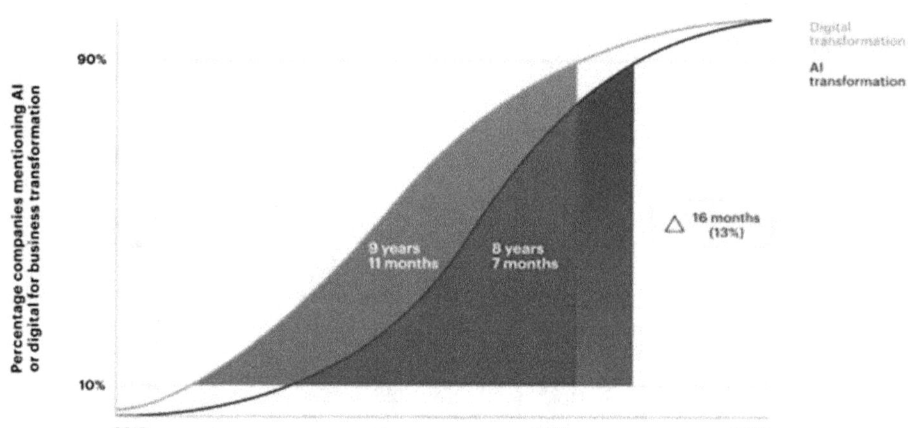

5.1 Robotic Surgery and Precision Medicine

AI-powered telemedicine systems have become essential to the medical tourism business, allowing for distant consultations and follow-up care. Patients can contact with healthcare specialists across boundaries to obtain expert guidance without having to travel extensively (Gretzel et al., 2020). AI algorithms help with diagnostic processes, ensuring that medical consultations conducted remotely maintain high levels of accuracy and effectiveness.

5.2 Remote Consultations and Telemedicine

AI helps to ensure the security and privacy of health data in the context of medical tourism (Alkhatib & Bayouq, 2021). Advanced encryption algorithms and AI-driven cybersecurity protections safeguard sensitive patient data, instilling faith in overseas patients concerned about the privacy of their medical records.

5.3 Health Data Security and Privacy

AI contributes significantly to ensuring the security and privacy of health data in the context of medical tourism (Ku & Chen, 2015). Advanced encryption algorithms and AI-driven cybersecurity measures protect sensitive patient information, fostering trust among international patients who are concerned about the privacy of their medical records.

Figure 2. Medical tourism market size, 2023 to 2032 (USD billion)
Source: https://www.precedenceresearch.com/medical-tourism-market

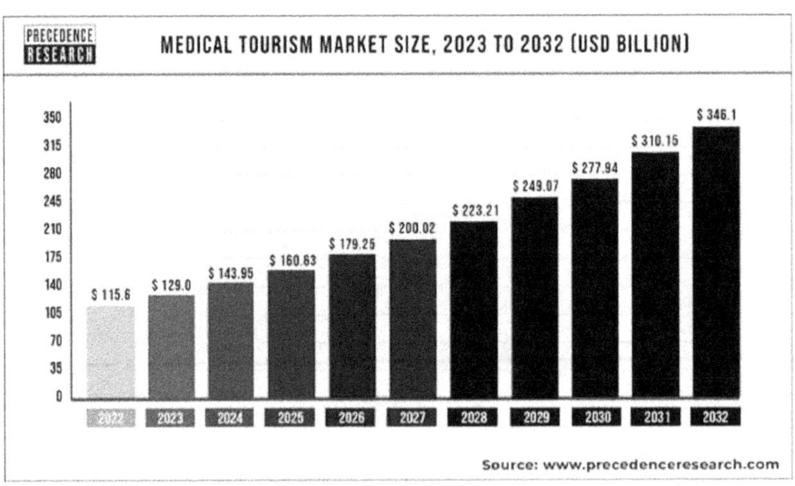

5.4 Enhanced Patient Experience

The application of AI and robotics in medical tourism goes beyond clinical processes to improve the overall patient experience. AI-powered chatbots offer real-time support with travel reservations, housing, and pre- and post-operative care. Robotics in healthcare service delivery, such as robotic nursing assistants, improve the comfort and well-being of international patients on their medical tourism journeys (Ku & Chen, 2015b). Countries and medical facilities that use AI and robotics have a competitive advantage in the global medical tourism sector. The availability of advanced technologies attracts people seeking cutting-edge medical treatments, establishing these locations as hubs for innovative and high-quality healthcare services. Furthermore, the integration of AI and robotics helps to reduce costs by streamlining operations and improving efficiencies, which leads to more competitive pricing for medical procedures.

6. IMPACT ON COMPETITIVE ADVANTAGE

The integration of artificial intelligence (AI) and robots into sectors has far-reaching ramifications for competitive advantage, radically altering how firms operate and compete in the global marketplace (Sahli, 2015). This section delves into the complex influence of AI and robotics on achieving and maintaining a competitive advantage, with a focus on how firms that strategically leverage these technologies can thrive in today's changing business climate.

6.1 Operational Efficiency and Cost Reduction

One of the key drivers of competitive advantage due to AI and robotics is operational efficiency. Routine jobs are automated using robotics and AI-driven algorithms, resulting in streamlined operations, fewer errors, and faster production (Masri et al., 2020). The ability to operate more efficiently results in cost savings, allowing organizations to charge competitive prices while retaining good profit margins. Organizations that incorporate AI and robotics into their workflows can gain economies of scale and operational excellence, putting them ahead of competitors that use traditional, manual procedures.

6.2 Innovation and Product Differentiation

AI drives innovation, enabling firms to develop unique goods and services that set them apart in the market. Machine learning algorithms examine large datasets, revealing patterns and insights that inspire creative problem-solving and innovation. AI allows for the development of cutting-edge solutions in industries such as healthcare, finance, and manufacturing, ranging from personalized medication to algorithmic trading methods. Businesses that innovate with AI and robots gain a competitive advantage by providing new value propositions, distinguishing themselves in an increasingly crowded market.

6.3 Agility and Adaptability

AI and robotics provide agility and adaptability that are important for achieving a competitive advantage in today's rapidly changing business market. These technologies offer real-time data analysis, allowing businesses to react quickly to market shifts, client preferences, and developing trends. Adaptive decision-making powered by AI algorithms enables firms to pivot strategically, capture opportunities, and overcome problems proactively. Organizations that rely on old, manual procedures, on the other hand, may struggle to match the speed and responsiveness of their technologically advanced rivals.

6.4 Customer Experience and Personalization

Customer experience is another area where AI and robotics provide a competitive advantage. Organizations that use these technologies can provide personalized and seamless interactions, ranging from AI-powered chatbots that assist clients in real time to robotics that improve service delivery (Tapar et al., 2017). The capacity to understand and anticipate customer needs using data-driven insights increases happiness and loyalty, resulting in a substantial competitive advantage. Businesses that focus the customer experience with AI and robotics get a strategic advantage in developing long-term connections and increasing repeat business (Vespestad et al., 2019). AI improves strategic decision-making processes by providing firms with predictive analytics and scenario modelling capabilities. This revolutionary ability to foresee market trends, identify risks, and optimize resource allocation promotes informed decision-making. Businesses that use AI for strategic insights gain a competitive advantage by making informed and data-driven decisions, reducing uncertainty, and preparing for long-term success.

AI and robots have far-reaching implications for competitive advantage, influencing operational efficiency, innovation, agility, customer experience, and strategic decision-making. Organizations that strategically adopt and integrate these technologies place themselves at the forefront of their respective sectors (González-Rodríguez et al., 2020a). As this chapter progresses, case studies and empirical evidence will provide light on how firms might negotiate these dynamics, employing AI and robotics to redefine industry standards and gain a competitive advantage in an era of technological upheaval.

7. CASE STUDIES: TRANSFORMATIVE IMPACT OF AI AND ROBOTICS ON COMPETITIVE ADVANTAGE

7.1 Manufacturing Industry: Robotics Revolution at XYZ Automotive

XYZ Automotive, a major player in the automotive manufacturing industry, launched a strategic drive to incorporate robotics into its production processes. Implementing robotic automation for duties ranging from assembly line operations to quality control resulted in a significant increase in operational efficiency. The organization saw a considerable reduction in manufacturing time, improved product quality, and a noticeable fall in error rates. XYZ Automotive's deliberate embrace of robotics not only increased its competitiveness, but also established the company as an industry leader in advanced production. Lessons from this instance stress the significance of a phased and well-planned implementation, constant monitoring, and upskilling the staff to collaborate with robots.

7.2. Healthcare Sector: Robotic-Assisted Surgeries at ABC Medical Centre

ABC Medical Centre accepted robotic-assisted procedures as a pillar of its commitment to offering cutting-edge healthcare services. The introduction of surgical robots improved precision and reduced invasiveness in a variety of medical operations, attracting patients worldwide who sought innovative and less invasive treatment choices. ABC Medical Centre's competitive edge went beyond technology breakthroughs to include improved patient outcomes, shorter recovery periods, and higher patient satisfaction. Key takeaways from this case include the value of investing in cutting-edge technologies, the requirement for complete training for medical practitioners to run these systems efficiently, and the importance of promoting such breakthroughs to overseas consumers.

7. 3 Financial Services: AI-Driven Analytics at DEF Bank

DEF Bank, a financial institution operating in a highly competitive sector, leveraged AI-driven data to acquire a strategic advantage. Machine learning algorithms were used to detect fraud in real time, provide tailored financial advice, and anticipate investment decisions. The deployment resulted in a significant reduction in fraudulent activity, increased customer interaction, and improved risk management. DEF Bank's proactive implementation of AI-driven analytics strengthened its position in the financial services business while also setting a standard for others. This instance demonstrates the need of having a strong data architecture, constantly adapting to emerging AI technologies, and focusing on producing customer-centric AI apps.

7.4 Logistics and Supply Chain: Drones and AI at LMN Logistics

LMN Logistics, a worldwide logistics and supply chain company, used drones and AI to streamline its operations. Drones with AI algorithms were used to track goods in real time, optimize routes, and operate warehouses more efficiently. This approach resulted in considerable delivery time reductions, increased inventory management accuracy, and overall supply chain efficiency. The strategic integration of AI and robots increased LMN Logistics' competitiveness while also establishing the company as an industry leader in harnessing modern technology for logistical excellence. The case emphasizes the necessity of data-driven decision-making, investment in developing technologies, and adaptation in the face of changing market conditions.

7.5 Tourism and Hospitality: AI-Powered Customer Service at PQR Resorts

PQR Resorts, a prominent player in the tourism and hospitality industry, implemented AI-powered customer service solutions to enhance the guest experience. Chatbots equipped with natural language processing capabilities were deployed to provide real-time assistance with reservations, inquiries, and personalized recommendations. This not only improved customer satisfaction but also allowed PQR Resorts to operate with a leaner workforce, focusing human resources on areas requiring personal touch. The case emphasizes the potential of AI in augmenting customer service, the importance of seamless integration into existing processes, and the need for a customer-centric approach in the hospitality sector.

8. DISCUSSION

8.1 Interpretation of Research Findings

The study findings given in this chapter highlight AI and robotics' substantial impact on redefining competitive advantage across multiple industries (González-

Rodríguez et al., 2020b). The interpretation of these findings reveals a common thread: strategic integration of AI and robotics results in increased operational efficiency, innovation, agility, and better customer experiences. From manufacturing to healthcare, banking, logistics, and tourism, firms that use these technologies deliberately position themselves at the forefront of their industry (Bi et al., 2020). The revolutionary nature of these technologies goes beyond simple automation; it includes a fundamental shift in how firms' function, make choices, and create value.

The findings are consistent with and enhance current research on the subject, offering a detailed exploration that includes both theoretical frameworks and real-world case studies. The literature continually highlights the potential of AI and robots to improve efficiency, stimulate innovation, and shape competitive tactics (Lim, 2020). The comparative study offered in this chapter expands on previous research by digging into industry-specific applications, providing insights into the strategic imperatives for firms seeking a competitive advantage. This discussion adds to our understanding of the complex relationships between AI, robots, and industry dynamics by synthesizing the research and using empirical evidence.

8.2 Implications

8.2.1 Implications for Industry Practitioners

The ramifications for industry practitioners are clear: strategic AI and robotics adoption is more than just a technological upgrade; it is a paradigm shift that has the potential to alter the competitive landscape (Cohen, 2014). Organizations that integrate these technologies proactively achieve a variety of benefits, including cost savings, increased innovation capabilities, higher customer satisfaction, and better strategic decision-making. Practitioners should see AI and robots as essential components of their strategic vision, cultivating a culture of continuous learning and modifying organizational structures to maximize the joint potential of humans and machines. Furthermore, the case studies offer practical insights into successful implementations, guiding industry practitioners through the intricacies of technological integration.

Despite the useful insights provided, this study has some drawbacks. To begin, the continuously growing nature of AI and robots means that the findings may not reflect the most recent breakthroughs in the area (Huang & Hsu, 2009). Furthermore, the case studies, while illustrative, are a snapshot of individual circumstances and may not fully represent the range of industry experiences. The study also focuses on successful deployments, which may miss problems and failures that provide useful lessons. Furthermore, the generalizability of findings across diverse cultural and legal contexts must be carefully considered. These constraints highlight the importance of ongoing study and a flexible approach to understanding the changing link between AI, robotics, and competitive advantage.

9. CONCLUSION

In conclusion, this chapter presents a detailed review of how Artificial Intelligence (AI) and robotics are redefining competitive advantage in several industries (Gohary et al., 2020). The findings support a consistent narrative: strategic use of these technologies alters operational processes, ignites creativity, and propels firms to the forefront of their industries. From manufacturing and healthcare to finance, logistics, and tourism, the integration of AI and robotics is emerging not only as an operational update, but also as a paradigm shift with far-reaching consequences. This chapter's contributions include its comprehensive exploration, conceptual framework, extensive case studies, and methodological approach. It reveals the multifaceted effects of AI and robotics, providing practical insights for industry practitioners and establishing the groundwork for future study. As the technological frontier evolves, the chapter proposes new routes of inquiry, focusing on ethical considerations, multidisciplinary studies, long-term impact investigations, global viewpoints, and human-AI collaboration optimization. It is the start of a voyage into a future influenced by technological disruption, in which the combination of AI and robotics reshapes industries and redefines the competitive landscape.

REFERENCES

Alkhatib, G., & Bayouq, S. T. (2021). A TAM-Based Model of Technological Factors Affecting Use of E-Tourism. *International Journal of Tourism and Hospitality Management in the Digital Age*, *5*(2), 50–67. doi:10.4018/IJTHMDA.20210701.oa1

Ayuningtyas, D., & Ariwibowo, D. A. (2020). The strategic role of information communication technology in succeeding medical tourism. *Enfermeria Clinica*, *30*, 170–173. doi:10.1016/j.enfcli.2020.06.039

Basheer, S., Farooq, S., Hassan, V., Malik, Y. M., & Reshi, M. A. (2023). *Augmented Reality and Virtual Reality in Cultural Heritage Tourism Enhancing Visitor Experiences*., doi:10.4018/978-1-6684-9957-3.ch002

Basheer, S., Walia, S., Farooq, S., Shah, M. A., & Mir, F. A. (2023). Exploring the metaverse: The future of tourism through ai and virtual reality. In *Influencer Marketing Applications Within the Metaverse* (pp. 195–204). IGI Global. doi:10.4018/978-1-6684-8898-0.ch012

Basheer, S., Walia, S., Mehraj, D., & Reshi, M. (2023). Creating Transformative Travel Unveiling the Link Between Meaningful Experiences and Sustainable Tourism. IGI Global. doi:10.4018/979-8-3693-0650-5.ch002

Bi, N., Yin, C. Y., & Chen, Y. (2020). A bittersweet experience! The effect of mixed emotions on business tourists' revisit intentions. *Journal of Travel & Tourism Marketing*, *37*(6), 695–710. doi:10.1080/10548408.2020.1795047

Boguszewicz-Kreft, M., Kuczamer-Kłopotowska, S., Kozłowski, A., Ayci, A., & Abuhashesh, M. (2020). The theory of planned behaviour in medical tourism: International comparison in the young consumer segment. *International Journal of Environmental Research and Public Health*, *17*(5), 1626. doi:10.3390/ijerph17051626 PMID:32138240

Bolger, R. K. (2021). Finding wholes in the metaverse: Posthuman mystics as agents of evolutionary contextualization. *Religions*, *12*(9), 768. doi:10.3390/rel12090768

Buhalis, D., & Karatay, N. (2022). Mixed Reality (MR) for Generation Z in Cultural Heritage Tourism Towards Metaverse. In Information and Communication Technologies in Tourism 2022. doi:10.1007/978-3-030-94751-4_2

Buhalis, D., Leung, D., & Lin, M. (2023). Metaverse as a disruptive technology revolutionising tourism management and marketing. In *Tourism Management* (Vol. 97). Elsevier Ltd. doi:10.1016/j.tourman.2023.104724

Camaréna, S. (2021). Engaging with artificial intelligence (AI) with a bottom-up approach for the purpose of sustainability: Victorian farmers market association, Melbourne Australia. *Sustainability (Basel)*, *13*(16), 9314. doi:10.3390/su13169314

Casillo, M., Clarizia, F., Colace, F., Lombardi, M., Pascale, F., & Santaniello, D. (2019). An approach for recommending contextualized services in e-tourism. *Information (Basel)*, *10*(5), 180. doi:10.3390/info10050180

Cohen, E. (2014). Recreational Hunting: Ethics, Experiences and Commoditization. *Tourism Recreation Research*, *39*(1), 3–17. doi:10.1080/02508281.2014.11081323

Doborjeh, Z., Hemmington, N., Doborjeh, M., & Kasabov, N. (2022). Artificial intelligence: A systematic review of methods and applications in hospitality and tourism. *International Journal of Contemporary Hospitality Management, 34*(3), 1154–1176. doi:10.1108/IJCHM-06-2021-0767

Farooq, S., Farooq, B., Basheer, S., & Walia, S. (2023). *Balancing Environmental Sustainability and Privacy Ethical Dilemmas in AI-Enabled Smart Cities.*, doi:10.4018/979-8-3693-0892-9.ch013

Filieri, R., D'Amico, E., Destefanis, A., Paolucci, E., & Raguseo, E. (2021). Artificial intelligence (AI) for tourism: An European-based study on successful AI tourism start-ups. *International Journal of Contemporary Hospitality Management, 33*(11), 4099–4125. doi:10.1108/IJCHM-02-2021-0220

Gohary, A., Pourazizi, L., Madani, F., & Chan, E. Y. (2020). Examining Iranian tourists' memorable experiences on destination satisfaction and behavioral intentions. In Current Issues in Tourism, 23(2). Routledge. doi:10.1080/13683500.2018.1560397

González-Rodríguez, M. R., Díaz-Fernández, M. C., & Pino-Mejías, M. Á. (2020a). The impact of virtual reality technology on tourists' experience: A textual data analysis. *Soft Computing, 24*(18), 13879–13892. doi:10.1007/s00500-020-04883-y

González-Rodríguez, M. R., Díaz-Fernández, M. C., & Pino-Mejías, M. Á. (2020b). The impact of virtual reality technology on tourists' experience: A textual data analysis. *Soft Computing, 24*(18), 13879–13892. doi:10.1007/s00500-020-04883-y

Gretzel, U., Fuchs, M., Baggio, R., Hoepken, W., Law, R., Neidhardt, J., Pesonen, J., Zanker, M., & Xiang, Z. (2020). e-Tourism beyond COVID-19: A call for transformative research. *Information Technology & Tourism, 22*(2), 187–203. doi:10.1007/s40558-020-00181-3

Gursoy, D., Malodia, S., & Dhir, A. (2022). The metaverse in the hospitality and tourism industry: An overview of current trends and future research directions. *Journal of Hospitality Marketing & Management, 31*(5), 527–534. doi:10.1080/19368623.2022.2072504

Hartwell, H. (2014). Medical Tourism: The Ethics, Regulation and Marketing of Health Mobility. *Tourism Management, 45*, 1–2. doi:10.1016/j.tourman.2014.04.001

Hassan, V. (n.d.). *The role of pro-poor tourism in the development of the local economy: A case study of Tripoli.* Research Gate. https://www.researchgate.net/publication/375869901

Heydari Fard, M., Sanayei, A., & Ansari, A. (2021). Determinants of Medical Tourists' Revisit and Recommend Intention. *International Journal of Hospitality & Tourism Administration*, *22*(4), 429–454. doi:10.1080/15256480.2019.1650688

Huang, S., & Hsu, C. H. C. (2009). Effects of travel motivation, past experience, perceived constraint, and attitude on revisit intention. *Journal of Travel Research*, *48*(1), 29–44. doi:10.1177/0047287508328793

Javed, A. R., Ahmed, W., Pandya, S., Maddikunta, P. K. R., Alazab, M., & Gadekallu, T. R. (2023). A Survey of Explainable Artificial Intelligence for Smart Cities. In Electronics (Switzerland), 12(4). MDPI. doi:10.3390/electronics12041020

Ku, E. C. S., & Der Chen, C. (2015a). Cultivating travellers' revisit intention to e-tourism service: The moderating effect of website interactivity. *Behaviour & Information Technology*, *34*(5), 465–478. doi:10.1080/0144929X.2014.978376

Ku, E. C. S., & Der Chen, C. (2015b). Cultivating travellers' revisit intention to e-tourism service: The moderating effect of website interactivity. *Behaviour & Information Technology*, *34*(5), 465–478. doi:10.1080/0144929X.2014.978376

Kye, B., Han, N., Kim, E., Park, Y., & Jo, S. (2021). Educational applications of metaverse: Possibilities and limitations. In Journal of Educational Evaluation for Health Professions, 18. doi:10.3352/jeehp.2021.18.32

Lim, J. E. (2020). Understanding the discrimination experienced by customers with disabilities in the tourism and hospitality industry: The case of Seoul in South Korea. *Sustainability (Basel)*, *12*(18), 7328. doi:10.3390/su12187328

Masri, N. W., You, J. J., Ruangkanjanases, A., Chen, S. C., & Pan, C. I. (2020). Assessing the effects of information system quality and relationship quality on continuance intention in e-tourism. *International Journal of Environmental Research and Public Health*, *17*(1), 174. Advance online publication. doi:10.3390/ijerph17010174 PMID:31881762

Mehraj, D., Qureshi, I. H., Singh, G., Nazir, N. A., basheer, S., & Nissa, V. (2023, December). basheer, S., & Nissa, V. u. (2023). Green marketing practices and green consumer behavior: Demographic differences among young consumers. *Business Strategy & Development*, *6*(4), 571–585. Advance online publication. doi:10.1002/bsd2.263

Mehraj, D., Ul Islam, M. I., Qureshi, I. H., Basheer, S., Baba, M. M., Nissa, V. u., & Asif Shah, M. (2023). Factors affecting entrepreneurial intention for sustainable tourism among the students of higher education institutions. In *Cogent Business and Management, 10*(3). Cogent OA. doi:10.1080/23311975.2023.2256484

Monaco, S., & Sacchi, G. (2023). Travelling the Metaverse: Potential Benefits and Main Challenges for Tourism Sectors and Research Applications. *Sustainability (Basel)*, *15*(4), 3348. doi:10.3390/su15043348

Noh, H.-K. (2022). Metaverse-related Issues in Tourism Using News Big Data. *Journal of Tourism and Leisure Research*, *34*(2), 151–166. doi:10.31336/JTLR.2022.2.34.2.151

Park, S., & Kim, S. (2022). Identifying World Types to Deliver Gameful Experiences for Sustainable Learning in the Metaverse. *Sustainability (Basel)*, *14*(3), 1361. doi:10.3390/su14031361

Park, S. M., & Kim, Y. G. (2022). A Metaverse: Taxonomy, Components, Applications, and Open Challenges. *IEEE Access : Practical Innovations, Open Solutions*, *10*, 4209–4251. doi:10.1109/ACCESS.2021.3140175

Prawiyogi, A. G., Purnama, S., Meria, L., & Giri, A. (2022). Smart Cities Using Machine Learning and Intelligent Applications. [ITALIC]. *International Transactions on Artificial Intelligence*, *1*(1), 102–116. doi:10.33050/italic.v1i1.204

Rokach, L. (2010). Ensemble-based classifiers. *Artificial Intelligence Review*, *33*(1–2), 1–39. doi:10.1007/s10462-009-9124-7

Sahli, A. (2015). Revisiting Perceived Risk and Trust in E-tourism Context: Toward an Extended Technology Acceptance Model. *Journal of Marketing Research & Case Studies*, 1–14. doi:10.5171/2015.516086

Subasinghe, M., Magalage, D., Amadoru, N., Amarathunga, L., Bhanupriya, N., & Wijekoon, J. L. (2020). Effectiveness of artificial intelligence, decentralized and distributed systems for prediction and secure channelling for Medical Tourism. *11th Annual IEEE Information Technology, Electronics and Mobile Communication Conference, IEMCON 2020*, (pp. 314–319). IEEE. 10.1109/IEMCON51383.2020.9284898

Tapar, A. V., Dhaigude, A. S., & Jawed, M. S. (2017). Customer experience-based satisfaction and behavioural intention in adventure tourism: Exploring the mediating role of commitment. *Tourism Recreation Research*, *42*(3), 344–355. doi:10.1080/02508281.2017.1303564

Um, T., Kim, H., Kim, H., Lee, J., Koo, C., & Chung, N. (2022). Travel Incheon as a Metaverse: Smart Tourism Cities Development Case in Korea. In Information and Communication Technologies in Tourism 2022. doi:10.1007/978-3-030-94751-4_20

Vespestad, M. K., Lindberg, F., & Mossberg, L. (2019). Value in tourist experiences: How nature-based experiential styles influence value in climbing. *Tourist Studies*, *19*(4), 453–474. doi:10.1177/1468797619837966

Vinuesa, R., Azizpour, H., Leite, I., Balaam, M., Dignum, V., Domisch, S., Felländer, A., Langhans, S. D., Tegmark, M., & Fuso Nerini, F. (2020). The role of artificial intelligence in achieving the Sustainable Development Goals. In Nature Communications, 11(1). doi:10.1038/s41467-019-14108-y

Wu, H. C., Li, T., & Li, M. Y. (2016). A Study of Behavioral Intentions, Patient Satisfaction, Perceived Value, Patient Trust and Experiential Quality for Medical Tourists. *Journal of Quality Assurance in Hospitality & Tourism*, *17*(2), 114–150. doi:10.1080/1528008X.2015.1042621

Yigitcanlar, T., Corchado, J. M., Mehmood, R., Li, R. Y. M., Mossberger, K., & Desouza, K. (2021). Responsible urban innovation with local government artificial intelligence (Ai): A conceptual framework and research agenda. *Journal of Open Innovation*, *7*(1), 1–16. doi:10.3390/joitmc7010071

Chapter 9
Artificial Intelligence-Enhanced Global Healthcare: The Future of Medical Tourism

Sheikh Najam mu Sahar
Lovely Professional University, India

Hafizullah Dar
https://orcid.org/0000-0003-2388-9474
Lovely Professional University, India

ABSTRACT

Health and tourism are benefiting from AI technology. This study examines medical tourism's positive outlook and emphasizes AI's role in improving the complete experience for foreign patients. Qualitative desk research and thorough review of contemporary research literature, including content analysis, identified AI and smart healthcare system (SHS) dimensions in medical tourism. The study found that AI is helping medical agencies connect physicians with prospective medical tourists via telehealth and virtual assistance. Results demonstrate that medical tourists utilize AI to research hospitals, treatments, check-ups, and consultations. AI secures medical data exchange and blockchain-based cashless transactions. Medical travelers find AI-enabled smart health devices useful for tracking health data. AI problems in medical tourism are highlighted in the study. The study concludes with insights on AI's transformative impact on medical tourism. The chapter helps comprehend the changing dynamics of AI, healthcare, AI's role in improving medical tourism and addresses its issues.

DOI: 10.4018/979-8-3693-2248-2.ch009

1. INTRODUCTION

The landscape of healthcare is in the midst of a profound transformation, with Artificial Intelligence (AI) emerging as a pivotal force reshaping the medical tourism industry. AI's capacity to enhance diagnostic accuracy, streamline patient care, and enable surgical innovation has created new opportunities for the global healthcare sector. According to Wong and Sa'aid Hazley (2020), technology has played a substantial role in shaping the healthcare business, and the growth of the medical tourism industry has been accelerated by the introduction of the latest technologies in the healthcare sector. Artificial Intelligence (AI), in particular, has been a transformative force in medical tourism, propelling the sector to new heights and advancements. AI applications in medical tourism have not only facilitated the delivery of quality treatments but have also streamlined the overall patient experience, marking a significant shift in the dynamics of the industry. The integration of Smart Healthcare Systems (SHS) leveraging Artificial Intelligence (AI) technologies plays a pivotal role in the healthcare ecosystem, especially in the context of monitoring medical tourists. This utilization of AI contributes to reducing treatment costs and streamlining information workflows. By incorporating advanced technologies, SHS enhances the overall efficiency, effectiveness, and cost-effectiveness of healthcare services, benefiting both providers and medical tourists.

With increased demand and expansion in medical tourism throughout the world, digital technologies under smart healthcare systems (SHS) are now offering a stronger boost to the sector. Healthcare providers, medical tourists, medical tourist facilitators (MTFs), the local community, and the government are the primary players in medical tourism and SHS leads to their network integration to benefit medical tourism (Dar & Kashyap, 2022). According to Akpojaro and Bello (2019), applications of Smart Healthcare Systems (SHS) including AI, play a crucial role in enhancing the quality of healthcare and medical tourism. These applications extend from fitness tracking to the remote monitoring of medical tourists within hospital settings. The integration of SHS technologies allows for continuous monitoring and tracking, contributing to improved healthcare outcomes and a more seamless and efficient experience for medical tourists. AI-enabled devices utilized in medical services and healthcare demonstrate the capability to surpass human physicians in making precise measurements for disease detection (Gärtner et al. (2020). This heightened accuracy is seen as a crucial factor in improving the overall customer experience. Further this accuracy and precision offered by AI technology contribute to a positive user experience by instilling confidence and fostering a better understanding of the medical information provided (Ameen et al. (2021). In addition to accessing information, the internet and mobile devices have practically and effortlessly connected medical tourists, healthcare providers, and other stakeholders, regardless of geography

(Nazir et al., 2019). Online websites are used to organize medical travel services. (Gada, 2019). AI and its applied technologies have helped to increase the number of medical tourists by increasing the number of personalized advertisements displayed in people's internet browsers. In addition, incoming medical tourists use advanced maps to travel a foreign country and obtain general information about the medical destination, facility, and fees (Senbekov et al.,2020). They use mobile health apps to get second views from healthcare persons. Online websites are used to organise medical travel services. (Gada, 2019).

The digitization of medical records has improved medical tourist care and safety by making it easier to understand and read medical tourist data, implying the important role of blockchain technology, an applied AI assisted tool. Medical tourists' medical records are accessible for follow-up care in their native country (Deloitte, 2018). As a result, cloud technology comes into the picture, assisting medical tourism destinations and stakeholders in welcoming more medical tourists from all over the world by establishing trust and trustworthiness. Cloud technology has aided the expansion of medical tourism, allowing medical tourists to seek the care they expect while having their medical data and information saved digitally (Shabarish & Shivashankar, 2019). Further AI health tools help in enhancing the patient's experience in Medical Tourism by improving interaction between health providers and patients, before and after medical travel, ensuring transparent access to information, and supporting the recovery process once the patient returns home following medical interventions (Goel, 2019).

This study seeks to examine the current state of AI integration in the medical tourism industry. Through an extensive literature it explores different dimensions and applications of AI in healthcare and medical tourism. AI coupled with other technological advancements has thoroughly been discussed. Further, this study identifies the Future aspects, challenges and ethical issues associated with AI in medical tourism as well.

2. LITERATURE REVIEW

2.1 Medical Tourism

Medical tourism, in health and medical tourism literature, refers to the phenomenon of medical tourists going from one country to another to get medical and related services and facilities (Wong & Hazley, 2020). Foreign people seek medical tourism owing to issues such as high treatment costs, inadequate quality, long wait times, weak health systems, limited treatment options, lack of insurance coverage, and lack of local knowledge (Collins et al., 2019). Medical tourism aims to offer 'cost-effective' private medical care for clients seeking surgical and specialized treatments in the tourist industry. Cosmetic surgery, orthopedic surgery, heart surgery and dental operations are among the most popular medical tourism procedures. Patients and their families are lured to the places for the convenience, greater quality, and/or reduced cost of care. As a result, the success of medical tourism depends on the overall health of a country's medical sector, as well as appropriate government regulations, patient safeguards, training standards, insurance frameworks, travel and visa restrictions, and other factors that influence the patient's experience and treatment outcomes (Ministry of Tourism, 2022). Although it conducts surgeries overseas, including dental treatments, reproductive therapy, cosmetic surgery, and essential interventions such as heart and organ transplants (Ayuningtyas & Ariwibowo, 2020). More than 130 countries are competing for a share of the global market. Top medical tourism destinations include India, Brunei, Cuba, Columbia, Hong Kong, Hungary, Jordan, Malaysia, Singapore, South Africa, Thailand, and the United States. According to Grandviewresearch, (2020), the worldwide medical tourism industry is expected to reach USD 207.9 billion by 2027, rising at a 21.1% CAGR.

According to Health-tourism, 2019, the medical tourism industry is expected to be worth USD 65-87.5 billion with medical tourists on average spending USD 3410 on every visit to a medical destination. Furthermore, this sector is a market of hundreds of hotels, hospitals, clinics, and it is expanding at a pace of 15-25%. Patients Beyond Borders, an international health travel data source, estimates that the worldwide medical tourism sector is valued between $65 billion and $87.5 billion. Medical tourism is fast expanding and becoming a global trend industry worth billions of dollars, while being a relatively minor specialty in the tourist market (Ghassemi et al., 2021). Medical tourism is gaining popularity in developing countries, after initially focusing on the developed nations. Developing countries aim to improve their health systems by boosting productivity and encouraging innovation, increasing foreign exchange gains, and driving GDP development (Rahman, 2019).

The majority of medical tourism markets come from affluent nations such as the United States, Western Europe, Great Britain, Canada, Australia, and the Middle East (Chandra, 2019). Medical tourists often search for destinations where they may relax during their trip while receiving affordable medical treatment. Thailand, India, Singapore, Hungary, Israel, Greece, Mexico, Dominican Republic, Peru, Jordan, Lithuania, Malaysia, South Africa, Argentina, Philippines, Brunei, Cuba, Columbia, Hong Kong, Hungary, Jordan, Costa Rica, Malaysia, and the United States are among the world's most popular medical tourism destinations(Chandra, 2019). As per Medical Tourism Magazine, (2021), The medical tourism market is expected to reach USD 160.8 billion by 2025. Hence, government health and tourist ministries throughout the world are increasingly focused on their attempts to grab their part of this rich sector (Hassan & Noaman, 2017).

2.2 Artificial Intelligence (AI)

As per Tater et al., (2022), AI is the intelligence displayed by machines which aids in the execution of various tasks using sentiment analysis and Natural Language Processing (NLP). Janiesch *et al.*, (2021), stated that Machine learning (an approach to achieve AI) and deep learning (branch of AI) together use massive datasets to train computers to mimic human intelligence. This technology enables the machines to learn from past data and information, make sense of it, and apply it to a variety of commercial tasks. As per Vishnoi et al., (2018), Artificial Intelligence (AI) refers to the natural predispositions, genetic inheritance or learned skill sets forming the core of individual personalities. It comprises of machine taught applications and uses computer programs to mimic human decision-making capabilities. AI systems incorporate human intelligence with business and marketing tasks including sales, budgeting, and decision-making to enhance consumer advantages (Advani, 2021). AI devices have shown the ability to replicate and even surpass human performance in areas such as learning, judgement, and emotion detection (Wright et al., 2018, Manyika et al., 2017). AI enables computers to simulate intelligent conduct with minimum human interaction, and it has been proved to outperform humans at some tasks. In 2017, it was claimed that deep neural networks (a branch of AI), has been used effectively to assess skin cancer photos with more accuracy than a dermatologist, as well as to identify diabetic retinopathy from retinal images (The Lancet, 2017).

2.3 AI in Healthcare

Artificial intelligence (AI), after captivating different sectors, is now being hailed as a technology that will help provide access and assistance for all to quality medical care, including the development and improvement of diagnostics, personalized medical care, the prevention of illnesses and the discovering new treatments (Parry et al., 2020). The advancement of AI is revolutionizing various aspects of human life as stated by Uzir et al., (2021) including healthcare. This advancement as researched by Pesapane et al., (2018) has resulted in a shift in areas of healthcare including health applications, big data storage, data gathering and analysis, reporting, follow-up scheduling. As per Tian et al., (2019), AI technology is increasingly transforming the health and tourism industries. According to (Parry et al., 2020), AI for health includes Machine learning (a core component of AI that provides systems with the ability to automatically learn and improve without being explicitly programmed), NLP(subfield of AI that helps computers understand, interpret and manipulate human language), speech recognition (text-to-speech and speech-to-text), image recognition and machine vision, expert systems (a computer system that emulates the decision-making ability of a human expert), robotics, and systems for planning, scheduling and optimization. According to Sundaravadivel et al., (2017), the implementation of AI methods play a pivotal role in predicting future illnesses and preventing pharmaceutical shortages. The advancement of AI and data science has resulted in many opportunities for hospital management. As researched by Yang et al., (2022), AI is being used in hospital management for hospital work-flow, Human resource, medical devices, In-Patient management (discharging patients, scheduling and assigning inpatient beds), medical imaging and diagnosis, drug discovery and precision medicine. This indicates that applying AI and data science appropriately is improving healthcare management and patient care.

2.4 Smart Healthcare System in Medical Tourism

Adoption of technology in health care is a critical predictor of productivity in this sector, and technology is usually regarded as the primary engine of long-term productivity improvements in the larger economy (Jorgenson, 2011). The integration of information and communication technology (ICT) has propelled conventional tourism into the era of smart tourism, and further the confluence of these technological services in healthcare and smart tourism has given rise to the notion of smart health tourism (Nazir *et al.*,2019). The concept of a Smart Healthcare System (SHS) is both conceptually and practically supported by the principles of smart tourism, as highlighted by Balandina et al. (2015). An SHS is essentially a network of wireless and interconnected wearable devices equipped with electronics, software, sensors and actuators facilitated by AI and connections to enable the collection, storage, and sharing of data, as outlined by Kelly et al. (2020). This interconnected system has the potential to facilitate proper diagnosis, consultation, and treatment of medical tourists, engage the local community, and enhance the overall efficiency of the medical destination. The integration of technology within the healthcare system reflects a broader trend towards creating smarter and more efficient solutions in the medical tourism industry.

The Smart Healthcare System (SHS), as described by Ghosh (2021) in Figure 1 operates in five stages with functions facilitated by Artificial Intelligence (AI). In the initial stage, users such as medical tourists, physicians, and family members register on the portal and enter and retrieve data from the database. In the second stage, devices, applications, and sensors receive data from users and transfer it to the third stage, where it is verified, pre-processed, and appraised for usable information. The fourth stage involves converting acquired data to digital representation and connecting devices to the internet. Finally, in the fifth step, data is checked, stored in data centers, monitored, and maintained, with additional activities planned to ensure the system's continued and efficient functioning.

Figure 1. SHS framework
Source: Ghosh (2021)
Source: DOI:10.1109/JIOT.2019.2941767

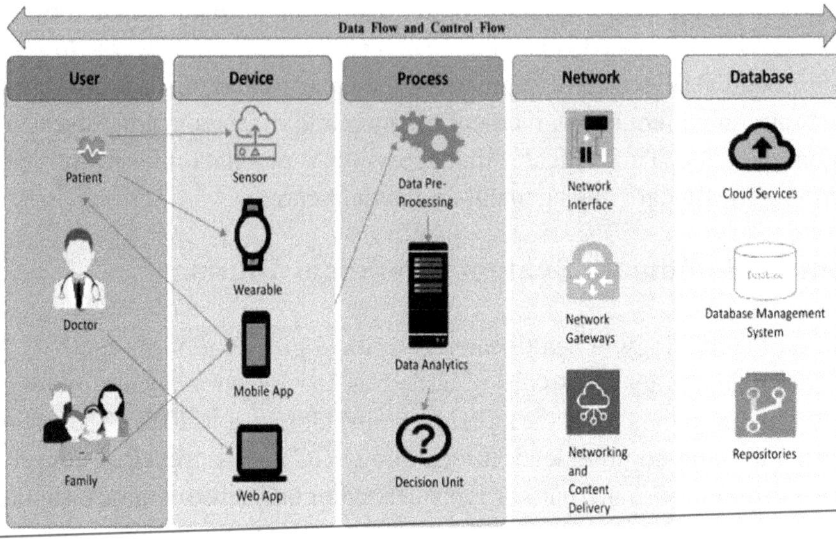

2.5 AI as an Application of SHS in Medical Tourism

According to Dar and Kashyap (2022), an evaluation of technology usage in Smart Healthcare Systems (SHS) reveals its significant impact on medical tourism and the network integration of various stakeholders to enhance the overall medical tourism experience. The study highlights several key technological applications within the context of expanding medical tourism through SHS. These applications include mobile health, telehealth, blockchain technology, cloud technology, the Internet of Medical Things (IOMT), virtual reality all backed and facilitated by AI. Each of these technologies plays a crucial role in shaping and advancing the landscape of medical tourism, offering new possibilities for improved healthcare services, communication, and coordination among stakeholders.

Deepthi and Bansal (2023) claim that the use of AI in medical tourism has considerably enhanced the client experience, resulting in widespread adoption of AI-powered chatbots and intelligent travel assistants. Existing research by Gretzel (2011) and Tussyadiah and Miller (2019) demonstrates that AI benefits customers by assisting them in finding more relevant information, increasing mobility, improving decision-making, and ultimately giving a better travel experience. Tian et al. (2019) emphasised the rapid growth of AI in both the health and tourism industries, opening up new opportunities for technology-enabled medical tourism services and personalised treatments. Uzir et al. (2021) emphasise that AI may play an important role in monitoring medical tourists, lowering treatment costs, and expediting the flow of medical tourist information.

This study delves into the optimistic outlook for medical tourism, emphasizing the crucial role of Artificial Intelligence (AI) in enhancing the overall experience for individuals seeking healthcare services beyond their native borders. The narrative discusses how AI contributes in creating a more enriching and streamlined experience for medical tourists in their quest for quality care internationally.

The research is qualitative in nature, employing a desk research approach. Data was gathered through an extensive literature review of recent research from reputable sources such as journals, reports, magazines, websites, and other publications. The literature review involved content analysis to identify key concepts related to AI and medical tourism. Specifically, the search focused on concepts like Artificial Intelligence, medical tourism, Robotics, smart healthcare, medical tourist, internet of medical things (IoMT), technologies in healthcare, blockchain technology, medical tourism facilitator, applications of technologies, smart technologies, smart tourism, mobile health, telehealth, online consultation, wearable health devices, etc. The purpose is to develop an integrated framework for the intersection of Artificial Intelligence and Medical tourism based on these identified concepts.

3. INTEGRATION, DIMENSIONS, AND EFFECTS OF AI IN MEDICAL TOURISM

3.1 AI and Mobile Health Internet

Medical tourists utilize digital technologies for navigation and obtaining information about Medical Tourist Facilitators (MFTs), medical service providers, and the destination country's healthcare offerings (Parekh et al., 2021). AI assists the users to access better and more relevant information, enhancing mobility, and ultimately contributing to an improved overall tourism experience (Tussyadiah and Miller,2019). Use of AI simplifies travel arrangements, aids tourists in making informed decisions about their future trips, and has a substantial impact on enhancing the overall tourist experience (Elsoud and Morsy, 2022). Moreover, automatic translation features of AI were found to facilitate tourists' navigation of the destination, enabling them to explore and engage in a wide range of activities. This suggests that personalized experiences and language translation, enabled by AI technology, contribute to a more enriched and seamless tourism experience for medical travellers (Benckendorff et al., 2019). These perspectives highlight the multifaceted positive influence of AI on various aspects of the tourism industry.

3.2 AI and Data Safety

Blockchain, a relatively new technology for safeguarding and maintaining medical data within the health and medical tourism, is described as a "digital and decentralized" ledger in which transactions are logged and connected in chronological order to generate tamper-proof records" (Rejeb et al., 2019). AI along with blockchain plays a crucial role in ensuring data security and facilitating safer payment transactions. It is an appropriate network for transferring and securing medical tourists' data, as well as improving the security and transparency of shared medical data (Balasubramanian et al., 2022). AI coupled with cloud technology allows healthcare practitioners to view the medical data of patients in real time. (Senbekov et al., 2020). These technologies contribute to minimizing challenges associated with bank intermediation, currency volatility, and transaction delays, medical record safety thereby enhancing the overall efficiency and security of financial transactions in the context of medical tourism (Basheer, Walia, Farooq, et al., 2023)..Medical tourist clinical history, diagnostic test scans, past medical conditions, electronic prescriptions, laboratory data, and discharge summaries can be accessed in digital formats in the form of (EHR) electronic health records (Premarathne et al., 2016). Adopting a standardized EHR enhances the capabilities of medical institutions in developing nations, which helps them to fetch more medical tourists from developed countries. Furthermore, EMR's can be used to coordinate and access medical tourist care across many hospitals and nations (Senbekov et al., 2020).

3.3 AI and Virtual Consultations

AI facilitates remote interactions between health experts and care seekers through telehealth and telemedicine apps (Luxton, 2016). The patients are able to consult with doctors or medical professionals virtually in a surreal environment using virtual reality (VR) technology. Consultations between doctors and patients are recorded, and the data, including images, speech, text, videos, clinical data, etc., is gathered, transferred, and stored on the blockchain. This data is then utilized for disease prediction and diagnosis through explainable artificial intelligence models (Ali et al., 2022).The use of AI enables the interactions between health professionals and care seekers at a distant, providing necessary treatment recommendations (Luxton, 2016), a very useful communication and interaction tool between health tourists and the healthcare professionals overseas (Farooq et al., 2023). This in future may reduce face-to-face appointments for health tourists, particularly for follow-up sessions after returning to their home country, and thus generate more convenient and costeffective medical experience (Wong & Hazley, 2020). Virtual Reality (VR) is also employed to allow medical travellers to virtually visit different potential hospitals before the visit and various locations, including their hometowns, providing a sense of familiarity after the visit and during the medical treatment (Wong and Hazley, 2020). Vibrant sights in virtual reality are also used to divert medical tourists during treatment. Additionally, VR is also utilized to inform and explain a medical tourist's treatment options.

3.4 AI and Health Monitoring

AI-enabled smartwatches help medical tourists to monitor and manage their own health. These gadgets employ artificial intelligence to give users with tools and features that promote personal health tracking, allowing individuals to actively monitor and manage their well-being. (Uzir et al., 2021). Internet of Medical Things (IOMT), an AI approach integrates healthcare information technology through interconnected physical infrastructure, medical devices, and software applications. This enables wireless medical equipment to securely connect to the internet, facilitating the swift and efficient handling of medical data (Pradhan et al., 2021). IoT-based health monitoring framework proves helpful for medical tourists in monitoring health status, detecting diseases, and providing online health services in the medical tourism sector (Jangra et al., 2021). The wearable and sensors forming parts of IOT are hailed to be beneficial to healthcare professionals with effective benefits (Junata & Tong, 2018). The usage of fitness-tracking bands, smart watches help in collection of data on patients' health status and communication with healthcare specialists across the world. While these technological gadgets may apply to any person, the utilization is prominent and feasible for health travelers who would want to elect to follow-up with their physicians in another nation remotely, since data may be delivered to them in real-time (Psiha & Vlamos, 2017).

A medical tourist uses all of the above-mentioned AI and applied utilities like blockchain, cloud, telehealth, VR, IOMT, prior to the medical travel, during the medical travel and post medical travel and treatment as explained by Dar and Kashyap (2022) through a triple staged framework.

3.5 Medical Tourist's Pre-Procedural Stage

Pre-procedural stage starts with the medical tourist's decision regarding MTF selection, the medical destination country and medical treatment providers such as hospitals, physicians (Parekh et al., 2020). Medical tourists meet virtually with the physician before making a final treatment decision in most of the cases. These decisions are influenced by MTF's websites offering different packages and other information related to medical tourism destinations and medical treatments. *Information management (IT) technology* offers required relevant information on the online portals. Before departure, interested medical tourists involve in online interactive videoconferencing sessions with the healthcare providers with the help of peer-to-peer availability of tele-health technology. At last, digital payments with the help of blockchain technology make transactions convenient and safe, reducing the conveniences of bank intermediation, and currency delays (Balasubramanian et al., 2022).

Figure 2. Triple staged framework for medical tourists
Dar and Kashyap (2022); Source: https://doi.org/10.1504/IJTP.2022.121982

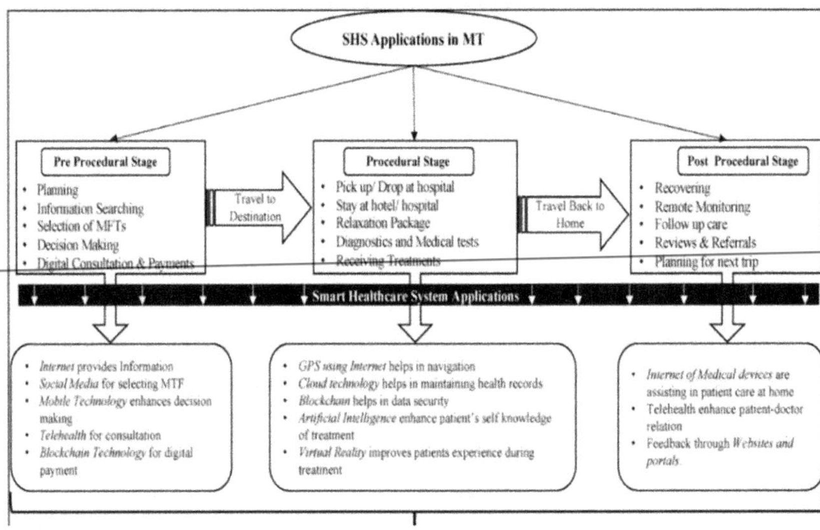

3.6 Medical Tourist's Procedural Stage

The procedure stage, which is the second stage, begins when the medical tourist arrives in the medical destination (Parekh et al., 2020). The MTF provides pick-up and drop-off facilities from the airport, and further *navigation* makes it convenient for the medical tourist to get to the hospital. Before the onset of the medical procedure, medical tourist undertakes required tests and further the healthcare providers do the arrangements for prerequisite medical reports and *cloud technology is* used to safely store the medical data of the tourist and later save it through *blockchain technology.* Smart contracts, backed by *blockchain technology* process medical insurance claims automatically after identifying the treatment completion, thus reducing paperwork (Tyan et al., 2021). Other technologies are used for treatment like AI is used to inform the medical tourists regarding the procedure's details, *VR technology* helps with distracting the medical tourists undergoing treatment to get some relief from pain. Further, after the treatment, the procedural stage includes post-operative care in the destination country itself which includes doctors' regular visits, nurse care, medical supplies, and relaxation packages (Balasubramanian et al., 2022).

3.7 Medical Tourist's Post-Procedural Stage

The medical tourist returns home in post procedural stage after recovery but is continuously receiving remote treatment, monitoring, and care therapy from the host or destination country or sometimes a new healthcare provider in the home country itself. With the help of *IoMT (internet of medical things)* medical tourists receive remote medical care from the home country (Kelly et al., 2020). These devices help in tracking and interacting of the medical tourists at home and doctors in the hospital (Dimitrov, 2016). Sometimes RPM (Remote medical tourist monitoring) device facilitates sending medical tourist's metrics like oxygen saturation or blood pressure to their physicians. Medical tourists further have tamper-proof and reliable accurate records of the received treatments and medications along with the contact information for the healthcare providers (Balasubramanian et al., 2022). Medical tourists who are in need of extensive postsurgical care or follow-up care and develop difficulties after returning back to home, use the interoperable IoMT facility to switch to a local healthcare provider (Vernon, 2021). Healthcare providers reward medical tourists through several technology assisted platforms, for trusting and sharing their experiences via ratings, reviews, or blogs (Deloitte, 2018). On the basis of the shared information, medical tourists sometimes, may receive offers and savings for their future trips or some personalized suggestions.

3.8 AI and Robotics in Healthcare and Medical Tourism

In the realm of medical tourism, AI-enabled robotics presents the potential for significant advancements (Yamin 2018). This includes improved diagnostics, less invasive surgery, higher precision in medical procedures, faster recovery times, and a reduction in waiting times for patients. Additionally, robotic nurses in hospitals, assist human nurses by performing tasks such as gathering supplies and delivering them to the rooms of medical tourists, transporting samples to labs, fetching items, and even cleaning rooms for new admissions (Tian et al., 2019). These applications showcase the evolving role of AI and robotics in enhancing both the efficiency and quality of healthcare services within the context of medical tourism. Figure 3, shows how AI and robotics are incorporated during procedural treatment stage of a medical tourist.

Figure 3. Procedural stage of medical tourist using AI and robotics
source: - https://doi.org/10.1504/IJTP.2022.121982

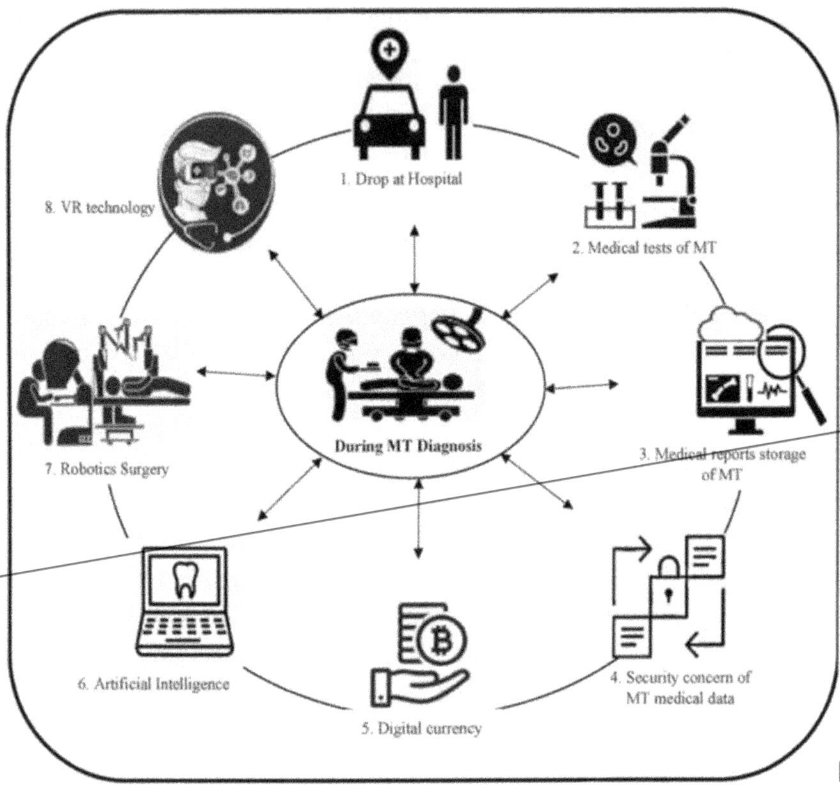

4. CHALLENGES IN AI BASED MEDICAL TOURISM

AI applications, coupled with technologies like big data and robotics, are anticipated to bring about transformational and disruptive changes in the healthcare and tourism sector. This impact is expected to extend across various domains, including hospitals and hospital management, pharmaceuticals, mental health and well-being, insurance, as well as predictive and preventive medicine. Despite the potential benefits, the integration of these applications introduces new risks and challenges. Addressing these concerns will necessitate the development of policy and institutional frameworks to guide the design and use of AI in healthcare. This reflects the importance of establishing governance structures to ensure responsible and ethical deployment of AI technologies in the evolving landscape of healthcare.

Uzir et al. (2021) argue that the impersonal nature of healthcare delivered by AI-enabled devices has the potential to replace in-person services, disrupting traditional expectations regarding physician-patient consultations and leading to trust issues. The shift toward AI in healthcare may raise concerns among users about the effectiveness and safety of AI technology in providing accurate diagnoses and prescriptions. Trust, being a dynamic and multi-faceted construct, plays a significant role in influencing customer satisfaction with the adoption and utilization of technology. This highlights the importance of addressing trust-related challenges in the implementation of AI in healthcare to ensure a positive reception and successful integration of these technologies.

As per a study by, Safeguarding citizens' health data emerges as a crucial responsibility for entities handling sensitive information for AI purposes. Healthcare organizations are confronted with the escalating challenges of cybersecurity, requiring robust responses. Policymakers, in turn, bear the responsibility of enacting laws that ensure meticulous governance and security arrangements for stored data. An illustrative example of the consequences of mishandling health data is the criticism faced by Google DeepMind's partnership with the Royal Free London NHS (National Health Service) Foundation Trust in 2017. The partnership faced severe backlash due to inappropriate sharing of confidential patient data and its utilization in an app called Streams, designed for alerting, diagnosing, and detecting acute kidney injury. This incident underscores the imperative need for stringent regulations and ethical practices in handling health-related data within the realm of AI applications in healthcare and medical tourism.

Algorithms inherently mirror the biases present in their training data, and AI tools often exhibit biases reflective of conditions prevalent in the high-income countries where they are developed. This stems from the fact that these algorithms necessitate millions of historical health datapoints, which may be lacking in low-resource settings, to generate accurate outputs tailored to the specific geography and population (USAID, 2019).This recognition underscores the importance of transparency and careful consideration in the development and application of AI tools, particularly in the healthcare domain. Another key issue with these software-controlled services is that a minor malware threat can interrupt the software programmes and operations of the service providers, resulting in trouble.

5. FUTURE OF AI BASED MEDICAL TOURISM

As per (Wong & Hazley, 2020), despite the existence of Artificial Intelligence (AI) for some time, its application in the healthcare industry is still in its early stages.. The incorporation of IR 4.0 technologies into the broader healthcare industry is poised to revolutionize patients' travel options and patterns, particularly for initial check-ups and follow-up stages. This transformation will become significant especially for older individuals and those facing mobility issues, making healthcare more accessible. According to Medical Tourism Magazine (2020), AI technology, being highly relevant and useful for health tourism, is expected to further enhance the processes of pre and post-operative care from both medical and customer service perspectives. This underscores the potential of AI to play a crucial role in improving various aspects of healthcare services within the context of medical tourism, contributing to enhanced patient care experiences and operational efficiency.

Wong and Hazley (2020) emphasized that the utilization of AI technology can further potentially free up face-to-face appointments for health tourists, especially during follow-up sessions after their return to their home country. This approach will create a more convenient and cost-effective healthcare experience. The availability of IR 4.0 technologies is also expected to enhance the virtual mobility of experts, including doctors, nurses, professors, and caregivers. This advancement contributes to the readiness of developing nations to welcome health tourists, thereby facilitating more accessible and efficient healthcare services.

Digitalized and connected healthcare has seen increased investment in recent years, with technology-enabled healthcare becoming more widespread. This trend is paving the way for the realization of the concept of "Smart Hospitals" (Frost and Sullivan, 2017), indicating a broader shift toward more advanced and interconnected healthcare systems As per (Lobo, 2020; Tim, 2018), the adoption of technologies in healthcare, including precision and timely applications, is generally seen as beneficial for patients. However, critics express concerns such as the perceived absence of empathy and a lack of trust in decision-making by robots. In the context of health tourism, additional concerns may arise, including issues related to data leakages, privacy matters, cybersecurity, and worries about over-commercialization (Medical Technology, 2020). These apprehensions highlight the importance of addressing ethical, privacy, and security considerations in the integration of technology into health tourism to ensure responsible and patient-centred practices.

6. DISCUSSION AND FINDINGS

The current section offers a thorough examination of the relationship that exists between medical tourism, a rapidly expanding industry, and AI technology. Through the use of desk research techniques and a qualitative research strategy, the study explores the complex ways that AI is changing the healthcare services that people seek out when traveling abroad. The portion's careful examination of the many facets of artificial intelligence and smart healthcare systems in relation to medical tourism is one of its strongest points. The study offers a comprehensive overview of how AI is transforming medical tourists' access to and experiences with healthcare services around the world by combining insights from recent literature. This Study explanation of the various ways AI is being used to ease medical tourists' travels is one of its main contributions. AI driven solutions have been demonstrated to improve accessibility and convenience for medical tourists. These range from telehealth platforms that link patients with physicians to chatbots and virtual assistants that provide real-time help. The chapter also emphasizes how AI helps to enable proactive healthcare measures, which guarantees better preparedness for possible health issues. Furthermore, the research illuminates the crucial function of artificial intelligence in tackling important issues in the medical tourism sector. AI helps allay worries about financial transactions and information security by enabling cashless transactions through blockchain technology, simplifying secure data transmission, and guaranteeing anonymity.

The Study does, however, also highlight a number of noteworthy difficulties and problems related to the application of AI in medical tourism. These include worries about the privacy of personal information, moral issues with AI-driven technology, and possible inequities in access to cutting-edge medical treatments. To sum up, this chapter provides insightful information about how artificial intelligence is changing the medical tourism industry. The study adds to a better knowledge of the changing dynamics at the nexus of AI, healthcare, and tourism by outlining the several ways that AI improves the medical tourism experience while resolving related difficulties.

7. CONCLUSION

The purpose of this study was to assess the applications of AI and smart healthcare systems in healthcare and medical tourism. It aimed to provide with a comprehensive understanding of medical tourism through the lens of AI and the potential for AI to improve patient outcomes, streamline healthcare processes, and attract a global clientele. A qualitative research approach was followed to conduct this study. According to the study results, AI applications in SHS are augmenting the medical tourism industry tremendously. Different types of Tourism and travel industries are taking advantage of AI in order to perform a variety of administrative and customer service tasks. For medical tourism the development of these technologies play a key role in supporting the medical agencies in connecting physicians and prospective medical tourists. The applications of AI not only enhances the overall efficiency of medical tourism destinations but also contributes to proactive measures in healthcare, ensuring better preparedness for potential health challenges and optimizing pharmaceutical resource management. In addition to make it more feasible and credible option for patients, these technologies respond to questions and provide valuable information to customers, and allows to reduce the time would be impossible for humans. The rise of technological agents, such as AI, virtual reality (VI), and robots, is creating new doors for tech-enabled medical tourist care and personalized treatment.

Results indicate that medical tourists use AI for getting information about medical tourist destinations, hospitals, treatments, check-ups, consultations and real time interactions with the medical fraternity. They find AI safe for sharing their medical data and records and also for doing cashless transactions. Further results implied that medical tourists use chatbots and virtual assistants to get an access to the hospitals, and required treatments. Medical tourists also find smart health wearables very useful in tracking and recording the health data.

Results also point out that a lack of explainability and trust in AI processes presents a significant hurdle, impeding the full trust in AI systems. As health-related data becomes more readily available and AI is increasingly used to analyse such data for medical purposes, a host of ethical, technical, and resource-related questions will need to be addressed. Issues related to quality, safety, governance, privacy, consent, and ownership pose ongoing challenges that are still not fully addressed in the deployment of AI in healthcare. Establishing comprehensive solutions and frameworks to address these challenges is crucial for ensuring the responsible and ethical use of AI in the medical domain.

The recent coronavirus pandemic brought global travel to a near standstill, emphasizing the need for the future of health tourism to evolve through advancements in technology, particularly in IR 4.0 technologies like AI. The landscape of healthcare is shifting from local accessibility to a more global perspective, transcending national borders, whether in physical presence or through virtual means. This transformation highlights the crucial role of technology, especially AI, in reshaping the way healthcare services are accessed and delivered across borders, reflecting the growing importance of innovative solutions in the face of global challenges like the pandemic.

In conclusion, the adoption of Artificial Intelligence (AI) technologies is seen as a transformative step that has the potential to enhance healthcare quality, reduce costs, improve transparency, increase accessibility, and provide greater flexibility among stakeholders. These improvements contribute to more positive impacts on and assimilation into the medical tourism destination, highlighting the broad-reaching benefits that AI can bring to the healthcare sector within the context of medical tourism.

REFERENCES

Abo-Elsoud, E., & Morsy, J. (2022). The Role of Artificial Intelligence in Improving Service and Strengthening Tourist Experience in Egypt. *ãÌáÉ ˜áíÉ ÇáÓíÇÍÉ æÇáÝäÇÏÞ. ÌÇãÚÉ ÇáãäÕæÑÉ, 2*(11), pp.821-855.

Advani, V. (2021). What is Artificial Intelligence? How does AI work, Types and Future of it. *mygreatlearning*.

Akpojaro, J., & Bello, R. W. (2019). Smart healthcare system: A primer. *International Journal for Innovative Research in Multidisciplinary Field, 5*(3), 143–149.

Ali, S., Abdullah, Armand, T. P. T., Athar, A., Hussain, A., Ali, M., Yaseen, M., Joo, M.-I., & Kim, H.-C. (2023). Metaverse in healthcare integrated with explainable ai and blockchain: Enabling immersiveness, ensuring trust, and providing patient data security. *Sensors (Basel), 23*(2), 565. doi:10.3390/s23020565 PMID:36679361

Ameen, N., Tarhini, A., Reppel, A., & Anand, A. (2021). Customer experiences in the age of artificial intelligence. *Computers in Human Behavior, 114*, 106548. doi:10.1016/j.chb.2020.106548 PMID:32905175

Ayuningtyas, D., & Ariwibowo, D. A. (2020). The strategic role of information communication technology in succeeding medical tourism. *Enfermeria Clinica, 30*, 170–173. doi:10.1016/j.enfcli.2020.06.039

Balasubramanian, S., Sethi, J. S., Ajayan, S., & Paris, C. M. (2022). An enabling framework for blockchain in tourism. *Information Technology & Tourism, 24*(2), 165–179. doi:10.1007/s40558-022-00229-6

Basheer, S., Farooq, S., Hassan, V., Malik, Y. M., & Reshi, M. A. (2023). *Augmented Reality and Virtual Reality in Cultural Heritage Tourism Enhancing Visitor Experiences*. doi:10.4018/978-1-6684-9957-3.ch002

Basheer, S., Walia, S., Farooq, S., Shah, M. A., & Mir, F. A. (2023). Exploring the metaverse: The future of tourism through ai and virtual reality. In *Influencer Marketing Applications Within the Metaverse* (pp. 195–204). IGI Global. doi:10.4018/978-1-6684-8898-0.ch012

Basheer, S., Walia, S., Mehraj, D., & Reshi, M. (2023). *Creating Transformative Travel Unveiling the Link Between Meaningful Experiences and Sustainable Tourism*. doi:10.4018/979-8-3693-0650-5.ch002

Benckendorff, P. J., Xiang, Z., & Sheldon, P. J. (2019). *Tourism information technology*. Cabi. doi:10.1079/9781786393432.0000

Collins, A., Medhekar, A., Wong, H. Y., & Cobanoglu, C. (2019). Factors influencing outbound medical travel from the USA. *Tourism Review*, *74*(3), 463–479. doi:10.1108/TR-06-2018-0083

Dar, H., & Kashyap, K. (2023). Smart Healthcare System (SHS): Medical Tourism Delivering, Consumption, and Elevating Tool in the Ages of Smart Technologies. *Tourism Planning & Development*, *20*(3), 397–415. doi:10.1080/21568316.2022.2109206

Deepthi, B., & Bansal, V. 2023. Applications of Artificial Intelligence (AI) in the Tourism Industry: A Futuristic Perspective. In Impact of Industry 4.0 on Sustainable Tourism: Perspectives, Challenges and Future (pp. 31-43). Emerald Publishing Limited.

Deloitte. (2018). *A journey towards smart health: The impact of digitalization on medical tourist experience*. Deloitte.

Dimitrov, D. V. (2016). Medical internet of things and big data in healthcare. *Healthcare Informatics Research*, *22*(3), 156–163. doi:10.4258/hir.2016.22.3.156 PMID:27525156

Farooq, S., Farooq, B., Basheer, S., & Walia, S. (2023). *Balancing Environmental Sustainability and Privacy Ethical Dilemmas in AI-Enabled Smart Cities*. doi:10.4018/979-8-3693-0892-9.ch013

Frost & Sullivan. (2017). *Future of smart hospitals – concept definition, application, and growth opportunities, K1FF-01*. Frost & Sullivan.

Gada, N. (2017). Four technological advancements that could change the medical tourism landscape as we know it. *Medical Travel Today*. https://medicaltraveltoday.com/four-technological-advancements-that-could-change-the- medicaltourism-landscape-as-we-know-it/

Gärtner, E., Pirinen, A., & Sminchisescu, C. (2020, April). Deep reinforcement learning for active human pose estimation. *Proceedings of the AAAI Conference on Artificial Intelligence*, *34*(07), 10835–10844. doi:10.1609/aaai.v34i07.6714

Ghasemi, M., Nejad, M. G., & Aghaei, I. (2021). Knowledge management orientation and operational performance relationship in medical tourism (overview of the model performance in the COVID- 19 pandemic and post-pandemic era). *Health Services Management Research*, *34*(4), 208–222. doi:10.1177/0951484820971438 PMID:33183088

Goel, P. (2019). *The power of digital health in medical tourism.* LinkedIn. https://www.linkedin.com/pulse/power-digital-health-medical-tourism-pramod-goel

Gretzel, U. (2011). Intelligent systems in tourism: A social science perspective. *Annals of Tourism Research, 38*(3), 757–779. doi:10.1016/j.annals.2011.04.014

Hassan, V., & Noaman, S. (2017). Relation between Tourism and Health: Case Study AIDS in Lebanon. In *Atiner Conference Paper Series.*

Jangra, P., & Gupta, M. (2021). Internet of Medical Things (IoMT) Based Framework for Smart Healthcare Tourism Sector. *International Journal of Hospitality and Tourism Systems*, 14.

Junata, M., & Tong, R. (2018). Wearable technology in medicine and health care: introduction. In R. Tong (Ed.), *Wearable Technology in Medicine and Health Care* (pp. 1–5). Academic Press. doi:10.1016/B978-0-12-811810-8.00001-4

Kelly, J. T., Campbell, K. L., Gong, E., & Scuffffham, P. (2020). The internet of things: Impact and implications for health care delivery. *Journal of Medical Internet Research, 22*(11), e20135. doi:10.2196/20135 PMID:33170132

Lancet, T.The Lancet. (2017). Artificial intelligence in health care: Within touching distance. *Lancet, 390*(10114), 2739. doi:10.1016/S0140-6736(17)31540-4 PMID:29303711

Lobo, F. (2020). *Industry 4.0 – manufacturing and the future of medical things.*

Luxton, D. D. (2016). An introduction to artificial intelligence in behavioral and mental health care. In D. D. Luxton (Ed.), *Artificial Intelligence in Behavioral and Mental Health Care* (pp. 1–26). Academic Press. doi:10.1016/B978-0-12-420248-1.00001-5

Manyika, J., Chui, M., Miremadi, M., Bughin, J., George, K., Willmott, P., & Dewhurst, M. (2017). A future that works: AI, automation, employment, and productivity. McKinsey Global Institute Research. *Tech. Rep, 60*, 1–135.

Medical Technology. (2020). *Medical tourism: how is digital tech reshaping the industry?* Medical Technology.

Medical Tourism Magazine (2020). Trends in healthcare digital revolution. *Medical Tourism Magazine.*

Medical Tourism Magazine. (2021). Will medical tourism survive the coronavirus pandemic? *Medical Tourism Magazine.*

Nazir, S., Ali, Y., Ullah, N., & García-Magariño, I. (2019). Internet of things for healthcare using effects of mobile computing: A systematic literature review. *Wireless Communications and Mobile Computing*, *2019*, 1–20. doi:10.1155/2019/5931315

Parekh, J., Jaffer, A., Bhanushali, U., & Shukla, S. (2021). Disintermediation in medical tourism through blockchain technology: An analysis using value-focused thinking approach. *Information Technology & Tourism*, *23*(1), 69–96. doi:10.1007/s40558-020-00180-4

Parry, C. M., & Aneja, U. (2020). *Artificial Intelligence for Healthcare: Insights from India.*

Pesapane, F., Volonté, C., Codari, M., & Sardanelli, F. (2018). Artificial intelligence as a medical device in radiology: Ethical and regulatory issues in Europe and the United States. *Insights Into Imaging*, *9*(5), 745–753. doi:10.1007/s13244-018-0645-y PMID:30112675

Pradhan, B., Bhattacharyya, S., & Pal, K. (2021). IoT-based applications in healthcare devices. *Journal of Healthcare Engineering*, *2021*, 1–18. PMID:33791084

Premarathne, U., Abuadbba, A., Alabdulatif, A., Khalil, I., Tari, Z., Zomaya, A., & Buyya, R. (2016). Hybrid cryptographic access control for cloud-based EHR systems. *IEEE Cloud Computing*, *3*(4), 58–64. doi:10.1109/MCC.2016.76

Rahman, R. (2019). The privatisation of healthcare system in Bangladesh. *International Journal of Health Care Quality Assurance*, *32*(1), 97–107. doi:10.1108/IJHCQA-11-2017-0217 PMID:30859884

RejebA.KeoghJ. G.TreiblmaierH. (2019). The impact of blockchain on medical tourism. Web2019 workshop on e-business. doi:10.2139/ssrn.3521841

Senbekov, M., Saliev, T., Bukeyeva, Z., Almabayeva, A., Zhanaliyeva, M., Aitenova, N., Toishibekov, Y., & Fakhradiyev, I. (2020). The recent progress and applications of digital technologies in healthcare: A review. *International Journal of Telemedicine and Applications*, *2020*, 1–18. doi:10.1155/2020/8830200 PMID:33343657

Shabarish, P. S., & Shivashankar, K. C. (2019). Impact of technology on medical tourism: A conceptual study. *International Journal of Commerce and Management Research*, *5*(3), 145–148.

Subasinghe, M., Magalage, D., Amadoru, N., Amarathunga, L., Bhanupriya, N., & Wijekoon, J. L. (2020, November). Effectiveness of artificial intelligence, decentralized and distributed systems for prediction and secure channelling for Medical Tourism. In *2020 11th IEEE Annual Information Technology, Electronics and Mobile Communication Conference (IEMCON)* (pp. 0314-0319). IEEE. 10.1109/IEMCON51383.2020.9284898

Sundaravadivel, P., Kougianos, E., Mohanty, S. P., & Ganapathiraju, M. K. (2017). Everything you wanted to know about smart health care: Evaluating the different technologies and components of the internet of things for better health. *IEEE Consumer Electronics Magazine*, *7*(1), 18–28. doi:10.1109/MCE.2017.2755378

Tater, U. R., Jain, A., & Tater, M. J. (2022). Artificial intelligence must be a revolutionary technique of marketing in medical-tourism. *2022 Indian Journal of Business Administration (IJBA).*

Tian, S., Yang, W., Le Grange, J. M., Wang, P., Huang, W., & Ye, Z. (2019). Smart healthcare: Making medical care more intelligent. *Global Health Journal (Amsterdam, Netherlands)*, *3*(3), 62–65. doi:10.1016/j.glohj.2019.07.001

Tim, W. (2018). *Health tourism today and into the future.*

Tussyadiah, I., & Miller, G. 2019. Perceived impacts of artificial intelligence and responses to positive behaviour change intervention. In *Information and Communication Technologies in Tourism 2019: Proceedings of the International Conference in Nicosia,* (pp. 359-370). Springer International Publishing. 10.1007/978-3-030-05940-8_28

Tyan, I., Plaza, A., & Yague, M. (2021). The benefifits of blockchain technology for medical tourism. *Sustainability (Basel)*, *13*(22), 12448–12412. doi:10.3390/su132212448

USAID. (2019). *Artificial Intelligence in Global Health: Defining a Collective Path Forward.* USAID. https://www.usaid.gov/cii/ai-in-global-health

Uzir, M. U. H., Al Halbusi, H., Lim, R., Jerin, I., Hamid, A. B. A., Ramayah, T., & Haque, A. (2021). Applied Artificial Intelligence and user satisfaction: Smartwatch usage for healthcare in Bangladesh during COVID-19. *Technology in Society*, *67*, 101780. doi:10.1016/j.techsoc.2021.101780 PMID:34697510

Vishnoi, S. K., Bagga, T. E. E. N. A., Sharma, A. A. R. U. S. H. I., & Wani, S. N. (2018). Artificial intelligence enabled marketing solutions: A review. Indian Journal of Economics & Business, 17(4), 167-177.

Wong, B. K. M., & Hazley, S. A. S. A. (2020). The future of health tourism in the industrial revolution 4.0 era. *Journal of Tourism Futures*, *7*(2), 267–272. doi:10.1108/JTF-01-2020-0006

Wright, S. A., & Schultz, A. E. (2018). The rising tide of artificial intelligence and business automation: Developing an ethical framework. *Business Horizons*, *61*(6), 823–832. doi:10.1016/j.bushor.2018.07.001

Yamin, M. (2018). IT applications in healthcare management: A survey. *International Journal of Information Technology : an Official Journal of Bharati Vidyapeeth's Institute of Computer Applications and Management*, *10*(4), 503–509. doi:10.1007/s41870-018-0203-3 PMID:32289102

Yang, Y., Siau, K., Xie, W., & Sun, Y. (2022). Smart health: Intelligent healthcare systems in the metaverse, artificial intelligence, and data science era. [JOEUC]. *Journal of Organizational and End User Computing*, *34*(1), 1–14. doi:10.4018/JOEUC.308814

Chapter 10
The Ultimate Guide to Medical Tourism Marketing:
A Comprehensive Systematic Review and Future Research Agenda

Prakash Singh
https://orcid.org/0000-0001-7307-1932
College of Administrative and Financial Sciences, Saudi Electronic University, Riyadh, Saudi Arabia

Viana Hassan
https://orcid.org/0000-0002-7372-5059
University of Malta, Malta

Santosh Dastane
https://orcid.org/0009-0009-7708-7718
ASM's Institute of Business Management and Research, India

ABSTRACT

The popularity of medical tourism has matured substantially over the last few decades due to escalating medical costs, advancements in medical technology, and reduced travel costs. The significance of medical marketing has thus become vital for hospitals to lure domestic and global patients, designate their brand, and boost awareness about life-saving treatments delivered by distinctive hospitals across the globe. Even though medical tourism has acquired growing interest, prior research has been limited in scope, particularly in studying the industry's marketing aspects. Because of this, comprehending the target market is vital. To address this need, this chapter aims to answer two research questions: RQ1, What key themes, study characteristics,

DOI: 10.4018/979-8-3693-2248-2.ch010

study contexts, and methodologies are investigated in medical tourism marketing research studies?; RQ2, What future research avenues in medical tourism marketing exist for scholars across diverse fields? This chapter thoroughly synthesizes existing medical tourism marketing literature and pinpoints future research opportunities.

1. INTRODUCTION

The medical tourism industry has undergone a rapid transition within the tourism industry since patients travel outside of specified cross-border care contracts to fetch medical benefits abroad. This practice is commonly paid for out-of-pocket and is a fast-growing sector (Ramirez de Arellano, 2007; Johnston et al., 2010; Crooks et al., 2011; Han & Hyun, 2015; Rahman et al., 2022) and is often referred to as "Medical tourism" (Johnston et al., 2010). The success of medical tourism relies on the effective communication of medical treatment opportunities, healthcare establishments, tourism prospects, travel arrangements, and destination countries to potential patients through diverse marketing strategies (Chinai & Goswami, 2007; Howze, 2007; Crooks et al., 2011). Furthermore, several countries globally are positioning themselves to attract visitors seeking accessible, affordable, and more readily available healthcare services. Consequently, effective marketing strategies have become increasingly substantial for medical tourism businesses to flourish in the competitive global medical tourism landscape (Heung et al., 2011; Connell, 2013; Han, 2013). Moreover, the fundamental concern for marketing practitioners in this medical tourism sector is attracting new medical travelers through suitable marketing practices and encouraging them to make repeat visits and avail of medical treatments via service efforts and effective marketing strategies (Han, 2013).

Previous studies have discovered that several factors, such as government policies and regulations, medical treatment costs, governmental funding, healthcare infrastructure, and community necessities, can significantly impact medical tourism marketing practices in terms of perceived value and revisit intention (Kim & Oh, 2021; Pighin et al., 2022; Rahman et al., 2022). Besides, the niche medical tourism market is expanding due to medical advancements and enhanced well-being, and hospitals and medical tour operators must embrace effective marketing practices to advertise their medical treatment services effectively (Cham et al., 2022; Saiprasert et al., 2022). Regardless, with the mounting number of medical establishments and world-class infrastructure in diverse countries globally, it can be difficult for medical tourists or patients to find the most promising contemporary hospitals and well-trained physicians who invariably deliver high-quality medical treatment services.

In this context, effective marketing practices are vital in publicizing to potential patients and medical tourists seeking medical treatment opportunities, tourism prospects, and additional pertinent information on medical tourism (Pighin et al., 2022; Abdul-Rahman et al., 2023). Hence, the survival and success of the medical tourism industry hinge on effective marketing strategies that prioritize broadcasting to potential patients and medical tourists about the available medical treatment opportunities and tourism possibilities. Medical establishments and practitioners can ensure survival and build an advantage in the increasingly competitive global medical tourism industry.

Medical tourism has boosted popularity over the last few decades as medical costs increase and medical technology and travel costs declined (Guru et al., 2022; Kim et al., 2022). Consequently, the significance of medical marketing has evolved as crucial for hospitals to lure domestic and global patients, market their brand, and boost awareness of life-saving treatments delivered by prominent hospitals worldwide (Das, 2017; Abdul-Rahman et al., 2023). Even though medical tourism has gained much attention, preceding studies have been limited, primarily in studying its marketing aspects (Jiang et al., 2022; Buckley, 2023); as an outcome, learning its target market is vital (Wang et al., 2022). To address this need, this chapter aims to answer two research questions (RQs): **RQ1**: What key themes, study characteristics, contexts, and methodologies are investigated in medical tourism marketing research studies? **RQ2**: What future research avenues in medical tourism marketing exist for scholars across diverse fields?

Through a systematic literature review, this study discovers answers to the stated research questions in a stringent, translucent, and reproducible manner (Elkhwesky et al., 2023). The Web of Science (WoS) electronic database is preferred for article extraction because of its selectivity compared with Scopus and other databases, making it vital to a systematic literature review approach (Gusenbauer & Haddaway, 2020; Singh et al., 2021).

Moreover, researchers and scholars in the medical tourism marketing discipline will benefit from this chapter's thorough synthesis of the research landscape in this field of study, irrespective of their disciplines. With the help of this chapter, businesses and marketers will gain a more profound acumen of the sophistication of medical tourism marketing, contributing to the body of knowledge and delivering insight into how to augment their medical tourism marketing strategies for reaching, targeting, and serving local and global patients. The book chapter is organized into five sections, which are as stated. Section 1 offers an overview of this research study. Section 2 presents the methodology used. Section 3 discusses our findings of the study. Section 4 presents the implications, limitations, and future research directions, and Section 5 outlines the concluding observations of the study.

2. METHODOLOGY

The rationale of any research study depends on the research method selected to align with the research questions. In management and business studies, systematic literature reviews (SLRs) have become the prevailing research methodology because of their ability to comprehend multidisciplinary research studies in detail (Paul et al., 2023; Tranfield et al., 2003). Moreover, researchers and scholars have used diverse methods, such as integrative reviews, bibliometric analysis, and the PRISMA protocol, to address their research questions based on the strengths and limitations of different approaches evaluated to determine the most suitable method (Cubric,2020; Williams Jr et al., 2021). By employing a quantitative approach, the SLR methodology fosters researchers' and scholars' understanding of the current state of research better and pinpoints areas for further study (Snyder, 2019). Accordingly, it is imperative to carefully select research methods and techniques to improve academic work and research impact. This study aims to provide valuable insights into medical tourism marketing by selecting the most suitable research methodology. The study adopted a Systematic Literature Review (SLR) approach, known for its thoroughness, transparency, and objectivity (Singh et al., 2023). The Web of Science (WoS) database was selected to locate articles on medical tourism marketing due to its reputation as a reputable and reliable source of high-quality academic literature (Guz & Rushchitsky, 2009).

Figure 1. Five-step process

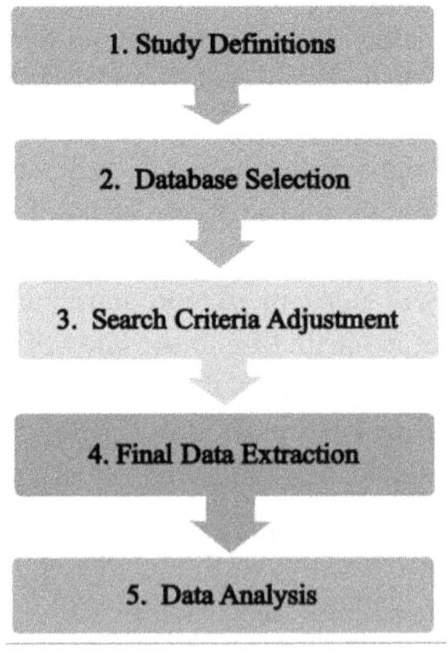

Figure 2. PRISMA flow diagram

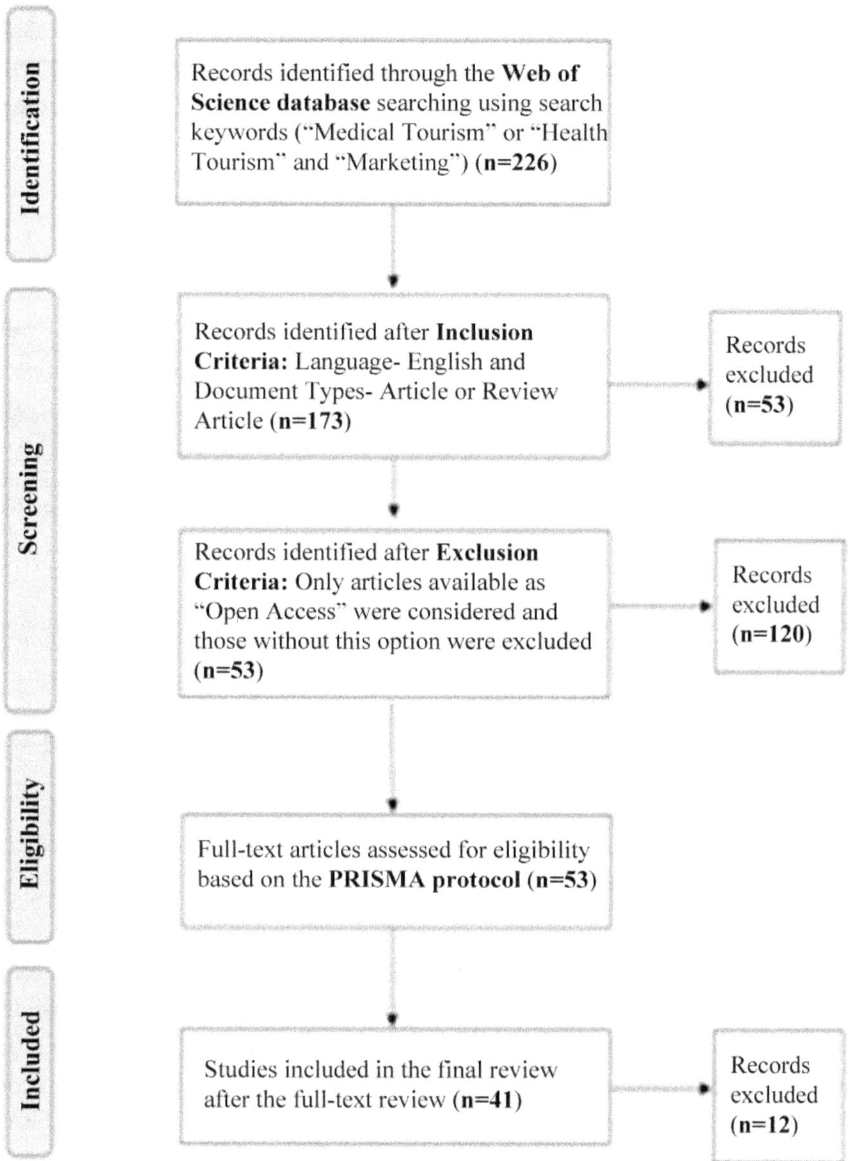

The WoS provides a wide range of topics and has a strict quality control system that ensures that only significant publications are indexed and included in this electronic database (Guz & Rushchitsky, 2009; Zhang et al., 2022; Singh et al., 2023). Therefore, a five-step process adapted from Chakma et al. (2021), as depicted in **Figure 1**, was followed to extract articles from the WoS electronic database.

Figure 2 presents the PRISMA flow diagram explaining the definitions of the study. A search was performed using the "Title-Abstract-Keywords" query field by searching articles on medical tourism marketing until December 31st, 2023, as the publication period. The search keywords used for extracting articles are "Medical Tourism," "Health Tourism," and "Marketing." The inclusion (review article as the document type and articles in English) and exclusion (articles not indexed under the open access option) criteria were applied to get the most eligible articles.

Table 1. Research protocol stepwise descriptions

Steps followed	*Descriptions of Research Protocol*
Step 1: Study Definitions	Figure 2 presents the PRISMA flow diagram explaining the definitions of the study.
Step 2: Database Selection	The Web of Science (WoS) electronic database was selected, and a search was performed using the "Title-Abstract-Keywords" query field by searching articles on medical tourism marketing until December 31st, 2023, as the publication period.
Step 3: Search Criteria Adjustment	The search keywords used for extracting articles are "Medical Tourism," "Health Tourism," and "Marketing." Inclusion Criteria- "Review Article" as the document type, and articles in English were included. Exclusion Criteria- Articles not indexed under the "Open access option" were excluded.
Step 4: Final Data Extraction	The final data list of 53 eligible articles was successfully exported to CSV format. Afterward, a thorough full-text review of all the eligible articles was performed, and 41 relevant articles that fully met the criteria were selected for final review.
Step 5: Data Analysis	The TCCM framework was used to present the study's findings.

Source: The Author.

Consequently, the final data list of 53 eligible articles was successfully exported to CSV format. Afterward, a thorough full-text review of all the eligible articles was performed, and 41 relevant articles that fully met the criteria were selected for final review. The research protocol descriptions are presented in **Table 1**. Furthermore, the TCCM framework (T-themes, C-Contexts, C-Characteristics, and M-methodologies) to present its findings (Paul & Rosado-Serrano, 2019). This framework fosters a structured and thorough understanding of the study data.

Table 2 provides the list of journals selected for the final review, including the number of articles selected from each journal. 41 articles on medical tourism marketing research discipline were published in 31 journals.

Table 2. List of journals selected for the final review

S. No.	*List of Journal selected for the final review*	*No. of the article (s) selected*
1	Anatolia-International Journal Of Tourism And Hospitality Research	01
2	Balneo Research Journal	01
3	BMC Medical Ethics	01
4	BMC Research Notes	01
5	Current Issues in Tourism	01
6	Developing World Bioethics	01
7	Entrepreneurial Business and Economics Review	01
8	European Business Review	01
9	Frontiers in Psychology	02
10	Globalization and Health	02
11	Health & Social Care in the Community	01
12	Health Services Management Research	01
13	Innovative Marketing	01
14	Inquiry-The Journal of Health Care Organization Provision and Financing	01
15	Interactive Journal of Medical Research	01
16	International Journal of Environmental Research and Public Health	03
17	International Journal of Reproductive Biomedicine	01
18	Journal of Ethnic and Migration Studies	01
19	Journal Of History Culture and Art Research	01
20	Jurnal The Messenger	01
21	Malaysian Journal of Medical Sciences	01
22	Managerial and Decision Economics	01
23	Marketing Intelligence & Planning	01
24	Plastic Surgery	01
25	Sage Open	01
26	Social Science & Medicine	01
27	Sustainability	07
28	Tehnicki Glasnik-Technical Journal	01
29	Tourism Culture & Communication	01
30	Tourism Management	01
31	Tourism Review	01
Total		41

Source: The Author.

3. RESULTS AND DISCUSSION

The researcher conducted a comprehensive bibliometric analysis to uncover the trends and significant characteristics of the bibliographic data. This involved analyzing publication patterns and identifying influential authors, journals, articles, etc. In addition, the researcher identified the most productive countries and organizations in the field. The researcher also delved deeper into the data by examining authors' keywords, co-occurrence of keywords, co-authorship, and citation analysis. The analysis provides valuable insights into the complexities of bibliographic data and its components.

3.1 Key Themes

The key themes investigated in this study are presented in **Figure 3.**

Figure 3. Medical tourism key themes

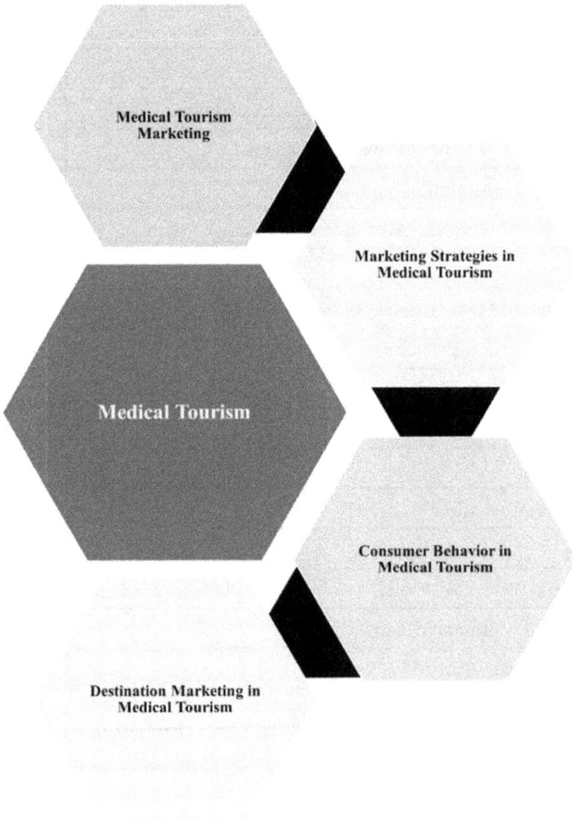

3.1.1 Medical Tourism

Medical tourism has emerged as a significant form that involves improving health as a vital journey component. According to Jagyasi (2008), medical tourism is an umbrella term that includes invasive medical procedures, unlike passive processes in wellness tourism. Habibi et al. (2022) describe medical tourism as traveling abroad for medical treatment or therapy. Medical services to foreign patients and tourist services were characterized by medical tourism (Aurelia-Felicia et al., 2014). Furthermore, Kazemha and Dehkordi (2017) explained that medical tourism was often referred to as tourism associated with treatment, and the increasing attention to medical tourism was reflected in the existing literature (Campra et al., 2022).

3.1.2 Medical Tourism Marketing

The provision of website information by medical clinics that were a study target needed more attention to patients and their family members, and the quality of information concerning cell therapies needed to be improved (Kashihara et al., 2016). This posed a significant challenge for medical tourism, where patients often rely on online information to make informed decisions about medical treatments in foreign countries. Consequently, word-of-mouth advice from family and friends has emerged as a prevalent source of information regarding surgery abroad (Robertson et al., 2022). The motivation for traveling abroad for medical treatment services, such as cosmetic procedures, was shaped by cost, post-operative care, customer service, and word-of-mouth recommendations, among the key factors influencing people's decisions (Robertson et al., 2022). The literature emphasized the importance of virtual marketing and health tourism websites in the development of the medical tourism industry (Shaygani et al., 2023). Many medical clinics have been using marketing techniques on their websites to attract patients' interest. However, the need for more unity in the global tourism market challenged the success of promotional actions. Expressly, the need for clear correlations between tourist attractions and the forms of tourism that should be the basis of brand elements necessitated the creation of specific marketing strategies for different consumer segments targeted through certain communication channels (Oltean et al., 2020).

3.1.3 Marketing Strategies in Medical Tourism

The competitiveness of medical tourism destinations heavily relied on their brand reputation, as Cham et al. (2020) noted. Push and pull marketing strategies were commonly employed to attract medical tourists, with a dialectical relationship between tourists, tour operators, and medical establishments playing a significant role. However, the associated costs and the satisfaction level were push factors, as pointed out by Horsfall (2020). In this context, effective marketing strategies were required to overcome several barriers. Furthermore, Turner (2012) suggested that marketing messages emphasizing timely access to medical care services could be instrumental in effective marketing strategies. However, building a brand with substantial equity was vital for differentiation and competitiveness, especially in today's highly competitive medical tourism industry, as emphasized by Shriedeh (2019). Therefore, creating a brand that stands out in the market became essential. Crooks et al. (2011) argued that marketing materials were crucial in informing potential patients about procedure options, treatment facilities, tourism opportunities, travel arrangements, and destination countries. Similarly, Shafii et al. (2022) suggested increasing the importance of advertising, implementing realistic advertising strategies, and developing adequate urban infrastructure and services to improve the demand for medical tourism. However, to target tourists effectively, it was crucial to understand their behaviors and intentions to revisit a particular destination, as highlighted by Abdul-Rahman et al. (2023). Hence, effective marketing communication strategies were essential to enable patients to comprehend the information they receive and make informed treatment decisions, as emphasized by Penney et al. (2011).

The study also identified loyalty and satisfaction as essential factors for describing consumer attitudes toward medical tourism in terms of relaxation, hospitality, and tourism, eventually resulting in flourishing tourism destinations. Tourism attitudes tend to be shaped by the intention to return to medical tourism destinations. As a result, destination marketers must refine their marketing strategies by paying close attention to the stimuli that foster tourists' loyalty and satisfaction (Kim et al., 2017; Yin et al., 2021). From a marketing perspective, people's appetite for looking suitable and fit inspires others to do the same. Medical tourists can be attracted to those medical treatment destinations that people they know or follow on social media visit. In this endeavor, social media marketing strategies can be a game changer in attracting and augmenting consumer attitudes toward medical tourism. In prior studies, the most influential channels were social media, word of mouth (WoM), and electronic word of mouth. Moreover, collaborating with social media influencers and celebrities can positively impact the marketing results of medical tourism destinations (Noaman & Montargot, 2022). To attract medical tourists, a medical tourism destination must establish its credibility and reputation as a safe and high-quality medical tourism destination. Specifically, more medical tourists may visit if a medical tourism destination is recognized for its safety and respect for everyone, especially children and women. However, combining medical treatment with other types of tourism could be an excellent selling point, should be explored more, and might become a key feature of marketing campaigns (Campra et al., 2022).

There were discussions of marketing strategies in prior studies stating that marketing strategies can be implemented in single businesses and by stakeholders in the medical tourism industry, including national medical establishments. Branding is all about awareness, and tourist destination brands help establish the unique identity of tourist destinations by differentiating them from their competitors (Mortezaei et al., 2021). An effective way to promote medical tourism is to use effective communication media, such as websites and online activities, to boost the destination image of medical tourism (Pessot et al., 2021). The combination of other tourist activities can support medical tourism. However, such diversification must be accompanied by corresponding marketing, which in all destinations needs to be improved in online marketing; it could be helpful if the marketing were diversified, focusing more on evidence-based studies and promoting wellness and health (Schmude et al., 2021). Although the concept of loyalty has been extensively studied in public marketing literature, brand loyalty in the form of tourist destinations has yet to be studied. It should be noted that loyalty increases the stability of the destination. Advantages of brand loyalty for destinations include lower marketing costs, increased influence of the travel business, and repetition of word-of-mouth advertising (Mortezaei et al., 2021).

Medical tourists can be attracted by forming cluster groups to manage medical tourism, aggressive marketing with business partners, and encouraging medical establishments and stakeholders to develop and create new products and services (Phuthong et al., 2022). A well-targeted medical tourism marketing campaign relies heavily on understanding travelers' behaviors and intentions to return to a specific medical destination (Abdul-Rahman et al., 2023). Although medical tourism faced challenges in branding the destination image of medical tourism, medical establishments did not consider these as important as improving a destination's competitiveness and increasing its capacity. It has been discovered that virtual marketing has proven to be the most effective way of promoting products. Studies show that a standard website with appropriate information for users can increase the likelihood of their willingness to buy products. Hence, there is no exception to medical tourism, which is among the most profitable and vital sectors within the healthcare industry (Mortezaei et al., 2021; Abdul-Rahman et al., 2023; Shaygani et al., 2023).

Medical establishments should use digital marketing, print media, and medical tourism fairs and exhibitions to promote medical infrastructure and services to potential medical tourists (Abdul-Rahman et al., 2023). Furthermore, it is crucial to correlate with the relevant resources of the destination to optimize the new type of tourism. Destination marketing strategies can be developed by harmonizing them (Aurelia-Felicia et al., 2014). However, marketing strategies emphasizing global medical tourism must be implemented across diverse countries to communicate via marketing campaigns. This may improve credibility, reputation, and trust among consumers and international bodies (Hartwell et al., 2018). Hence, marketing strategies to provide adequate information for medical tourists on the medical establishments' websites are critical components of attracting them and turning them into loyal customers, implementing the changes in medical tourism based on emerging marketing trends (Shaygani et al., 2023).

3.1.4 Consumer Behavior in Medical Tourism

Effectively managing the variables that impact consumer behavior and marketing was crucial for retaining customers in the medical tourism industry (Kim et al., 2017). It was essential to understand the significant role psychological factors played in shaping tourists' intentions and behavior (Li & Wen, 2022). This, in turn, influenced not only tourists but also their friends and family members, indicating the significant role of behavioral attitudes in determining health-related preferences (Li et al., 2022). It was vital to disclose risk information to help tourists make informed decisions and shape their intentions towards medical tourism (Yin et al., 2021). Moreover, value co-creation was also crucial in shaping tourists' behavior as they sought to experience a better quality of life through medical treatment services (Majeed et al., 2020). Word of mouth (WoM) remained a vital acquisition channel for medical tourism businesses, indicating the significance of personal recommendations from satisfied customers (Taheri et al., 2021). However, electronic Word of mouth (eWOM) and social media communication significantly impacted medical tourists' overall behavior (Noaman & Montargot, 2023).

3.1.5 Destination Marketing in Medical Tourism

Natural resources played a pivotal role in enhancing the wellness of individuals, thereby contributing to the development and sustainability of medical tourism destinations. These medical tourist destinations were often marketed as value-for-money opportunities due to natural attractions augmenting tourist experiences (Pessot et al., 2021). In medical tourism, touristic experiences have assumed great significance and formed an integral part of the destinations' marketing strategies (Hartwell et al., 2018). Furthermore, the brand equity of a health tourism destination impacts the value of the services and experiences offered at that destination. Brand equity refers to the assets or liabilities associated with the brand name and symbol (Mortezaei et al., 2021). Natural attractions could be leveraged as a powerful marketing tool in promoting medical tourism destinations (Schmude et al., 2021). Consequently, Innovative services and products, improved transportation, and leveraged rich cultural traditions that appeal to tourists enhanced medical tourism destinations' competitiveness. They tended to attract more medical tourists from diverse geographical locations (Phuthong et al., 2022).

Figure 4. Geographical context

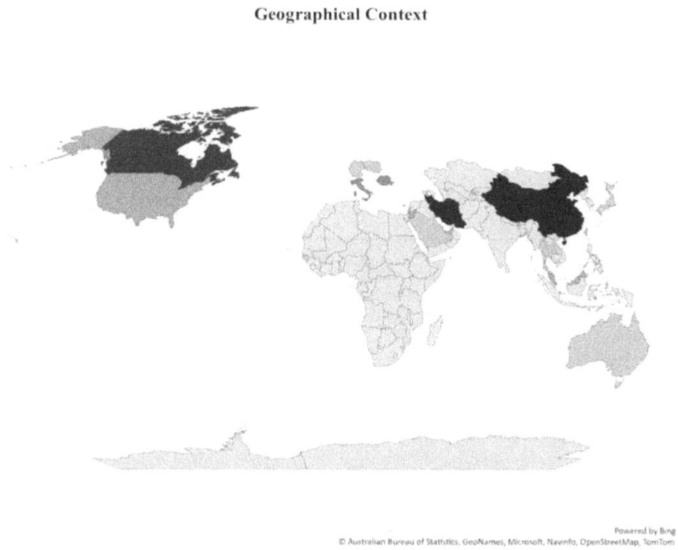

3.2 Key Contexts

The data shown in **Figure 4** and **Table 3** furnishes an overview of the geographical distribution of the contexts investigated in a study on medical tourism marketing. The study examined the frequency of countries where medical tourists search for various medical procedures. As per the data, the countries with the highest frequency of searches were China (6), Iran (6), and Canada (5), followed by Italy (3), Romania (3), and England (2). While the data nourishes valuable insights into the preferences of medical tourists, the analysis is specific to studying the underlying reasons for the favor of certain countries.

The study only surrounds the search patterns of medical tourists, but it does not delve into the characteristics that make certain countries more popular than others. For instance, what factors make China and Iran more prevalent destinations for medical tourists than other countries? Answering such questions can help medical establishments and tourism professionals better understand their target audience's priorities and design their marketing campaigns accordingly. This can significantly affect the marketing strategies of medical tourism establishments. Thus, there is a clear requirement for further research to bridge this gap and furnish a more in-depth knowledge of the characteristics that impact the priorities of medical tourists for typical geographical contexts.

Table 3. Geographical contexts investigated in the study

Country	Frequency
Australia	1
Canada	5
China	6
England	2
Germany	1
Iran	6
Italy	3
Japan	1
Jordan	2
Malaysia	2
Poland	1
Romania	3
Saudi Arabia	1
Scotland	1
Taiwan	1
Thailand	1
United Arab Emirates	2
United States of America	2

Source: The Author.

3.3 Characteristics: Publication Trends

The upswing in publications earmarked to medical tourism marketing studies in recent years can be justified by several factors. According to data presented in **Figure 5**, there has been an incremental upsurge in the number of articles published, with only 11 published up until 2017. Since then, there has been a substantial surge in articles published, with 30 published between 2018 and 2023.

Figure 5. Publication trend

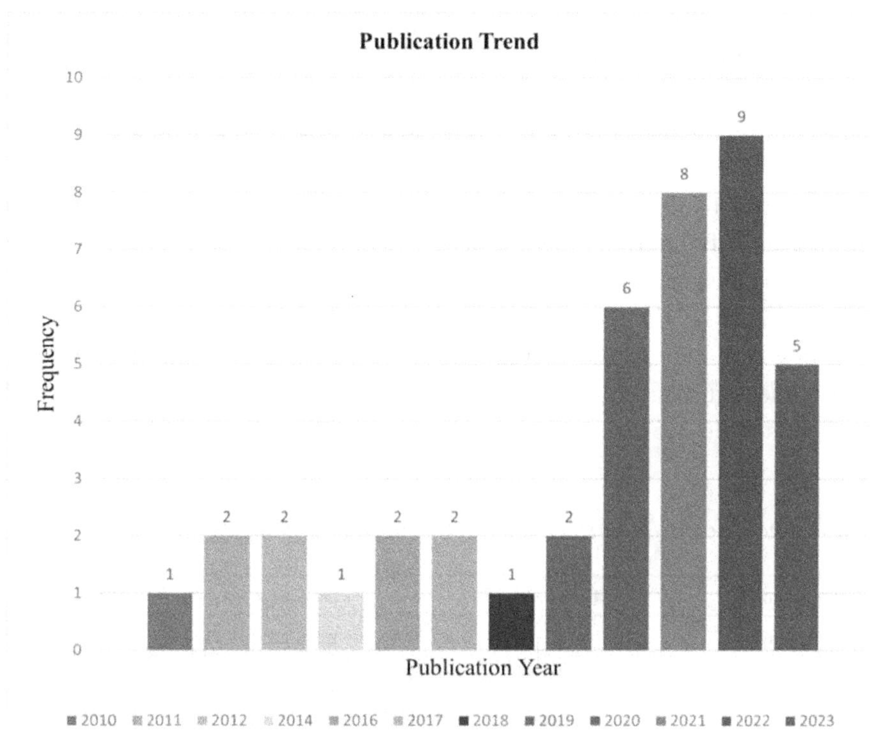

This proliferation can be attributed to several factors. For instance, with the emergence of innovative technologies and marketing strategies, there has been a requirement for contemporized studies that can furnish insights into the best practices and trends in medical tourism marketing. Moreover, the advancement of medical tourism as a global industry has devised a complementary demand for study that furnishes insights into the industry's dynamics, challenges, and opportunities. However, the mounting interest of researchers and scholars in medical tourism and marketing has led to more research and publications concentrating on the convergence of these two research domains. The interdisciplinary disposition of the medical tourism discipline has created a need for research that can draw from multiple domains, such as marketing, healthcare, and tourism.

3.4 Key Methodologies

The most used research methods in medical tourism marketing research are quantitative and qualitative research, content analysis, and case studies. Quantitative research accounts for 46.34% of studies reviewed, while qualitative research accounts for

21.95%. Content analysis and case study research are also generally used. The least widely used research methods are bibliometric analysis, conceptual research, and systematic literature review approaches. **Table 4** depicts all methodologies pinpointed in this study. The trend of using quantitative and qualitative research methods, content analysis, and case study research in medical tourism marketing research is justified by the need for a thorough acquaintance with the complex and multidisciplinary disposition. The focus on empirical data is apparent in prior studies' prevailing usage of quantitative research. However, employing qualitative research approaches and case study methods underlines the significance of comprehending the diverse nature of the medical tourism marketing discipline. Further research studies using different methods might provide valuable insights into the overall behavior of medical tourists, empowering policymakers, medical service providers, and marketers to devise effective marketing strategies.

Table 4. Methodologies investigated in the published articles

Methodology	Frequency	Percentage Exemplary studies
Bibliometric analysis	03	07.32% Campra et al. (2022)
Case Study	01	02.44% Schmude et al. (2021)
Conceptual	03	07.32% Dini and Pencarelli (2021)
Content analysis	04	09.76% Crooks et al. (2011)
Qualitative research	09	21.95% Horsfall (2020)
Quantitative research	19	46.34% Majed et al. (2020); Taheri et al. (2021)
Systematic Literature Review approach	02	04.87% Hartwell et al. (2018)

Note: The stated frequencies are based on 103 articles from the final sample size.
Source: The Author.

4. IMPLICATIONS, LIMITATIONS, AND FUTURE RESEARCH AGENDA

4.1 Theoretical Implications

A systematic literature review study on medical tourism marketing and its usefulness to business and academic communities offers a helpful opportunity to augment the conceptual understanding of this rapidly growing discipline. Theoretically, the study raises the conceptual knowledge of medical tourism marketing scholars and researchers and further guides future research. The study can also assist in pinpointing research gaps in the literature where further study is vital. This can encourage scholars and researchers to identify research problems and frame related research questions to uncover their answers. Moreover, by identifying key research themes related to medical tourism, such as medical tourism marketing, consumer behavior in medical tourism, and marketing strategies related to medical tourism destinations, the study can contribute to a more profound knowledge of the medical tourism industry and marketing.

4.2 Practical Implications

The study presents practical insights into medical tourism marketing. Firstly, medical establishments and marketing professionals may consider developing well-targeted medical tourism marketing campaigns using diverse marketing channels such as electronic, print, and social media. Secondly, medical establishments can utilize electronic media, print media, medical tourism fairs, and exhibitions to attract medical tourists. Thirdly, they can utilize virtual marketing to promote medical services to potential tourists. Fourthly, social media marketing strategies can be implemented to augment consumer attitudes toward medical tourism by emphasizing brand building and employing social media platforms. This study spotlights vital marketing implications for medical tourism establishments and marketing professionals. Furthermore, It is also essential to understand the behaviors and intentions of medical tourists. Additionally, word-of-mouth referrals from family, friends, and relatives have emerged as a significant source of information for potential medical tourists. Similarly, medical cluster groups can be started to operate medical tourism, and stakeholders can be motivated to devise new medical treatment products and services to attract potential medical tourists and build brand loyalty. Moreover, medical establishments and marketing professionals must embrace these traditional and digital marketing strategies to effectively promote medical establishments' infrastructure, reputation, and medical treatment services to potential medical tourists while building brand awareness and loyalty. This study attempts to support medical establishments and medical tourism marketing professionals in devising suitable marketing strategies to conduct effective marketing campaigns for attracting and retaining medical tourists. Hence, leveraging these marketing implications, medical tourism establishments and marketing professionals can enhance their performance and build a competitive advantage in this rapidly growing medical tourism industry by attracting medical tourists.

4.2 Limitations and Future Research Agenda

While this study provides valuable insights into the concept of medical tourism marketing, it is essential to acknowledge its specific limitations. Firstly, this study's systematic literature review approach integrated the PRISMA protocol and the TCCM framework. While this approach effectively extracts, reviews, and analyzes eligible articles, alternative research approaches may yield different results and insights. Future researchers may benefit from exploring alternative approaches, such as meta-analyses, structured literature reviews, and integrative literature reviews. Moreover, the inclusion criterion was limited to the "Article" document type, which may have excluded relevant studies published in other formats. Additionally, this study only selected articles with open-access availability, which may have resulted in a limited pool of eligible articles. Lastly, this study only explored articles from the Web of Science database, which may have resulted in a partial outlook on medical tourism marketing. Future research initiatives could explore other databases, such as Google Scholar, Scopus, or ScienceDirect, to comprehensively understand this emerging research domain.

The study proposes six future research questions (FRQ), namely, **FRQ1.** How can medical tourism destinations effectively market their distinctive medical services and establishments to entice medical tourists or patients? **FRQ2.** What are the most effective marketing strategies for boosting medical tourism destinations to potential medical tourists or patients? **FRQ3.** How does consumer behavior differ across medical tourism destinations? **FRQ4.** What role do destination marketing institutions play in fostering medical tourism destinations? **FRQ5.** In what way do diverse countries' cultural, economic, and political contexts influence the medical tourism industry's marketing strategies and consumer behavior? **FRQ6.** What research approaches are most suitable for exploring the effectiveness of medical tourism marketing strategies? These future research questions offer an opportunity to investigate diverse characteristics of medical tourism marketing that can further assist scholars and researchers in identifying new approaches, understanding the role of destination marketing institutions, designing adequate studies, identifying consumer behavior trends, and adapting marketing strategies to diverse contexts. Moreover, this can provide a marketing tool as a starting point for businesses to attract more medical tourists.

5. CONCLUSION

The tourism industry has expanded the range of services and experiences that contribute positively to people's well-being (Dini & Pencarelli, 2021). In this context, the growth of the Internet has enabled medical facilities in countries around the world to market their services to a receptive audience of international patients (Johnston et al., 2012). With the growth of the tourism industry, medical tourism has become one of the fastest-growing sectors worldwide, contributing to countries' sustainable development and economic dynamism (Ghasemi et al., 2021). The present study has employed a systematic literature review approach and carried the study on medical tourism marketing to augment the conceptual understandings of academic and business communities. The Web of Science database was selected to extract relevant articles using suitable search keywords and the TCCM framework; the study's key findings have been presented. However, it was found that the promotion of the healthcare travel industry heavily relies on tourism sources and healthcare medical infrastructures, as Dang et al. (2020) noted. Thus, for a successful medical venture, understanding the target market is paramount, as highlighted by Wang et al. (2023). In this regard, medical establishments' websites can be instrumental in attracting and convincing patients to visit medical tourism, as Hartman (2016) suggested. However, Bulatovic and Iankova (2021) highlighted several barriers to medical tourism, such as high costs, lack of marketing activities, and insufficient collaboration between medical and tourism service providers. Several factors motivate people to prefer medical care services in a foreign country, including the reputation of medical practitioners and facilities, hospital accreditation, and expenses (Al Adwan, 2020). It is essential to note that the favorable behaviors of tourists towards a destination's marketing stimuli can promote the tourism business globally (Majeed et al., 2020). Overall, it is evident that a holistic approach that considers various factors, including natural resources, touristic experiences, brand equity, and government support, is critical to the success and sustainability of medical tourism destinations. Governments must adopt good practices that support and promote creating a favorable image for medical tourism destinations in their respective countries (Dryglas et al., 2019). To promote a country as a medical tourism destination, advertising strategies highlight its affordability and value for money, which further helps to create a positive consumer attitude (Zain et al., 2022).

Competing Interests

The authors of this publication declare there are no competing interests.

Funding

This research received no specific grant from any funding agency in the public, commercial, or not-for-profit sectors. Funding for this research was covered by the author(s) of the article.

REFERENCES

Abdul-Rahman, M. N., Hassan, T. H., Abdou, A. H., Abdelmoaty, M. A., Saleh, M. I., & Salem, A. E. (2023). Responding to tourists' intentions to revisit medical destinations in the post-COVID-19 era through the promotion of their clinical trust and well-being. *Sustainability (Basel)*, *15*(3), 2399. doi:10.3390/su15032399

Al Adwan, A. (2020). The impact of motivation factors and intention to adopt Jordan as a destination for medical tourism in the Middle East. *Innovative Marketing*, *16*(2), 146–158. doi:10.21511/im.16(2).2020.11

Al-Azzam, A. F. M. (2016). A study of the impact of marketing mix for attracting medical tourism in Jordan. *International Journal of Marketing Studies*, *8*(1), 139–149. doi:10.5539/ijms.v8n1p139

Aurelia-Felicia, S., Andreea, B., & Ion, P. (2014). Clarifications Regarding Medical Tourism Destination Marketing. *Balneo Research Journal*, *5*(3), 84–98. doi:10.12680/balneo.2014.1068

Buckley, R. (2023). Tourism and mental health: Foundations, frameworks, and futures. *Journal of Travel Research*, *62*(1), 3–20. doi:10.1177/00472875221087669

Bulatovic, I., & Iankova, K. (2021). Barriers to medical tourism development in the United Arab Emirates (UAE). *International Journal of Environmental Research and Public Health*, *18*(3), 1365. doi:10.3390/ijerph18031365 PMID:33540935

Campra, M., Riva, P., Oricchio, G., & Brescia, V. (2022). Bibliometrix analysis of medical tourism. *Health Services Management Research*, *35*(3), 172–188. doi:10.1177/09514848211011738 PMID:33957813

Chakma, R., Paul, J., & Dhir, S. (2021). Organizational ambidexterity: A review and research agenda. *IEEE Transactions on Engineering Management*, 1–17. doi:10.1109/TEM.2021.3114609

Cham, T. H., Cheng, B. L., Low, M. P., & Cheok, J. B. C. (2020). Brand image as the competitive edge for hospitals in medical tourism. *European Business Review*, *33*(1). doi:10.1108/EBR-10-2019-0269

Cham, T. H., Lim, Y. M., & Sigala, M. (2022). Marketing and social influences, hospital branding, and medical tourists' behavioural intention: Before-and after-service consumption perspective. *International Journal of Tourism Research*, *24*(1), 140–157. doi:10.1002/jtr.2489

Chinai, R., & Goswami, R. (2007). Medical visas mark growth of Indian medical tourism. *Bulletin of the World Health Organization*, *85*(3), 164–165. doi:10.2471/BLT.07.010307 PMID:17486202

Connell, J. (2013). Contemporary medical tourism: Conceptualisation, culture and commodification. *Tourism Management*, *34*, 1–13. doi:10.1016/j.tourman.2012.05.009

Crooks, V. A., Turner, L., Snyder, J., Johnston, R., & Kingsbury, P. (2011). Promoting medical tourism to India: Messages, images, and the marketing of international patient travel. *Social Science & Medicine*, *72*(5), 726–732. doi:10.1016/j.socscimed.2010.12.022 PMID:21310519

Cubric, M. (2020). Drivers, barriers and social considerations for AI adoption in business and management: A tertiary study. *Technology in Society*, *62*, 101257. doi:10.1016/j.techsoc.2020.101257

Dang, H. S., Nguyen, T. M. T., Wang, C. N., Day, J. D., & Dang, T. M. H. (2020). Grey system theory in the study of medical tourism industry and its economic impact. *International Journal of Environmental Research and Public Health*, *17*(3), 961. doi:10.3390/ijerph17030961 PMID:32033177

Das, R. (2017). Medical marketing in promoting medical tourism in India. *ELK Asia Pacific Journal of Marketing and Retail Management*, *8*(3), 1–10.

Dini, M., & Pencarelli, T. (2021). Wellness tourism and the components of its offer system: A holistic perspective. *Tourism Review*, *77*(2), 394–412. doi:10.1108/TR-08-2020-0373

Dryglas, D., & Lubowiecki-Vikuk, A. (2019). The Attractiveness of Poland as a Medical Tourism Destination from the Perspective of German and British Consumers. *Entrepreneurial Business and Economics Review*, *7*(2), 45–62. doi:10.15678/EBER.2019.070203

Elkhwesky, Z., Castañeda-García, J. A., Abuelhassan, A. E., & Tag-Eldeen, A. (2023). A systematic and critical review of restaurants' business performance: Future directions for theory and practice. *Tourism and Hospitality Research*, *23*(3), 441–458. doi:10.1177/14673584221104983

Ghasemi, P., Mehdiabadi, A., Spulbar, C., & Birau, R. (2021). Ranking of sustainable medical tourism destinations in Iran: An integrated approach using fuzzy SWARA-PROMETHEE. *Sustainability (Basel)*, *13*(2), 683. doi:10.3390/su13020683

Guru, S., Sinha, A., & Kautish, P. (2022). Determinants of medical tourism: Application of Fuzzy Analytical Hierarchical Process. *International Journal of Emerging Markets*, *18*(11), 4819–4842. doi:10.1108/IJOEM-08-2021-1173

Gusenbauer, M., & Haddaway, N. R. (2020). Which academic search systems are suitable for systematic reviews or meta-analyses? Evaluating retrieval qualities of Google Scholar, PubMed, and 26 other resources. *Research Synthesis Methods*, *11*(2), 181–217. doi:10.1002/jrsm.1378 PMID:31614060

Guz, A. N., & Rushchitsky, J. J. (2009). Scopus: A system for the evaluation of scientific journals. *International Applied Mechanics*, *45*(4), 351–362. doi:10.1007/s10778-009-0189-4

Habibi, A., Mousavi, M., Jamali, S. M., & Ale Ebrahim, N. (2022). A bibliometric study of medical tourism. *Anatolia*, *33*(3), 415–425. doi:10.1080/13032917.2021.1954042

Han, H. (2013). The healthcare hotel: Distinctive attributes for international medical travelers. *Tourism Management*, *36*, 257–268. doi:10.1016/j.tourman.2012.11.016

Han, H., & Hyun, S. S. (2015). Customer retention in the medical tourism industry: Impact of quality, satisfaction, trust, and price reasonableness. *Tourism Management*, *46*, 20–29. doi:10.1016/j.tourman.2014.06.003

Hartman, A. E. (2016). Transparency of hope: Ethical issues in marketing cross border reproductive care: A multi-region content analysis of clinic websites. *Marketing Intelligence & Planning*, *34*(7), 943–963. doi:10.1108/MIP-06-2015-0121

Hartwell, H., Fyall, A., Willis, C., Page, S., Ladkin, A., & Hemingway, A. (2018). Progress in tourism and destination wellbeing research. *Current Issues in Tourism*, *21*(16), 1830–1892. doi:10.1080/13683500.2016.1223609

Heung, V. C., Kucukusta, D., & Song, H. (2011). Medical tourism development in Hong Kong: An assessment of the barriers. *Tourism Management*, *32*(5), 995–1005. doi:10.1016/j.tourman.2010.08.012

Horsfall, D. (2020). Medical tourism from the UK to Poland: How the market masks migration. *Journal of Ethnic and Migration Studies*, *46*(20), 4211–4229. doi:10.1080/1369183X.2019.1597470

Howze, K. S. (2006). Medical tourism: Symptom or cure. *Ga. L. Rev.*, *41*, 1013.

Jiang, L., Wu, H., & Song, Y. (2022). Diversified demand for health tourism matters: From a perspective of the intra-industry trade. *Social Science & Medicine*, *293*, 114630. doi:10.1016/j.socscimed.2021.114630 PMID:34954675

Johnston, R., Crooks, V. A., & Snyder, J. (2012). "I didn't even know what I was looking for": A qualitative study of the decision-making processes of Canadian medical tourists. *Globalization and Health*, *8*(1), 1–12. doi:10.1186/1744-8603-8-23 PMID:22769723

Johnston, R., Crooks, V. A., Snyder, J., & Kingsbury, P. (2010). What is known about the effects of medical tourism in destination and departure countries? A scoping review. *International Journal for Equity in Health*, *9*(1), 1–13. doi:10.1186/1475-9276-9-24 PMID:21047433

Kashihara, H., Nakayama, T., Hatta, T., Takahashi, N., & Fujita, M. (2016). Evaluating the quality of website information of private-practice clinics offering cell therapies in Japan. *Interactive Journal of Medical Research*, *5*(2), e5479. doi:10.2196/ijmr.5479 PMID:27222494

Kazemha, A. S., & Dehkordi, K. S. (2017). The Principles of Designing Hospital Hotel with the Approach of Health Tourism in Kish Island. *Tarih Kültür ve Sanat Arastirmalari Dergisi*, *6*(1), 515–531. doi:10.7596/taksad.v6i1.758

Kim, H. L., & Hyun, S. S. (2022). The future of medical tourism for individuals' health and well-being: A case study of the relationship improvement between the UAE (United Arab Emirates) and South Korea. *International Journal of Environmental Research and Public Health*, *19*(9), 5735. doi:10.3390/ijerph19095735 PMID:35565130

Kim, M., Koo, D. W., Shin, D. J., & Lee, S. M. (2017). From Servicescape to loyalty in the medical tourism industry: A medical Clinic's service perspective. *Inquiry*, *54*, 0046958017746546. doi:10.1177/0046958017746546 PMID:29233057

Kim, W. O. N. J. O. N. G., & Oh, J. (2021). A Study on the Antecedents of Revisit Intention for Korean Medical Tourism: Focusing on Chinese Medical Tourists. *ICIC Express Lett. Part B Appl. Int. J. Res. Surv*, *12*, 741–750. doi:10.24507/icicelb.12.08.741

Li, Y., Han, Q., & Wen, T. (2022). Effect of Frugality and Cognition on Forest Health Tourism Intention–A Mediating Effect Analysis Based on Multigroup Comparison. *Frontiers in Psychology*, *13*, 844628. doi:10.3389/fpsyg.2022.844628 PMID:35572268

Li, Y., & Wen, T. (2022). Impact of cognition and social trust on forest-based health tourism intention during COVID-19. *Sustainability (Basel)*, *15*(1), 714. doi:10.3390/su15010714

Majeed, S., Zhou, Z., Lu, C., & Ramkissoon, H. (2020). Online tourism information and tourist behavior: A structural equation modeling analysis based on a self-administered survey. *Frontiers in Psychology*, *11*, 599. doi:10.3389/fpsyg.2020.00599 PMID:32373008

Majeed, S., Zhou, Z., & Ramkissoon, H. (2020). Beauty and elegance: Value co-creation in cosmetic surgery tourism. *SAGE Open*, *10*(2), 2158244020932538. doi:10.1177/2158244020932538

Mortezaei, G., Otaghvar, H. A., Vazifehdoost, H., Saeedi, P., & Pegheh, A. (2021). Prioritization of Brand Equity Components Using an Integrated Structural Equation Modelling and Fuzzy AHP. *Tehnički glasnik*, *15*(3), 357-365. doi:10.31803/tg-20210410214620

Noaman, S., & Montargot, N. (2023). Understanding Interactions Between Providers of Cosmetic Tourism and Tourists: The Lebanon Experience. *Tourism, Culture & Communication*, *23*(1), 45–60. doi:10.3727/109830422X16510695152055

Oltean, F. D., Gabor, M. R., Stăncioiu, A. F., Kardos, M., Kiss, M., & Marinescu, R. C. (2020). Aspects of marketing in dental tourism—Factor of sustainable development in Romania. *Sustainability (Basel)*, *12*(10), 4320. doi:10.3390/su12104320

Paul, J., Alhassan, I., Binsaif, N., & Singh, P. (2023). Digital entrepreneurship research: A systematic review. *Journal of Business Research*, *156*, 113507. doi:10.1016/j.jbusres.2022.113507

Paul, J., & Rosado-Serrano, A. (2019). Gradual internationalization vs born-global/international new venture models: A review and research agenda. *International Marketing Review*, *36*(6), 830–858. doi:10.1108/IMR-10-2018-0280

Penney, K., Snyder, J., Crooks, V. A., & Johnston, R. (2011). Risk communication and informed consent in the medical tourism industry: A thematic content analysis of Canadian broker websites. *BMC Medical Ethics*, *12*(1), 1–9. doi:10.1186/1472-6939-12-17 PMID:21943392

Pessot, E., Spoladore, D., Zangiacomi, A., & Sacco, M. (2021). Natural resources in health tourism: A systematic literature review. *Sustainability (Basel)*, *13*(5), 2661. doi:10.3390/su13052661

Phuthong, T., Anuntavoranich, P., Chandrachai, A., & Piromsopa, K. (2022). Developing and validating an assessment scale to measure the competitiveness of wellness destinations. *Sustainability (Basel)*, *14*(7), 4152. doi:10.3390/su14074152

Pighin, M., Alvarez-Risco, A., Del-Aguila-Arcentales, S., Rojas-Osorio, M., & Yáñez, J. A. (2022). Factors of the Revisit Intention of Patients in the Primary Health Care System in Argentina. *Sustainability (Basel), 14*(20), 13021. doi:10.3390/su142013021

Popa, A. L., Ţarcă, N. N., Sasu, D. V., Bodog, S. A., Roşca, R. D., & Tarcza, T. M. (2022). Exploring Marketing Insights for Healthcare: Trends and Perspectives Based on Literature Investigation. *Sustainability (Basel), 14*(17), 10499. doi:10.3390/su141710499

Rahman, I., Martin, D. S., & Liu, S. (2022). Outbound medical tourists: The interplay of perceived quality, length of stay, group-size, post-visit destination image and revisit intention. *PLoS One, 17*(5), e0267755. doi:10.1371/journal.pone.0267755 PMID:35536776

Ramírez de Arellano, A. B. (2011). Medical tourism in the Caribbean. *Signs (Chicago, Ill.), 36*(2), 289–297. doi:10.1086/655908 PMID:21114073

Robertson, E. M., Moorman, S. W., & Korus, L. J. (2022). Why do canadians travel abroad for cosmetic surgery? A qualitative analysis on motivations for cosmetic surgery tourism. *Plastic Surgery (Oakville, Ont.), 30*(4), 353–359. doi:10.1177/22925503211019607 PMID:36212104

Saiprasert, W., Xu, B., & Tavitiyaman, P. (2022). The Relationships Among Perceived Medical Quality, Well-Being Perception, And Behavioral Intention: A Comparison Between Domestic And Overseas Medical Destinations. *Journal of Quality Assurance in Hospitality & Tourism, 23*(4), 1011–1036. doi:10.1080/1528008X.2021.1943598

Schmude, J., Pillmayer, M., Witting, M., & Corradini, P. (2021). Geography Matters, But horizontal ellipsis Evolving Success Factors for Nature-Oriented Health Tourism within Selected Alpine Destinations. *International Journal of Environmental Research and Public Health*, (10). Advance online publication. doi:10.3390/ijerph18105389 PMID:34070110

Shafii, M., Madadizadeh, F., Askari, R., Zarezadeh, M., Abdoli, A. M., & Bahariniya, S. (2022). Investigating medical tourism development indicators in Yazd Reproductive Sciences Institute: A cross-sectional study. *International Journal of Reproductive BioMedicine, 20*(12), 1039. 10.18502%2Fijrm.v20i12.12565 PMID:36819203

Shaygani, F., Jahangiri, S., Roozegar, M. H. Z., Kavosi, Z., & Marzaleh, M. A. (2023). Investigating the quality of Iranian hospitals' websites and their association to the Province's share of medical tourism. *BMC Research Notes, 16*(1), 332. doi:10.1186/s13104-023-06619-1 PMID:37964374

Shriedeh, F. B. (2019). Strategic factors for building brand equity: Jordan medical tourism. *Jurnal The Messenger*, *11*(1A), 121–137. doi:10.26623/themessenger.v11i1A.821

Singh, P., Arora, L., & Choudhry, A. (2023). Consumer Behavior in the Service Industry: An Integrative Literature Review and Research Agenda. *Sustainability (Basel)*, *15*(1), 250. doi:10.3390/su15010250

Singh, V. K., Singh, P., Karmakar, M., Leta, J., & Mayr, P. (2021). The journal coverage of Web of Science, Scopus and Dimensions: A comparative analysis. *Scientometrics*, *126*(6), 5113–5142. doi:10.1007/s11192-021-03948-5

Smith, E., Behrmann, J., Martin, C., & Williams-Jones, B. (2010). Reproductive tourism in Argentina: Clinic accreditation and its implications for consumers, health professionals and policy makers. *Developing World Bioethics*, *10*(2), 59–69. doi:10.1111/j.1471-8847.2009.00256.x PMID:19508291

Snyder, H. (2019). Literature review as a research methodology: An overview and guidelines. *Journal of Business Research*, *104*, 333–339. doi:10.1016/j.jbusres.2019.07.039

Taheri, B., Chalmers, D., Wilson, J., & Arshed, N. (2021). Would you really recommend it? Antecedents of word-of-mouth in medical tourism. *Tourism Management*, *83*, 104209. doi:10.1016/j.tourman.2020.104209

Tranfield, D., Denyer, D., & Smart, P. (2003). Towards a methodology for developing evidence-informed management knowledge by means of systematic review. *British Journal of Management*, *14*(3), 207–222. doi:10.1111/1467-8551.00375

Turner, L. (2012). Beyond "medical tourism": Canadian companies marketing medical travel. *Globalization and Health*, *8*(1), 16. doi:10.1186/1744-8603-8-16 PMID:22703873

Wang, H., Ghasemi, M., Ghadiri Nejad, M., & Khandan, A. S. (2023). Assessing the Potential Growth of Iran's Hospitals with Regard to the Sustainable Management of Medical Tourism. *Health & Social Care in the Community*, *2023*, 1–12. Advance online publication. doi:10.1155/2023/8734482

Wang, H., Ghasemi, M., Ghadiri Nejad, M., & Khandan, A. S. (2023). Assessing the Potential Growth of Iran's Hospitals with Regard to the Sustainable Management of Medical Tourism. *Health & Social Care in the Community*, *2023*, 1–12. Advance online publication. doi:10.1155/2023/8734482

Williams, R. I. Jr, Clark, L. A., Clark, W. R., & Raffo, D. M. (2021). Re-examining systematic literature review in management research: Additional benefits and execution protocols. *European Management Journal*, *39*(4), 521–533. doi:10.1016/j.emj.2020.09.007

Yin, J., Chen, Y., & Ji, Y. (2021). Effect of the event strength of the coronavirus disease (COVID-19) on potential online organic agricultural product consumption and rural health tourism opportunities. *Managerial and Decision Economics*, *42*(5), 1156–1171. doi:10.1002/mde.3298 PMID:34149119

Zain, N. A. M., Connell, J., Zahari, M. S. M., & Hanafiah, M. H. (2022). Intra-Regional Medical Tourism Demand in Malaysia: A Qualitative Study of Indonesian Medical Tourists' Rationale and Preferences. *The Malaysian Journal of Medical Sciences : MJMS*, *29*(2), 138–156. doi:10.21315/mjms2022.29.2.13 PMID:35528809

Zhang, M., Ding, H., Naumceska, M., & Zhang, Y. (2022). Virtual reality technology as an educational and intervention tool for children with autism spectrum disorder: Current perspectives and future directions. *Behavioral Sciences (Basel, Switzerland)*, *12*(5), 138. doi:10.3390/bs12050138 PMID:35621435

Chapter 11
The Indian Medical Tourism Industry's Repercussions From AI and Robotics

Birendra Kishore Roy
https://orcid.org/0000-0001-8202-5137
C.V. Raman Global University, Bhubaneswar, India

Viveka Nand Sharma
https://orcid.org/0000-0002-2807-105X
International Society for Hospitality Education, India

ABSTRACT

The proposed chapter delves into the transformative repercussions of Robotics and Artificial Intelligence (AI) on the burgeoning Indian Medical Travel and Tourism Sector. As India emerges as a global hub for medical tourism, the integration of innovative technologies promises to reshape patient experiences and healthcare delivery. The chapter meticulously investigates the expansion of the medical travel sector in India, the role of AI in healthcare, the impact of robotics, and the intersection of these technologies in the context of medical tourism. It not only highlights the positive outcomes and success stories but also addresses the challenges and ethical concerns related to healthcare robots and artificial intelligence. Furthermore, the chapter provides a forecast for the future of these technologies in India's wellness sector, offering recommendations for stakeholders to navigate the evolving landscape successfully.

DOI: 10.4018/979-8-3693-2248-2.ch011

1. INTRODUCTION

In this era of unmatched technological advancement, robotics and artificial intelligence (AI) are becoming more and more integrated, revolutionizing several industries (Sharma, Chauhan, & Tyagi, 2023). Of them, the healthcare industry is one that particularly stands out as a field where new advances could transform conventional methods and improve patient care in general (Singh, Sharma, & Bhadula, Automated Deep Learning based Disease Prediction Using Skin Health Records: Issues Challenges and Future Directions, 2022). One area of healthcare that has seen significant transformation is medical tourism, a growing sector that draws patients looking for high-quality, reasonably priced care from all over the world (Pandey, Sharma, & Matta, 2021). This study explores the complex effects of robots and artificial intelligence on the medical tourism market in India, looking at how these technologies may affect healthcare delivery, the economy, and the sector's general environment (Vitthal, Subhash, Sharma, & Ramachandran, 2015).

India has a rich cultural history and is swiftly rising to the top of the global medical tourism rankings. Modern healthcare facilities, highly skilled healthcare professionals, and reasonable prices have made India a popular destination for overseas patients requiring medical care, from intricate surgeries to wellness services (Chen & Decary, 2020). The rising integration of technology and neural networks (or AI) throughout healthcare procedures presents both potential and challenges for India's healthcare sector (Li, Asch, & Shah, 2020) Robotics and computational intelligence have revolutionized the medical industry, affecting not just the experience but also the concept and delivery of healthcare services. With its unparalleled capacity to scrutinize extensive datasets and discern patterns, artificial intelligence presents unparalleled prospects for customized healthcare, precise diagnosis, and optimal treatment planning. Conversely, robots are revolutionizing surgical techniques, increasing accuracy, minimizing invasiveness, and shortening recovery periods (Khedkar, Gandhi, Shinde, & Subramanian, 2020). The decisions made by medical tourists who want to strike a compromise between affordability and high-quality medical care could be influenced by these technological developments, which could raise the standard of healthcare services overall (Wen & Huang, 2022).

This study seeks to cast light on the complex ties present between the application of AI and robotics and the sector of medical tourism in India. We want to offer insights into how new technologies are affecting patient expectations, medical tourism patterns, and the financial aspects of the Indian healthcare industry (Tariq, Gill, & Hussain, 2023) by analyzing the changing landscape (Crooks, Turner, Snyder, Johnston, & Kingsbury, 2011).

1.1. Indian Medical Tourism's Historical Context: A historical understanding of the Indian tourism for medicine business is necessary to fully understand the current situation (De la Hoz-Correa, Muñoz-Leiva, & Bakucz, 2018). Due to its well-known medical facilities, highly qualified medical staff, and reasonably priced treatments, India has seen a consistent influx of medical tourists over the past few decades. A framework to assess the impacts of robots and computational intelligence is provided by the historical backdrop (Reddy, Fox, & Purohit, 2019).

1.2. AI's Ascent in Indian Healthcare: AI's revolutionary introduction into healthcare services has provided answers to enduring problems (Singh, Sharma, & Mohan, Industry 4.0 Internet of Medical Things Enabled Cost Effective Secure Smart Patient Care Medicine Pouch, 2023). AI applications are becoming a more important part of the medical scene, from treatment planning to diagnostic tools. Initiatives like personalized medicine, predictive analytics, and AI-driven diagnostic tools are becoming more popular in India (Sharma & Sharma, 2022). In-depth discussions of AI's specific uses in the Indian healthcare system and an examination of how these developments are influencing medical travel are provided in this section (Diwakar, Singh, & Shankar, 2021).

1.3. The Effect of Robotic Surgery on Medical Travel: Robotics has transformed surgery by giving doctors access to high-precision instruments that improve their skills. Patients are now looking for robotic surgery because of its minimally invasive treatments and quick recovery periods (Rahman, Khatun, Sami, & Uzzaman, 2022). This section will analyze robotic surgery's impact on the medical tourism market in India by looking at how these innovations in technology are drawing patients looking for cutting-edge surgical procedures (Kansal, Ranjan, Sinha, Tiwari, & Wickramasinghe, 2021).

1.4. Experience and Expectations of the Patient: The patient experience is becoming a critical component as more and more medical tourists look for value outside of conventional boundaries (Santosh & Gaur, 2022). A more patient-centric approach is facilitated by robotically assisted procedures and AI-driven tailored healthcare. This section will examine how the employment of these technologies is affecting patient desires and encounters in the framework of Indian healthcare tourism (Sousa, Nunes, do Nascimento, & Chakraborty, 2023).

1.5. Economic Consequences and Industry Structure: The use of AI and robots has significant effects on the financial aspects of the medical tourism sector, even beyond the domain of healthcare delivery (Gaur, Afaq, Singh, & Dwivedi, 2021). This section will look at how these technologies affect cost structures, market competitiveness, and the general sustainability of the Indian medical tourism industry (Dar & Kashyap, 2023).

1.6. Difficulties and Ethical Issues: Healthcare AI and robotics hold great promise, but there are many obstacles and moral dilemmas to be resolved (Wright & Zascerinska, 2023). With an emphasis on data privacy, responsibility, and human contact in healthcare, the following subsection will critically evaluate the potential hazards and ethical conundrums involved with the incorporation of technological advances in the Indian healthcare tourism business (Baran & Karaca, 2023).

1.7. India's Increasing Medical Travel Industry

1.8. An outline of how the medical tourism industry in India has developed and grown:

India has been a well-known center for medical tourism, with significant expansion and advancement in the field throughout the previous ten years (Wang, et al., 2022). The nation's medical tourism sector has grown quickly, drawing in worldwide patients seeking excellent wellness services at a significantly affordable price than in developed nations. This rise is attributed to multiple sources (Hassan & Bellos, COVID-19: Reshaping Medical Tourism through Artificial Intelligence (AI) and Robotics, 2022). First off, India has excellent hospitals and medical facilities, as well as a strong healthcare infrastructure. India's reputation as a dependable location for medical care has been greatly aided by accredited facilities with cutting-edge equipment and knowledgeable medical staff. For medical disciplines like organ transplants, orthopedics, cardiac surgery, and cosmetic surgery, the nation has gained a reputation on a global scale (Khaksarshahmirzadi, et al., Evaluation of the use of robotic surgery with remote simulation for use in the field of Medical Tourism and Pro-Tourism development, 2022).

Furthermore, the regulatory framework in India has facilitated the expansion of medical tourism. Foreign patients now feel more confident because of the existence of hospitals with international accreditation and the observance of global medical standards (Shaikh, Dar, & Sofi, 2022). The number of overseas patients has increased in part because medical professionals who speak English are more readily available and because obtaining medical visas is easier (AYKIN, et al., 2023). Furthermore, medical tourism has become easier to arrange because of partnerships between foreign insurance companies and Indian healthcare facilities. These collaborations guarantee patients a smooth experience with affordable treatment plans and all-inclusive medical care (Bauer, 2023).

1.9. Determine the key characteristics that make India a popular choice for medical travel: India is a desirable location for medical tourism due to several factors. The main consideration is cost-effectiveness since medical care in India is far less expensive than in developed nations (Mbunge & Muchemwa, 2022). The quality of healthcare services is not jeopardized by this cost advantage, either, as numerous hospitals maintain worldwide accreditations and certifications (Navaz, Serhani, El Kassabi, Al-Qirim, & Ismail, 2021). The nation's varied cultural history and tourism hotspots add to their allure. Indian holistic medicine, which incorporates ancient therapies like yoga and Ayurveda into conventional medical care, is frequently a source of comfort for patients (Mishra & Singh, 2023) Moreover, the high doctor-to-patient ratio in India guarantees individualized attention and care. Patient results are further improved by the availability of state-of-the-art medical equipment and creative treatments (Crooks, Turner, Snyder, Johnston, & Kingsbury, 2011). When compared to their native countries, patients frequently value shorter wait periods for treatments and consultations (Sedik, Hammad, Abd El-Samie, Gupta, & Abd El-Latif, 2021).

1.10. The benefits of medical tourism for the economy and healthcare: India's economy is significantly impacted by medical tourism. By supporting the healthcare industry, generating job opportunities, and bolstering ancillary industries like hospitality and tourism, the flow of overseas patients boosts economic growth (Liaqat, Akhunzada, Shaikh, Giannetsos, & Jan, 2020). The money received from medical tourism goes toward funding cutting-edge medical technologies, upholding the standard of healthcare services, and developing infrastructure. From the standpoint of healthcare, medical tourism encourages communication and cooperation between medical specialists in India and abroad (Magaña-Espinoza, et al., 2014). By fostering a culture of innovation and constant improvement in medical practices, this exposure to other countries improves the abilities of local practitioners (Ghubaish, et al., 2020). Furthermore, the increased patient volume supports R&D projects that promote improvements in medical technology and science (Li & Carayon, 2021).

2. THE ROLE OF AI IN HEALTHCARE:

2.1. An Explanation for Artificial Intelligence in Medical Tourism and Healthcare: The practice of people traveling to other nations seeking medical care and healthcare services is known as medical tourism. For medical tourism, artificial intelligence (AI) is essential to improving the effectiveness and caliber of healthcare services (Ning, et al., 2020). AI benefits the medical tourism industry in several ways. First of all, it helps determine which medical locations are best suited for particular therapies (Chinzei, 2019). AI systems evaluate a ton of information, such as success rates, patient ratings, and cost-effectiveness, to suggest the best medical facilities for people traveling overseas for treatment. Additionally, AI makes it easier for medical tourists to receive individualized healthcare (Schäfer, Stewart, & Pott, 2019). AI can foresee patient requirements, preferences, and possible health hazards using predictive analytics, resulting in a customized and patient-centered approach (Locsin & Ito, 2018).

This degree of customization raises medical tourists' general level of happiness and helps healthcare providers' standing in the international marketplace. AI also plays a key role in cultural sensitivity and language translation. Artificial intelligence (AI)-powered translation systems assist in closing the communication gaps that frequently occur in medical tourism, enabling efficient communication between medical professionals and patients from different backgrounds in languages and cultures enhancing operational effectiveness for healthcare facilities that serve medical tourists, which has positive economic effects. The workflow is streamlined by the automation of billing, scheduling, and administrative chores, which lowers the overall expense of healthcare services. This can thereby lower the cost of medical travel and draw in a wider patient base (Bhaidkar & Goswami, 2017). Not only does the use of AI in medical tourism improve healthcare quality, but it also helps the economies of nations that market themselves as centers of medical tourism. These nations may draw in more foreign patients by using cutting-edge AI technologies, which will boost earnings and generate jobs in the healthcare industry (Su, Samani, Yang, & Fernando, 2018).

2.2. The Numerous Uses of AI in Patient Care, Diagnosis, and Therapy: Through the use of analytics, machine learning, and other cutting-edge technologies, artificial intelligence (AI) is transforming the handling of patients, diagnosis, and therapy. Personalized medicine is one of the main uses for AI, where it is used to create treatment plans that are specific to each patient by analyzing genetic, clinical, and lifestyle data (Nkurunziza, Udahemuka, Dusenge, & Umutesi, 2022). Artificial intelligence (AI)-)-powered chatbots and virtual assistants give immediate, customized assistance with medical care. They can arrange appointments, provide emotional support, and deliver medication information. This improves patients' engagement with healthcare services and their overall experience (Grespan, Fiorini, & Colucci, 2019).

AI systems are quite good at diagnosing diseases by evaluating data from medical imaging tests like CT, MRI, and X-rays. These systems help medical personnel make more accurate and quick diagnoses by detecting patterns and anomalies with high accuracy. This lowers the possibility of a false positive as well as enhances patient outcomes. Furthermore, through predictive modeling, AI makes a substantial contribution to therapy (Zarei, Feiz, Maleki Minbashrazgah, & Maleki, 2020). Through the examination of patient data and treatment results, artificial intelligence (AI) can forecast which treatments will work best for particular ailments. This enables medical professionals to select the most appropriate and customized course of treatment for their patients. AI's application in patient care, diagnosis, and therapy reduces costs in an economic sense. Healthcare personnel may concentrate on more difficult tasks by having less work to do thanks to AI-driven automation, which lets

AI systems take care of mundane and administrative duties. Healthcare facilities can save money on operations thanks to this efficiency (Dahl & Kamel Boulos, 2013).

2.3. AI-Driven Innovations in Telemedicine and Healthcare Management: AI-enabled telemedicine has revolutionized healthcare service delivery, increasing accessibility and efficiency. Artificial intelligence (AI)-driven virtual health platforms provide remote consultations, diagnosis, and monitoring, thereby reducing geographical obstacles and enhancing healthcare accessibility, especially in underprivileged or distant areas. AI improves staff scheduling, inventory control, and resource allocation in hospital administration. Healthcare facilities can better prepare for peak times and avoid resource shortages by using predictive analytics to forecast patient admission rates. Overall patient care and satisfaction have increased as a result of this proactive strategy (Rohokale, Prasad, & Prasad, A cooperative Internet of Things (IoT) for rural healthcare monitoring and control, 2020). Economic savings are substantial when AI is included in healthcare administration and telemedicine. A more economical healthcare system is the result of patient travel expenses being decreased, resources being used effectively, and procedures being streamlined. Furthermore, early detection and response can save expensive problems and hospital stays because of AI's preventive skills in healthcare management (Rohokale, Prasad, & Prasad, A cooperative Internet of Things (IoT) for rural healthcare monitoring and control, 2011).

4. ROBOTICS IN MEDICAL TOURISM:

4.1. Robotics in Healthcare, including Surgical Robotics and Rehabilitation Robotics: The use of robotics in healthcare is revolutionizing the field, improving patient outcomes, and fostering the expansion of medical travel. For example, surgical robots enable precise and minimally invasive operations (Oberoi & Kansra, 2019). When proficient surgeons command robots to carry out intricate procedures with great precision, recovery times are shortened and results are generally enhanced. Rehabilitation robotics helps patients rehabilitate by giving them focused exercises and encouragement as they receive physical therapy. Because these robotics developments offer cutting-edge technologies for a range of surgical procedures, they not only improve healthcare quality but also increase the appeal of medical tourism locations (Pang, Tian, & Chen, 2014).

4.2. The Integration of Robotic Surgery in Indian Hospitals Catering to Medical Tourists: Indian medical tourism hospitals have made robotic surgery one of their main offerings. In Indian hospitals, the well-liked robotic surgical platform da Vinci Surgical System is frequently utilized for operations like gynecological, urological, and cardiac surgery. Robotic technologies' accuracy and dexterity lead to fewer complications, faster healing times, and shorter hospital stays. International patients looking for cutting-edge, top-notch medical care are drawn to this. The incorporation of robotic surgery into Indian healthcare facilities can further augment the nation's standing as a premier medical tourism destination, hence stimulating economic growth via amplified medical tourism revenue (Hongal & Kshirsagar, 2023).

4.3. AI-Driven Innovations in Telemedicine and Healthcare Management: Travelers seeking medical tourism are drawn to locales because Artificial Intelligence (AI) has transformed telemedicine and healthcare administration. AI-powered virtual health platforms in telemedicine enable remote consultations, diagnosis, and monitoring, giving patients a smooth and customized experience (Sikdar & Guha, 2020). By analyzing patient data, AI algorithms enable predictive modeling for the administration of healthcare. By assisting with staff scheduling, inventory management, and resource allocation, this predictive capacity helps healthcare facilities operate more efficiently. Medical tourism locations become more competitive and financially viable when AI is integrated into telemedicine and healthcare management. This not only enhances overall healthcare services but also improves patient care and reduces costs (Kumar, Kumar, & Attri, 2021).

5. AI AND ROBOTICS' EFFECTS ON MEDICAL TOURISM:

AI and robotics are transforming medical tourism in India by providing both local and foreign patients with cutting-edge healthcare experiences that can change their lives. The improvement of the entire medical tourism experience and the growing draw of foreign patients looking for robotic surgery as a minimally invasive technique are two important factors influencing this paradigm shift (Tehseen, et al., 2021).

5.1. Advancements in robots and artificial intelligence for medical travel in India:

5.1.1. Improving India's Medical Tourism Experience:

5.1.1.1. Personalized Medical Supplies: AI is proving to be invaluable in customizing healthcare solutions to meet the demands of each patient. To provide individualized treatment regimens, sophisticated algorithms examine genetic information, medical history, and patient data. This degree of personalization guarantees the best possible care for patients in the shortest amount of time, making medical travel enjoyable (Quazi, Saha, & Singh, 2022).

5.1.1.2. Effective Diagnosis and Therapy: Medical assessments are being completed more quickly and accurately thanks to AI-powered diagnostic technologies.

Accurate and prompt diagnosis results in more effective treatment regimens, which cut down on the total amount of time needed for medical operations. This is advantageous to patients as well as increasing India's appeal as a medical travel destination (Sony, Antony, & McDermott, 2023).

5.1.1.3. Online and Virtual Medical Consultations: Patients can have remote consultations with medical providers thanks to AI-powered virtual health consultations. For patients from other countries, being able to speak with Indian physicians before booking travel is quite helpful. Medical tourism is now easier to access and more convenient thanks to virtual consultations (Saxena, et al., 2023).

5.1.1.4. Analytics Predictive for Healthcare Administration: AI makes it possible to forecast possible health problems using data analysis. This proactive approach to managing healthcare aims to avoid complications, lessen the need for lengthy treatments, and provide medical tourists with a more comprehensive and preventive healthcare experience (Kwon, et al., 2022).

5.2. Robotic Surgery Drawing Patients from Abroad:

5.2.1. International Patients Drawn to Robotic Surgery: The growing integration of there is an increase in qualified workers who are trained in AI and robots in Indian healthcare, providing international patients with access to a pool of expertise. Leading Indian hospitals and healthcare institutions are obtaining international accreditations and adhering to global quality standards, ensuring world-class medical care (Medhekar & Wong, 2020). The symbiotic relationship between AI-driven medical technologies and the Indian healthcare sector is set to propel the country to new heights in medical tourism. The allure of robotic surgery, coupled with its precision, faster recovery times, cost-effectiveness, availability of skilled professionals, and adherence to international quality standards, positions India as a frontrunner in the global medical tourism landscape (Wang, Su, Zhang, & Li, 2021).

5.2.1.1. Procedures with Minimal Invasiveness: Artificial intelligence (AI)--driven robotic surgery has made tremendous progress in the realm of minimally invasive surgeries. Robotic systems' accuracy and dexterity enable surgeons to execute intricate procedures with tiny incisions, which shorten recovery periods and minimize discomfort and scarring. Modern robotic surgery is attracting more and more foreign patients to India in search of these advantages (Saxena, et al.).

5.2.1.2. Knowledge and Creativity: India is become a center for knowledge on robotic surgery. Patients from throughout the world seeking cutting-edge medical treatments are drawn to surgeons versed in the most recent robotic technology. Robotics integration improves surgical accuracy and highlights India's dedication to medical innovation, which attracts individuals looking for cutting-edge healthcare solutions (Kumar, Sharma, & Dutot, 2023).

5.2.1.3. Economical Solutions: Robotic surgery can be more economical while still delivering high-quality medical care. Because robotic treatments are more efficient, patients often recover more quickly and spend less time in the hospital overall. Additionally, because of this financial benefit, patients from outside are drawn to receive excellent healthcare at a far lower cost than they would back home (Li, Yin, Qiu, & Bai, 2021).

6. FACILITIES BY THE OTHER NEIGHBORING COUNTRIES TO THE TOURIST IN TERMS OF HEALTH TOURISM AND MEDICAL TOURISM

Each state in India promotes health tourism in a particular way, which helps the business flourish in its own special way (Gyu Ko, 2011). A number of states have used tactics to draw in foreign patients looking for top-notch medical care after realizing the potential of medical tourism. Reputable medical centers like Maharashtra, Tamil Nadu, and Karnataka have aggressively marketed their cutting edge medical facilities, cutting edge technology, and highly qualified medical staff. These states market themselves as top destinations for medical tourists by utilizing their current infrastructure and knowledge (Aydin & Karamehmet, 2017).

Furthermore, state governments work with private healthcare professionals to provide customized treatment plans that include wellness initiatives and possibilities for recovery following treatment (Rai, Medical Tourism in Kolkata, Eastern India, 2019). Moreover, programs like expedited visa applications, special medical tourism desks, and advertising campaigns are launched to facilitate the travel of patients from abroad (Timothy & Nyaupane, 2022). Particularly the southern states have developed into centers for medical tourism; Kerala, with its reputation for providing holistic healthcare, draws patients looking for alternative therapies and wellness programs (Hosain, Amin, Rehman, & Koirala, Know thy neighbors: The status of cardiac surgery in the South Asian countries around India, 2017). As the aforementioned research title highlights, these states play a major role in the growth of the Indian medical tourism business, which is currently dealing with the disruptive impact of AI and robotics, by actively advertising their healthcare skills (Binoy, 2017).

Internationally recognized medical institutes and hospitals, including the All India Institute of Medical Sciences (AIIMS), Apollo Hospitals, Fortis Healthcare, and Manipal Hospitals, have a significant impact on India's health tourism sector. These hospitals draw patients from outside because of their cutting-edge facilities, cutting-edge medical equipment, and highly qualified medical staff (Hosain, Amin, Rehman, & Koirala, Know thy neighbors: The status of cardiac surgery in the South Asian countries around India, 2017). Government agencies that work with state health departments to create policy and carry out activities, such as the Ministry of Health and Family Welfare, also aggressively encourage health tourism. International patients benefit from the favorable atmosphere created by groups such as the Medical Tourism Promotion Board (MTPB), which also helps with visa applications, works with healthcare professionals, and hosts promotional events. These establishments play a pivotal role in molding India's image as a leading medical tourism destination, shaped by the amalgamation of artificial intelligence and robotics (Rai & Rai, Medical tourism: An introduction, 2019).

The Indian Medical Tourism Industry has experienced significant transformations and challenges due to the integration of AI and robotics. Major hospitals and clinics have adopted advanced technologies like AI-driven diagnostic tools, robotic-assisted surgeries, and smart healthcare systems to improve patient experience and medical outcomes. However, this shift raises concerns about the potential displacement of human jobs in the sector. Debates are ongoing to find a balance between technological advancements and maintaining a skilled workforce. India, a hub for medical treatments, is now facing competition from neighboring countries like Nepal, Bhutan, Bangladesh, Sri Lanka, Maldives, Thailand, Singapore, and Malaysia, who offer high-quality medical services at competitive prices (Akoijam & Khan, 2020).

With its stunning scenery and reasonably priced medical services, Nepal has begun to draw patients looking for a tranquil, all-encompassing setting for healing. Bhutan, which places a strong focus on overall health, has deliberately branded himself as a site for healthcare-seeking tourists looking for cutting-edge care as well as a peaceful, enlightening spiritual experience. By providing specialist medical care with a dash of hospitality, Bangladesh, Sri Lanka, and the Maldives have been focusing on using their close proximity and shared cultural heritage to entice medical tourists (Moghavvemi,, et al., 2017).

Malaysia, Singapore, and Thailand have made significant investments in state-of-the-art hospitals, advanced equipment, and highly qualified medical personnel (Connell, 2013). These nations are marketing themselves as locations for advanced medical procedures, offering a blend of top-notch medical care and exciting travel opportunities. Thailand, for example, has capitalized on its well-known friendliness and tourism attractions to evolve into a hotspot for healthcare professionals seeking wellness treatments and cosmetic surgery. Patients looking for cutting-edge therapies and individualized healthcare services increasingly choose Singapore due to its reputation for competence in the fields of research and innovation. Malaysia is becoming more and more popular as a destination for medical tourism across a range of specializations thanks to its cutting-edge medical infrastructure and varied cultural offers (Revathi & Rani, 2019).Stakeholders must reevaluate and modify their strategy because the competitive landscape presents a challenge to the Indian Health and Medical Tourism Industry (Manhas, 2015). The surrounding nations have purposefully positioned themselves as alternatives, stressing unique selling propositions like cultural experiences, natural beauty, and professional healthcare services, even though India still provides a wide range of medical treatments at reasonable costs (Ganguli & Ebrahim, 2017). To ensure that technological breakthroughs do not impair the human dimension and empathy that are essential to the patient experience, the Indian healthcare sector needs to address concerns about the ethical usage of robotics and AI in healthcare (Hassan & Bellos, COVID-19: Reshaping Medical Tourism through Artificial Intelligence (AI) and Robotics, 2022).

7. INDIA'S MEDICAL TOURISM SECTOR AND OTHER SURROUNDING NATIONS:

India (Tripathi, 2023) has always been a top medical tourism destination, with a variety of services that make it more appealing than other nearby nations. India is superior in the following areas when it comes to offering better medical tourism facilities:

7.1. Wide Range of Specialties: From orthopedics and heart surgery to cutting-edge treatments for cancer and organ transplants, India offers a wide variety of medical specialties. The nation is home to top-notch hospitals and specialty clinics with highly qualified medical staff in a range of specializations (Nagar, Singh, & Jaglan, 2021).

7.2. Advanced Medical Infrastructure: India has made large investments to construct cutting-edge medical facilities. Big cities like Bangalore, Mumbai, and Delhi are home to state-of-the-art hospitals with state-of-the-art facilities, cutting-edge medical technology, and international standards compliance. This facility guarantees that patients get top-notch medical attention (Manjunath, 2022).

7.3. Cost-Effective Healthcare: The affordability of medical travel to India is one of its main benefits. Without sacrificing quality, medical care in India is significantly less expensive than in many Western nations. For patients looking for top-notch medical care at a fraction of the price they could pay in their native nations, this makes India a desirable option (Amiri & Safariolyaei, 2017).

7.4. Accredited Hospitals and International Standards: International organizations have accredited numerous hospitals in India, guaranteeing that they meet high standards of quality. These accreditations, from organizations like National Accreditation Board for Hospitals & Healthcare Providers (NABH) and Joint Commission International (JCI), show that the medical facilities meet or surpass international standards for patient safety and care (Paul & Kulshreshtha, 2019).

7.5. English-Speaking Healthcare Professionals: The majority of medical personnel in India speak English fluently, which facilitates communication between foreign patients and physicians, nurses, and support personnel (GAUTAM & BHATTA, 2020). With regard to diagnosis, treatment planning, and post-operative care, this linguistic advantage guarantees clear communication and improves the patient experience overall (Kaushal & Srivastava, 2021).

7.6. Holistic Healthcare Services: India is well known for its comprehensive conception of healthcare, which blends modern techniques with age-old methods such as Ayurveda, naturopathy, and yoga. This integrative approach attracts patients seeking both specialized medical procedures and experiences related to overall wellness and well-being (Shukla, Singh, & Saxena, 2019).

8. CHALLENGES AND ETHICAL CONSIDERATIONS

8.1. Challenges And Ethical Quandaries Raised By Robotics and AI Advances in Healthcare and Medical Travel:

Numerous difficulties and moral dilemmas are brought up by the use of machine learning and artificial intelligence in healthcare and medical travel. The moral dilemmas raised by the application of these technologies to delicate medical processes and treatments are among the most important ones. For example, the use of powered AI surgical robots raises concerns regarding accountability and ethical duty in the event of mistakes or failures. When an apparatus is involved, figuring out who is at fault becomes complicated (Ivanov & Webster). This problem is made even more difficult in a medical tourism context where different legal jurisdictions may apply. Furthermore, there is a chance that using AI to diagnose and cure patients could dehumanize medical care. Although AI algorithms can greatly improve diagnosis accuracy, there is a delicate balance to be struck between increased efficiency and possible reduction of manual interaction in patient care (Zeng, Chen, & Lew, 2020). An important ethical concern is making sure that the use of AI and robotics for medical purposes does not jeopardize the fundamental components of empathy and individualized care. Policies that guarantee reasonable access to AI-enhanced medical services are particularly important in the realm of health tourism, as the use of sophisticated artificial intellect and robotic systems in healthcare may result in a situation during which only those who are capable of paying for these state-of-the-art treatments benefit, dividing society into those with and without financial means. These issues of justice and fairness raise ethical concerns (Yoon & Lee, 2018).

8.2. Issues Related To Data Privacy, Liability, And Transparency:

Large volumes of private information about patients must be collected and processed to integrate AI and robotics into tourism for medical reasons and healthcare. Significant problems with data security and privacy result from this. The ethical requirement to protect patient data becomes critical since any security lapse might have serious repercussions for people and undermine public confidence in AI-powered healthcare systems (Sandberg, 2017). Concerns about medical data ownership and control also surface. Concerns about data ownership emerge because AI algorithms need large datasets for training and development. These datasets are created during medical tourism operations. To address these ethical problems, transparent and broad data ownership rules are necessary. This is because patients, healthcare practitioners, and AI developers may have competing interests in this respect (Dang, Nguyen, Wang, Day, & Dang, 2020).

Liability concerns pose an additional ethical dilemma. It can be difficult to determine who is responsible for robotic or AI blunders that occur during medical procedures. It is imperative to establish unambiguous boundaries between responsibilities and liabilities to safeguard patients and provide them with channels of appeal in the event of unfavorable consequences. A delicate balance between promoting robots and AI research and putting in place accountability systems that put patient safety first is required due to ethical concerns (Genç, 2023). For deploying AI algorithms, it's imperative to keep in mind the moral precept of transparency. Some sophisticated AI models are "black boxes," which makes it difficult to comprehend the decision-making process. Ensuring that consumers as well as healthcare practitioners can understand and appreciate any choices made by AI systems represents a priority in the healthcare industry, where transparency is essential for making educated decisions. It can be difficult to strike the correct ethical balance between preserving secret algorithms and offering transparency (Hassan & Bellos, COVID-19: Reshaping Medical Tourism through Artificial Intelligence (AI) and Robotics, 2022).

8.3. Requirement for Guidelines and Regulatory Frameworks:

Strong regulatory frameworks and rules must be established to address the difficulties and moral dilemmas associated with the incorporation of AI and robotics into medical tourism and healthcare. A lack of responsibility, insufficient safeguards, and inconsistent standards are just a few of the problems that can arise from unclear legislation (Chakraborty, Edirippulige, & Ilavarasan, 2023). Policies must consider the ethical concerns brought up by healthcare robotics and artificial intelligence. This means setting standards for the development and use of algorithms for AI and ensuring that patient welfare, safety, and privacy are given first consideration. The regulatory environment ought to be flexible enough to keep up with the quick advances in AI technology without sacrificing its moral underpinnings (De La Hoz-Correa & Muñoz-Leiva, 2023).

As more clients travel abroad for therapies, synchronizing legislative requirements becomes crucial in the context of medical tourism. The necessity of international cooperation to create uniform regulations that safeguard patients' rights wherever they may be found is one ethical factor (Vargo, Zhu, Benwell, & Yan, 2021). This entails tackling problems like international data transfer, uniform AI training for medical personnel, and a generally acknowledged paradigm for moral AI in healthcare. Furthermore, ethical activities should be discouraged and responsible innovation should be promoted by regulatory frameworks. To maintain regulations that are both effective and relevant, regulators must keep up with technology advancements and actively collaborate with stakeholders to strike a balance between the need for invention and ethical considerations (Mintz & Brodie, 2019).

9. FUTURE PROSPECTS AND RECOMMENDATIONS

9.1. Perspectives On Artificial Intelligence And Automation's Future Throughout The Indian Medical Travel Sector:

In recent times, the nation's medical tourism industry has expanded dramatically, drawing patients from all over the world who are looking for reasonably priced, high-quality healthcare services. Technology is advancing, especially in the areas of robotics and artificial intelligence (AI), and these developments bode well for the Indian medical tourism sector (Chen & Decary, 2020). These developments possess the potential to fundamentally transform the provision of healthcare enhance patient outcomes, and greatly accelerate the industry's expansion (Andrikopoulou, 2023). AI and robotics in healthcare can improve efficiency and precision by streamlining a variety of procedures, from diagnosis to therapy. By utilizing these technologies, India can become a more popular medical tourism destination, especially in those parts of the nation where the industry contributes significantly to the national economy (Hassan & Bellos, COVID-19: Reshaping Medical Tourism through Artificial Intelligence (AI) and Robotics, 2022).

AI-powered diagnostic solutions can shorten the duration that patients spend in the phase of diagnosis by enabling faster and more accurate diagnoses. This improves patient satisfaction while also hastening the course of treatment. Furthermore, patients seeking cutting-edge medical solutions are drawn to robotically assisted surgeries due to their accuracy and minimally intrusive nature (Alowais, et al., 2023). AI can also be a key component of predictive analytics, which aids healthcare professionals in forecasting illness outbreaks, effectively allocating resources, and developing individualized treatment programs (Khaksarshahmirzadi, et al., Evaluation of the use of robotic surgery with remote simulation for use in the field of Medical Tourism and Pro-Tourism development, 2022). Concerning medical tourism, this predictive ability might enhance the preparedness of healthcare facilities to fulfill the particular needs of international patients (Jiang, et al., 2017). To fully realize the promise of artificial intelligence and robots in the medical tourism industry, stakeholders in India must collaborate to address challenges such as regulatory constraints, data privacy concerns, and a lack of qualified workers. To guarantee the effective integration and application of new technologies, infrastructure investments, and ongoing training initiatives will also be essential (Hussain, Malik, Halim, & Ali, 2014).

9.2. Suggestions for Healthcare Professionals, Government Officials, and Interested Parties Regarding Optimizing The Advantages Of These Technologies:

9.2.1. Healthcare Providers: Modern AI and robotics technology should be purchased by healthcare practitioners to improve their capacity for diagnosis and treatment. This entails implementing robotic surgery equipment, AI-powered diagnostic tools, and telemedicine platforms to provide remote consultations. Furthermore, it is crucial to promote a culture of ongoing education and training among medical practitioners. Institutions and technology vendors should work together to create specific training programs that give medical personnel the know-how to efficiently use and operate these cutting-edge technologies. When it comes to patient engagement, healthcare practitioners can employ chatbots and virtual assistants driven by AI to instantly deliver foreign patients' information, assistance, and direction, guaranteeing a smooth and customized experience throughout their treatment (Ulnicane, 2022).

9.2.2. Policymakers: When it comes to fostering an atmosphere that allows automation and contemporary technologies will be used in the medical travel industry. Policymakers are essential. They ought to endeavor to put in place precise legal frameworks that deal with matters of data protection, medical ethics, and AI application standards. Healthcare providers may be encouraged to invest in these technologies through financial incentives and subsidies. To pinpoint the most important areas where financial assistance and clear regulations may foster innovation and growth, policymakers should consult with industry experts. Knowledge transmission and technology transfer can also be facilitated by international collaboration and partnerships. To share best practices and encourage innovation in healthcare, policymakers should actively support and encourage efforts that bring together specialists from other nations (Enderwick & Nagar, 2011).

9.2.3. Stakeholders: Stakeholders should work together to develop integrated solutions that address the various demands of medical tourists, including service providers for technology, insurance providers, and travel agents. For instance, insurance providers can create customized plans to include robotic-assisted and AI procedures, increasing their accessibility to patients from throughout the world. Putting money into marketing and public awareness initiatives is essential to establishing India as a center for cutting-edge medical care. By banding together, stakeholders may develop a cohesive brand image that highlights the nation's dedication to patient-centered care and state-of-the-art medical technology (Cath, 2018).

10. CONCLUSION RECOMMENDATION

The amalgamation of robotics and artificial intelligence (AI) is bringing about a paradigm shift in the Indian medical tourism sector. Economic viability, high-quality medical care, legal frameworks, and partnerships between medical facilities and foreign entities are some of the drivers propelling this industry's expansion. Predictive analytics, diagnostic tools, and customized care are examples of AI-driven breakthroughs that contributed to better patient outcomes and shifted the focus to the patient. The effectiveness and accessibility of medical treatments have also improved because of the applications of artificial intelligence in online medical care and healthcare administration, especially in impoverished areas. In India, the use of robotics in surgical procedures, in particular, has transformed the provision of healthcare services by providing individualized medical supplies, efficient diagnosis and treatment, and virtual online consultations. Unfortunately, issues with data privacy, liability, openness, and the requirement for precise rules and legal frameworks become ethical issues. It's critical to strike an appropriate equilibrium between innovation and ethical considerations, which calls for cooperation from stakeholders, legislators, and healthcare professionals. With the potential to improve productivity, accuracy, and patient satisfaction, artificial intelligence (AI) and robotics have bright futures in the Indian medical tourism sector. To fully realize these benefits, stakeholders should work together to build integrated solutions, legislators should establish precise legal structures, and medical facilities should invest in cutting-edge technology.

10.1. Recommendation

It's clear that robotics and artificial intelligence (AI) together have the power to completely transform India's medical travel industry. The report offers insightful information about the development factors, historical background, and salient features that contribute to India's appeal as a medical travel destination. It looks at how robotics and artificial intelligence can improve surgical techniques, patient outcomes, and the standard of healthcare as a whole. Alongside these technological breakthroughs, the piece also discusses the ethical and economic implications. Given the revolutionary potential of artificial intelligence (AI) and robotics to improve the medical tourism sector, it is advised that stakeholders—including legislators, healthcare providers, and business professionals—work together closely. Healthcare professionals should support a culture of lifelong learning while investing in cutting-edge technologies like AI-powered diagnostic tools and robotic surgery equipment. To facilitate the blending of AI and robotics, policymakers are essential in developing clear legal frameworks, providing financial incentives, and encouraging international cooperation. To position India as a top destination for state-of-the-art healthcare, industry stakeholders should collaborate to create integrated solutions, such as personalized insurance plans and joint marketing campaigns. Overall, the suggestions place a strong emphasis on using a comprehensive strategy to fully utilize robotics and artificial intelligence, guaranteeing a smooth integration that improves patient outcomes and strengthens India's standing in the world of medical tourism.

11. REFERENCES

Akoijam, S. L., & Khan, T. (2020). Prospects and Challenges of Medical Tourism: Evidences From Manipur. *Global Developments in Healthcare and Medical Tourism*, 265-276.

Alowais, S. A., Alghamdi, S. S., Alsuhebany, N., Alqahtani, T., Alshaya, A. I., Almohareb, S. N., & Albekairy, A. M. (2023). Revolutionizing healthcare: The role of artificial intelligence in clinical practice. *BMC Medical Education*, *23*(1), 689. doi:10.1186/s12909-023-04698-z PMID:37740191

Amiri, M., & Safariolyaei, N. (2017). Medical tourism in India: Current scenario. *International Journal of Management and Social Sciences*, *5*(3), 106–111.

Andrikopoulou, E. (2023). The rise of AI in telehealth. In *Emerging Practices in Telehealth* (pp. 183–207). Academic Press. doi:10.1016/B978-0-443-15980-0.00011-9

Aydin, G., & Karamehmet, B. (2017). Factors affecting health tourism and international health-care facility choice. *International Journal of Pharmaceutical and Healthcare Marketing*, *11*(1), 16–36. doi:10.1108/IJPHM-05-2015-0018

Aykin, Ö., Uluhan, F., Gümüş, İ., Çabuk, Ş., Bozbayir, U., Duran, V., Özbay, C., Özbay, B., Gündoğmuş, Ş. E., Türkegün Şengül, M., & Övey, İ. S.AYKIN. (2023). Artificial Intelligence And Telemedicine Applications In Health Tourism Marketing. *Eurasian Journal of Health Technology Assessment*, *7*(2), 134–149. doi:10.52148/ehta.1396111

Baran, Z., & Karaca, S. (2023). Next-Generation Technologies in Health Tourism. In *Global Perspectives on the Opportunities and Future Directions of Health Tourism* (pp. 138–164). IGI Global., doi:10.4018/978-1-6684-6692-6.ch006

Bauer, I. L. (2023). Robots in travel clinics: Building on tourism's use of technology and robots for infection control during a pandemic. *Tropical Diseases, Travel Medicine and Vaccines*, *9*(1), 10. doi:10.1186/s40794-023-00197-7 PMID:37525269

Bhaidkar, A., & Goswami, A. (2017). Medical tourism in India and the Health policy of the NDA government: An overview. *Aweshkar Research Journal, 22*(1), 82-89. Retrieved from https://web.s.ebscohost.com/abstract?direct=true&profile=ehost&scope=site&authtype=crawler&jrnl=09741119&AN=124266520&h=o5zzf875if8NBE19DOCDYQTUt0I6VtEIctiyrSzrHUWnMxQvKTBcmrN5Q2WgXdJLxbWjvkbqiVHf%2bzSh9shFIQ%3d%3d&crl=c&resultNs=AdminWebAuth&resultLocal=

Binoy, T. A. (2017). Dynamics of medical tourism in India-A microscopic analysis. *Asia Pacific Journal of Research*, *1*(58), 161–167.

Cath, C. (2018). Governing artificial intelligence: Ethical, legal and technical opportunities and challenges. *Philosophical Transactions. Series A, Mathematical, Physical, and Engineering Sciences*, *376*(2133), 20180080. doi:10.1098/rsta.2018.0080 PMID:30322996

Chakraborty, I., Edirippulige, S., & Ilavarasan, P. V. (2023). The role of telehealth startups in healthcare service delivery: A systematic review. *International Journal of Medical Informatics*, *105048*, 105048. Advance online publication. doi:10.1016/j.ijmedinf.2023.105048 PMID:36963322

Chen, M., & Decary, M. (2020). Artificial intelligence in healthcare: An essential guide for health leaders. In *Healthcare management forum* (pp. 10–18). SAGE Publications. doi:10.1177/0840470419873123

Chinzei, K. (2019). Safety of surgical robots and IEC 80601-2-77: The first international standard for surgical robots. *Acta Polytechnica Hungarica*, *16*(8), 171–184.

Connell, J. (2013). Contemporary medical tourism: Conceptualisation, culture and commodification. *Tourism Management*, *34*, 1–13. doi:10.1016/j.tourman.2012.05.009

Crooks, V. A., Turner, L., Snyder, J., Johnston, R., & Kingsbury, P. (2011). Promoting medical tourism to India: Messages, images, and the marketing of international patient travel. *Social Science & Medicine*, *72*(5), 726–732. doi:10.1016/j.socscimed.2010.12.022 PMID:21310519

Dahl, T. S., & Kamel Boulos, M. N. (2013). Robots in health and social care: A complementary technology to home care and telehealthcare? *Robotics (Basel, Switzerland)*, *3*(1), 1–21. doi:10.3390/robotics3010001

Dang, H. S., Nguyen, T. T., Wang, C. N., Day, J. D., & Dang, T. H. (2020). Grey system theory in the study of medical tourism industry and its economic impact. *International Journal of Environmental Research and Public Health*, *17*(3), 961. doi:10.3390/ijerph17030961 PMID:32033177

Dar, H., & Kashyap, K. (2023). Smart Healthcare System (SHS): Medical Tourism Delivering, Consumption, and Elevating Tool in the Ages of Smart Technologies. *Tourism Planning & Development*, *20*(3), 397–415. doi:10.1080/21568316.2022.2109206

De La Hoz-Correa, A., & Muñoz-Leiva, F. (2023). The role of information sources and image on the intention to visit a medical tourism destination: A cross-cultural analysis. *Journal of Travel & Tourism Marketing*, *36*(2), 204–219. doi:10.1080/10548408.2018.1507865

De la Hoz-Correa, A., Muñoz-Leiva, F., & Bakucz, M. (2018). Past themes and future trends in medical tourism research: A co-word analysis. *Tourism Management*, *65*, 200–211. doi:10.1016/j.tourman.2017.10.001

Diwakar, M., Singh, P., & Shankar, A. (2021). Multi-modal medical image fusion framework using co-occurrence filter and local extrema in NSST domain. *Biomedical Signal Processing and Control*, *68*, 102788. doi:10.1016/j.bspc.2021.102788

Enderwick, P., & Nagar, S. (2011). The competitive challenge of emerging markets: The case of medical tourism. *International Journal of Emerging Markets*, *6*(4), 329–350. doi:10.1108/17468801111170347

Ganguli, S., & Ebrahim, A. H. (2017). A qualitative analysis of Singapore's medical tourism competitiveness. *Tourism Management Perspectives*, *21*, 74–84. doi:10.1016/j.tmp.2016.12.002

Gaur, L., Afaq, A., Singh, G., & Dwivedi, Y. K. (2021). Role of artificial intelligence and robotics to foster the touchless travel during a pandemic: A review and research agenda. *International Journal of Contemporary Hospitality Management*, *33*(11), 4079–4098. doi:10.1108/IJCHM-11-2020-1246

GAUTAM, P., & BHATTA, K. (2020). Medical Tourism in India: Possibilities and problems of alternative medical treatment. *International Journal of Health Management and Tourism*, *5*(3), 181–207.

Genç, R. (2023). Risks and benefits of medical tourism in terms of quality of life: Physical, economic, psychological and social aspects. In Handbook of Tourism and Quality-of-Life Research II: Enhancing the Lives of Tourists, Residents of Host Communities and Service Providers (pp. 181-196). Springer International Publishing.

Ghubaish, A., Salman, T., Zolanvari, M., Unal, D., Al-Ali, A., & Jain, R. (2020). Recent advances in the internet-of-medical-things (IoMT) systems security. *IEEE Internet of Things Journal*, *8*(11), 8707–8718. doi:10.1109/JIOT.2020.3045653

Grespan, L., Fiorini, P., & Colucci, G. (2019). *The route to patient safety in robotic surgery*. Springer International Publishing. doi:10.1007/978-3-030-03020-9

Gyu Ko, T. (2011). Medical tourism system model. *International Journal of Tourism Sciences*, *11*(1), 17–51. doi:10.1080/15980634.2011.11434634

Hassan, V., & Bellos, G. (2022). COVID-19: Reshaping Medical Tourism through Artificial Intelligence (AI) and Robotics. *Athens Journal of Tourism*, 77-98. doi:https://www.athensjournals.gr/tourism/2022-9-2-2-Hassan.pdf

Hassan, V., & Bellos, G. (2022). COVID-19: Reshaping Medical Tourism through Artificial Intelligence (AI) and Robotics. *Athens Journal of Tourism*, *9*(2), 77–98. doi:10.30958/ajt.9-2-2

Hassan, V., & Bellos, G. (2022). COVID-19: Reshaping Medical Tourism through Artificial Intelligence (AI) and Robotics. *Athens Journal of Tourism*, 77-98.

Hongal, P., & Kshirsagar, Y. I. (2023). Healthcare management in India: Issues, challenges and prospects. *International Journal of Engineering and Management Research*, *13*(2), 73–82. doi:10.31033/ijemr.13.2.11

Hosain, N., Amin, F., Rehman, S., & Koirala, B. (2017). Know thy neighbors: The status of cardiac surgery in the South Asian countries around India. *Indian Heart Journal*, *69*(6), 790–796. doi:10.1016/j.ihj.2017.09.219 PMID:29174261

Hosain, N., Amin, F., Rehman, S., & Koirala, B. (2017). Know thy neighbors: The status of cardiac surgery in the South Asian countries around India. *Indian Heart Journal*, *69*(6), 790–796. doi:10.1016/j.ihj.2017.09.219 PMID:29174261

Hussain, A., Malik, A., Halim, M. U., & Ali, A. M. (2014). The use of robotics in surgery: A review. *International Journal of Clinical Practice*, *68*(11), 1376–1382. doi:10.1111/ijcp.12492 PMID:25283250

Ivanov, S., & Webster, C. (n.d.). *Robots, artificial intelligence, and service automation in travel, tourism and hospitality*. Emerald Publishing Limited. doi:10.1108/9781787566873

Jiang, F., Jiang, Y., Zhi, H., Dong, Y., Li, H., Ma, S., Wang, Y., Dong, Q., Shen, H., & Wang, Y. (2017). Artificial intelligence in healthcare: Past, present and future. *Stroke and Vascular Neurology*, *2*(4), 230–243. Advance online publication. doi:10.1136/svn-2017-000101 PMID:29507784

Kansal, V., Ranjan, R., Sinha, S., Tiwari, R., & Wickramasinghe, N. (2021). *Healthcare and Knowledge Management for Society 5.0: Trends, Issues, and Innovations*. CRC Press. doi:10.1201/9781003168638

Kaushal, V., & Srivastava, V. (2021). Hospitality and tourism industry amid COVID-19 pandemic: Perspectives on challenges and learnings from India. *International Journal of Hospitality Management*, *92*, 102707. doi:10.1016/j.ijhm.2020.102707 PMID:33024348

Khaksarshahmirzadi, E., Saner, T., Attaeyan, A., Abdollahi, S., Mohammadi, M., & Heydari, S. (2022). Evaluation of the use of robotic surgery with remote simulation for use in the field of Medical Tourism and Pro-Tourism development. *Journal of Simulation and Analysis of Novel Technologies in Mechanical Engineering*, *14*(1), 47–59. 20.1001.1.27834441.2022.14.1.5.5

Khaksarshahmirzadi, E., Saner, T., Attaeyan, A., Abdollahi, S., Mohammadi, M., & Heydari, S. (2022). Evaluation of the use of robotic surgery with remote simulation for use in the field of Medical Tourism and Pro-Tourism development. *Journal of Simulation and Analysis of Novel Technologies in Mechanical Engineering*, *14*(1), 47–59. 20.1001.1.27834441.2022.14.1.5.5

Khedkar, S., Gandhi, P., Shinde, G., & Subramanian, V. (2020). Deep learning and explainable AI in healthcare using EHR. In Deep learning techniques for biomedical and health informatics (pp. 129-148). doi:10.1007/978-3-030-33966-1_7

Kumar, P., Sharma, S. K., & Dutot, V. (2023). Artificial intelligence (AI)-enabled CRM capability in healthcare: The impact on service innovation. *International Journal of Information Management*, *69*, 102598. doi:10.1016/j.ijinfomgt.2022.102598

Kumar, S., Kumar, V., & Attri, K. (2021). Impact of artificial intelligence and service robots in tourism and hospitality sector: current use & future trends. *Administrative Development: A Journal of HIPA, 8*, 59-83.

Kwon, H., An, S., Lee, H. Y., Cha, W. C., Kim, C., Cho, M., & Kong, H. J. (2022). Review of smart hospital services in real healthcare environments. *Healthcare Informatics Research*, *28*(1), 3–15. doi:10.4258/hir.2022.28.1.3 PMID:35172086

Li, J., & Carayon, P. (2021). Health Care 4.0: A vision for smart and connected health care. *IISE Transactions on Healthcare Systems Engineering*, *11*(3), 171–180. doi:10.1080/24725579.2021.1884627 PMID:34497970

Li, M., Yin, D., Qiu, H., & Bai, B. (2021). A systematic review of AI technology-based service encounters: Implications for hospitality and tourism operations. *International Journal of Hospitality Management*, *95*, 102930. doi:10.1016/j.ijhm.2021.102930

Li, R. C., Asch, S. M., & Shah, N. H. (2020). Developing a delivery science for artificial intelligence in healthcare. *NPJ Digital Medicine*, *3*(1), 107. doi:10.1038/s41746-020-00318-y PMID:32885053

Liaqat, S., Akhunzada, A., Shaikh, F. S., Giannetsos, A., & Jan, M. A. (2020). SDN orchestration to combat evolving cyber threats in Internet of Medical Things (IoMT). *Computer Communications*, *160*, 697–705. doi:10.1016/j.comcom.2020.07.006

Locsin, R. C., & Ito, H. (2018). Can humanoid nurse robots replace human nurses. *Journal of Nursing (Luton, England)*, *5*(1), 1–6. doi:10.7243/2056-9157-5-1

Magaña-Espinoza, P., Aquino-Santos, R., Cárdenas-Benítez, N., Aguilar-Velasco, J., Buenrostro-Segura, C., Edwards-Block, A., & Medina-Cass, A. (2014). WiSPH: A wireless sensor network-based home care m. *Sensors (Basel)*, *14*(4), 7096–7119. doi:10.3390/s140407096 PMID:24759112

Manhas, P. S. (2015). Marketing analysis of medical tourism in India. *Enlightening Tourism. A Pathmaking Journal, 5*(1), 1-39.

Manjunath, N. M., & Panda, K. (2022). Evolution of Cancer Patient Health Care in Odisha. *Indian Journal of Surgical Oncology*, *13*(1), 81–82. doi:10.1007/s13193-022-01661-0 PMID:36691494

Mbunge, E., & Muchemwa, B. (2022). Towards emotive sensory Web in virtual health care: Trends, technologies, challenges and ethical issues. *Sensors International*, *3*, 100134. doi:10.1016/j.sintl.2021.100134

Medhekar, A., & Wong, H. Y. (2020). Medical travellers' perspective on factors affecting medical tourism to India. *Asia Pacific Journal of Tourism Research*, *25*(12), 1295–1310. doi:10.1080/10941665.2020.1837893

Mintz, Y., & Brodie, R. (2019). Introduction to artificial intelligence in medicine. *Minimally Invasive Therapy & Allied Technologies*, *28*(2), 73–81. doi:10.1080/13645706.2019.1575882 PMID:30810430

Mishra, P., & Singh, G. (2023). Internet of medical things healthcare for sustainable smart cities: Current status and future prospects. *Applied Sciences (Basel, Switzerland)*, *13*(15), 8869. doi:10.3390/app13158869

Moghavvemi, S., Ormond, M., Musa, G., Isa, C. R., Thirumoorthi, T., Mustapha, M., & Chandy, J. J. (2017). Connecting with prospective medical tourists online: A cross-sectional analysis of private hospital websites promoting medical tourism in India, Malaysia and Thailand. *Tourism Management*, *58*, 154–163. doi:10.1016/j.tourman.2016.10.010

Nagar, R., Singh, Y., & Jaglan, V. (2021). A review on machine learning applications in medical tourism. *Fourth International Conference on Computational Intelligence and Communication Technologies (CCICT)* (pp. 208-215). IEEE. 10.1109/CCICT53244.2021.00048

Navaz, A. N., Serhani, M. A., El Kassabi, H. T., Al-Qirim, N., & Ismail, H. (2021). Trends, technologies, and key challenges in smart and connected healthcare. *IEEE Access : Practical Innovations, Open Solutions*, *9*, 74044–74067. doi:10.1109/ACCESS.2021.3079217 PMID:34812394

Ning, Z., Dong, P., Wang, X., Hu, X., Guo, L., Hu, B., & Kwok, R. Y. (2020). Mobile edge computing enabled 5G health monitoring for Internet of medical things: A decentralized game theoretic approach. *IEEE Journal on Selected Areas in Communications*, *1*(2), 463–478. doi:10.1109/JSAC.2020.3020645

Nkurunziza, J. M., Udahemuka, J. C., Dusenge, J. B., & Umutesi, F. (2022). Overview of Trending Medical Technologies. *Global Clinical Engineering Journal*, *16*(46), 6–46. doi:10.31354/globalce.v4i3.142

Oberoi, S., & Kansra, P. (2019). Factors influencing medical tourism in India: A critical review. *SAMVAD*, *17*, 9–16. doi:10.53739/samvad%2F2019%2Fv17%2F140676

Pandey, C., Sharma, S., & Matta, P. (2021). Body Sensor Network Architectures in Healthcare Internet-of-Things (HIoT): A Survey. *6th International Conference on Communication and Electronics Systems* (pp. 494-499). ICCES. 10.1109/ICCES51350.2021.9489205

Pang, Z., Tian, J., & Chen, Q. (2014). Intelligent packaging and intelligent medicine box for medication management towards the Internet-of-Things. *16th international conference on advanced communication technology* (pp. 352-360). IEEE.

Paul, S., & Kulshreshtha, S. K. (2019). *Global Developments in Healthcare and Medical Tourism*. IGI Global.

Quazi, S., Saha, P. R., & Singh, M. K. (2022). Applications of Artificial Intelligence in Healthcare. *Journal of Experimental Biology and Agricultural Sciences*, *10*(1), 211–226. doi:10.18006/2022.10(1).211.226

Rahman, M. M., Khatun, F., Sami, S. I., & Uzzaman, A. (2022). The evolving roles and impacts of 5G enabled technologies in healthcare: The world epidemic COVID-19 issues. *Array (New York, N.Y.)*, *14*, 100178. doi:10.1016/j.array.2022.100178 PMID:35571870

Rai, A. (2019). *Medical Tourism in Kolkata, Eastern India*. Springer International Publishing., doi:10.1007/978-3-319-73272-5

Rai, A., & Rai, A. (2019). Medical tourism: An introduction. *Medical Tourism in Kolkata, Eastern India*, 1-41.

Reddy, S., Fox, J., & Purohit, M. P. (2019). Artificial intelligence-enabled healthcare delivery. *Journal of the Royal Society of Medicine*, *112*(1), 22–28. doi:10.1177/0141076818815510 PMID:30507284

Revathi, J., & Rani, S. J. (2019). Analysis of Medical Tourism and its Economic Impact. *Indian Journal of Public Health Research & Development*, *10*(1), 21. doi:10.5958/0976-5506.2019.00006.8

Rohokale, V. M., Prasad, N. R., & Prasad, R. (2011). A cooperative Internet of Things (IoT) for rural healthcare monitoring and control. *2nd international conference on wireless communication, vehicular technology, information theory and aerospace & electronic systems technology (Wireless VITAE)* (pp. 1-6). IEEE.

Rohokale, V. M., Prasad, N. R., & Prasad, R. (2020). A cooperative Internet of Things (IoT) for rural healthcare monitoring and control. *2nd international conference on wireless communication, vehicular technology, information theory and aerospace & electronic systems technology (Wireless VITAE)* (pp. 1-6). IEEE.

Sandberg, D. S. (2017). Medical tourism: An emerging global healthcare ind. *International Journal of Healthcare Management, 10*(4), 281–288. doi:10.1080/20479700.2017.1296213

Santosh, K. C., & Gaur, L. (2022). *Artificial intelligence and machine learning in public healthcare: Opportunities and societal impact*. Springer Nature.

Saxena, P. D., MayI, K., Arun, R., Kumar, S. S., Mishra, B. R., & Praveen, K. B. (n.d.). Impact of Artificial Intelligence on Healthcare Informatics: Opportunities and Challenges. *Journal of Informatics Education and Research, 3*(2).

Saxena, P. D., Mayi, K., Arun, R., Kumar, S. S., Mishra, B. R., & Praveen, K. B. (2023). Impact of Artificial Intelligence on Healthcare Informatics: Opportunities and Challenges. *Journal of Informatics Education and Research, 3*(2).

Schäfer, M. B., Stewart, K. W., & Pott, P. P. (2019). Industrial robots for teleoperated surgery–a systematic review of existing approaches. *Current Directions in Biomedical Engineering, 153-156*(1), 153–156. Advance online publication. doi:10.1515/cdbme-2019-0039

Sedik, A., Hammad, M., Abd El-Samie, F. E., Gupta, B. B., & Abd El-Latif, A. A. (2021). Efficient deep learning approach for augmented detection of Coronavirus disease. *Neural Computing & Applications*, •••, 1–18. doi:10.1007/s00521-020-05410-8 PMID:33487885

Shaikh, T. A., Dar, T. R., & Sofi, S. (2022). A data-centric artificial intelligent and extended reality technology in smart healthcare systems. *Social Network Analysis and Mining, 12*(1), 122. doi:10.1007/s13278-022-00888-7 PMID:36065420

Sharma, I., & Sharma, S. (2022). Blockchain Enabled Biometric Security in Intemet-of-Medical-Things (IoMT) Devices. *2022 International Conference on Augmented Intelligence and Sustainable Systems (ICAISS)* (pp. 971-979). IEEE. 10.1109/ICAISS55157.2022.10010716

Sharma, S., Chauhan, Y., & Tyagi, R. (2023). Artificial Intelligence based Applications in Medical Tourism. *2023 14th International Conference on Computing Communication and Networking Technologies (ICCCNT)* (pp. 1-6). IEEE. 10.1109/ICCCNT56998.2023.10308020

Shukla, R., Singh, M., & Saxena, S. (2019). Consumer perception of hospitality services in JCI accredited hospitals at Delhi–NCR: An exploratory research on growth of medical tourism. *Humanities & Social Sciences Reviews*, 2395-6518.

Sikdar, S., & Guha, S. (2020). Advancements of healthcare technologies: Paradigm towards smart healthcare systems. *Recent trends in image and signal processing in computer vision*, 113-132. doi:10.1007/978-981-15-2740-1_9

Singh, S., Sharma, S., & Bhadula, S. (2022). Automated Deep Learning based Disease Prediction Using Skin Health Records: Issues Challenges and Future Directions. *International Conference on Electronics and Renewable Systems (ICEARS)* (pp. 638-643). ICEARS. 10.1109/ICEARS53579.2022.9752422

Singh, S., Sharma, S., & Mohan, S. (2023). Industry 4.0 Internet of Medical Things Enabled Cost Effective Secure Smart Patient Care Medicine Pouch. In *New Horizons for Industry 4.0 in Modern Business* (pp. 149-170). International Publishing. Retrieved from https://link.springer.com/chapter/10.1007/978-3-031-20443-2_7

Sony, M., Antony, J., & McDermott, O. (2023). The impact of healthcare 4.0 on the healthcare service quality: A systematic literature review. *Hospital Topics*, *101*(4), 288–304. doi:10.1080/00185868.2022.2048220 PMID:35324390

Sousa, M. J., Nunes, F. G., do Nascimento, G., & Chakraborty, C. (2023). *Future Health Scenarios: AI and Digital Technologies in Global Healthcare Systems*. Springer Nature. doi:10.1201/9781003227892

Su, C. Y., Samani, H., Yang, C. Y., & Fernando, O. N. (2018). Doctor robot with physical examination for skin disease diagnosis and telemedicine application. *International Conference on System Science and Engineering (ICSSE)* (pp. 1-6). 10.1109/ICSSE.2018.8520218

Tariq, A., Gill, A. Y., & Hussain, H. K. (2023). Evaluating the Potential of Artificial Intelligence in Orthopedic Surgery for Value-based Healthcare. *International Journal of Multidisciplinary Sciences and Arts, 2*(1), 27-35. doi:https://jurnal.itscience.org/index.php/ijmdsa/article/view/2394

Tehseen, K. Y., Imran, N., Qazi, A. S., Ahmad, W., Arslan, F., Ijaz, A., & Ali, M. (2021). Transformative Effects of COVID-19 on Global Econo. *International J. on Emerging Technologies*, *12*(2), 66–76.

Timothy, D. J., & Nyaupane, G. P. (2022). Conclusion: The Experience of Himalayan Tourism. In Tourism and Development in the Himalaya (pp. 308-323). Routledge.

Tripathi, R. P. (2023). Harnessing the Potential of AI and ML in Medical Tourism: Challenges, Opportunities, and Ethical Implications. *3rd International Conference on Technological Advancements in Computational Sciences (ICTACS)* (pp. 1068-1072). IEEE. 10.1109/ICTACS59847.2023.10390358

Ulnicane, I. (2022). Artificial Intelligence in the European Union: Policy, ethics and regulation. In The Routledge handbook of European integrations. Taylor & Francis. doi:10.4324/9780429262081-19

Vargo, D., Zhu, L., Benwell, B., & Yan, Z. (2021). Digital technology use during COVID-19 pandemic: A rapid review. *Human Behavior and Emerging Technologies*, *3*(1), 13–24. doi:10.1002/hbe2.242

Vitthal, P. C., Subhash, A. R., Sharma, B. R., & Ramachandran, M. (2015). Emerging trends and future prospects of medical tourism in India. *Journal of pharmaceutical sciences and research, 7*(5), 248. Retrieved from https://www.proquest.com/openview/60f1635cc6f0bf652e264ee5040645a9/1?cbl=54977&pq-origsite=gscholar&parentSessionId=CSnA2me%2FMJ7z5eQxqemQnO4b7ALEsYCy0qTaXqhL0Kw%3D

Wang, K., Kong, H., Bu, N., Xiao, H., Qiu, X., & Li, J. (2022). AI in health tourism: Developing a measurement scale. *Asia Pacific Journal of Tourism Research*, *27*(9), 954–966. doi:10.1080/10941665.2022.2142620

Wang, Q., Su, M., Zhang, M., & Li, R. (2021). Integrating digital technologies and public health to fight Covid-19 pandemic: Key technologies, applications, challenges and outlook of digital healthcare. *International Journal of Environmental Research and Public Health*, *18*(11), 6053. doi:10.3390/ijerph18116053 PMID:34199831

Wen, Z., & Huang, H. (2022). The potential for artificial intelligence in healthcare. *Journal of Commercial Biotechnology*, *27*(4). Advance online publication. doi:10.5912/jcb1327

Wright, D. W., & Zascerinska, S. (2023). Becoming immortal: Future wellness and medical tourism markets. *Journal of Tourism Futures*, *9*(2), 168–195. doi:10.1108/JTF-05-2021-0119

Yoon, S. N., & Lee, D. (2018). Artificial intelligence and robots in healthcare: What are the success factors for technology-based service encounters? *International Journal of Healthcare Management.*

Zarei, A., Feiz, D., Maleki Minbashrazgah, M., & Maleki, F. (2020). Factors influencing selection of medical tourism destinations: A special niche market. *International Journal of Healthcare Management, 13*(sup 1), 192-198. doi:10.1080/20479700.2018.1492764

Zeng, Z., Chen, P. J., & Lew, A. A. (2020). From high-touch to high-tech: COVID-19 drives robotics adoption. *Tourism Geographies*, *22*(3), 724–734. doi:10.1080/14616688.2020.1762118

Chapter 12
The Role of Augmented Reality (AR) and Virtual Reality (VR) in Medical Tourism

Sheezan Farooq
Islamic University of Science and Technology, India

Murtaza Reshi
Lovely Professional University, India

Bazillah Farooq
University of Kashmir, India

ABSTRACT

The revolutionary impact of virtual reality (VR) and augmented reality (AR) on medical tourism is examined in this article. AR and VR are transforming the patient experience and changing the face of healthcare tourism destinations as technology advances. The abstract explores the use of AR and VR in medical tourism, emphasizing how they can be used for patient education, pre-treatment consultations, and post-treatment care. Patients can virtually visit medical facilities, comprehend intricate medical processes, and participate in therapeutic virtual reality interventions through immersive experiences, which improves their decision-making process and general well-being. The study also looks at case studies from top medical tourism locations that highlight effective AR and VR technology applications. The study addresses how these technologies could draw in a worldwide patient population by providing a special fusion of state-of-the-art medical treatment with innovative experiential elements. The incorporation of AR and VR into medical tourism offers new prospects for improved patient involvement and comprehensive healthcare services as these technologies continue to develop.

DOI: 10.4018/979-8-3693-2248-2.ch012

1. INTRODUCTION

The healthcare industry is undergoing a significant shift, with artificial intelligence (AI) playing a key role in changing the medical tourism sector. The ability of AI to improve surgical innovation, expedite patient care, and improve diagnostic accuracy has opened up new possibilities for the global healthcare industry. Wong and Sa'aid Hazley (2020) assert that technology has significantly shaped the healthcare industry and that the adoption of cutting-edge technologies in the field has sped up the expansion of the medical tourism sector. In particular, artificial intelligence (AI) has revolutionized the medical tourism industry, taking it to new heights and advancing it significantly. The use of AI in medical tourism has significantly changed the dynamics of the sector by streamlining the patient experience overall and facilitating the provision of high-quality treatments (Farooq et al., 2023). The key component of the healthcare ecosystem is the integration of Smart Healthcare Systems (SHS) that make use of Artificial Intelligence (AI) technology, particularly when it comes to tracking medical tourists. By using AI, treatment costs can be decreased and information workflows can be streamlined. SHS improves the total efficacy, efficiency, and affordability of healthcare services by integrating cutting-edge technologies, which benefits medical tourists as well as providers. Applications of Smart Healthcare Systems (SHS), including AI, are critical to improving medical tourism and healthcare quality, claim Akpojaro and Bello (2019). These uses range from remote medical tourist monitoring in hospital environments to fitness tracking (Basheer, Walia, Mehraj, et al., 2023). The incorporation of SHS technologies facilitates ongoing surveillance and monitoring, which enhances healthcare results and makes the experience of medical tourists smoother and more effective.

This study explores the promising future of medical tourism, highlighting the critical role that artificial intelligence (AI) plays in improving the overall experience for those who travel abroad to receive medical care. The story talks about how medical tourists seeking high-quality care abroad might have a more enjoyable and efficient experience thanks to AI. Through several sections, this study continues to investigate different aspects of AI. A comprehensive literature review, an analysis of AI technology applications, the integration of AI in healthcare and medical tourism, and a strategic analysis of smart health technologies for the progress of medical tourism are just a few of the notable study topics that are discussed in detail. Each main section is supported by in-depth subsections that offer a thorough and in-depth examination of the various facets of the subject.

2. OBJECTIVES OF THE STUDY

1. Analyze how AI is currently being used in the medical tourism sector.
2. Examine how AI could improve surgical processes, individualized healthcare, and patient experiences in the context of medical tourism.
3. Examine how medical tourism is changing and how healthcare providers are implementing AI-driven solutions.

3. LITERATURE REVIEW

3.1 Medical Journey

In the literature on health and medical tourism, the term "medical tourism" describes the practice of individuals traveling abroad for medical treatment and related services (Wong & Hazley, 2020). Due to problems including excessive treatment costs, poor quality, lengthy wait times, shoddy health systems, constrained treatment alternatives, lack of insurance coverage, and ignorance of the area, foreigners are turning to medical tourism (Collins et al., 2019). The goal of medical tourism is to provide "cost-effective" private medical care to patients looking for specialized and surgical care while traveling. Despite this, it performs surgeries abroad for dental care, reproductive therapy, cosmetic surgery, and necessary procedures including organ and heart transplants (Ayuningtyas & Ariwibowo, 2020). The medical tourism business is anticipated to be worth USD 65-87.5 billion, with medical travelers spending an average of USD 3410 on each visit to a medical location, according to a survey of about 20 million medical tourists globally (Health-tourism.com, 2019). In addition, this market, which comprises hundreds of hotels, clinics, and hospitals, is growing at a rate of 15–25% annually. According to data from Patients Beyond Borders, a global health travel data source, the global medical tourism market is worth between $65 billion and $87.5 billion.

Despite being a relatively modest specialization in the travel industry, medical tourism is rapidly growing and becoming a billion-dollar global trend (Ghassemi et al., 2021). The medical tourism market is anticipated to reach USD 160.8 billion by 2025, according to Medical Tourism Magazine (2021). As a result, government agencies in charge of tourism and health around the globe are concentrating more on trying to take a piece of this lucrative market (Hassan & Noaman, 2017).

3.2 Towards Artificial Intelligence (AI)

Tater et al. (2022) defines artificial intelligence (AI) as the intelligence exhibited by machines that facilitates the completion of diverse tasks through the use of sentiment analysis and natural language processing (NLP). According to (Basheer, Walia, Farooq, et al., 2023) machine learning—a method for achieving artificial intelligence—and deep learning—a subset of AI—cooperately employ enormous datasets to teach computers to resemble human intellect. Thanks to this technology, the machines can interpret and learn from historical data and information, which can then be used for a range of business applications. According to Vishnoi et al. (2018), the term artificial intelligence (AI) describes the innate tendencies, inherited genetic makeup, or acquired skill sets that make up each person's unique personality. It mimics human decision-making abilities using computer algorithms and includes machine-learned applications. To improve consumer advantages, artificial intelligence (AI) solutions combine human intelligence with business and marketing responsibilities like budgeting, sales, and decision-making (Advani, 2021). In domains like emotion recognition, learning, and decision-making, artificial intelligence (AI) devices have demonstrated the capacity to equal or even exceed human performance (Wright et al., 2018, Manyika et al., 2017). Artificial Intelligence (AI) has been shown to surpass humans on certain tasks and allows computers to mimic intelligent behavior with little to no human intervention. It was reported in 2017 (The Lancet, 2017) that deep neural networks, a subfield of artificial intelligence, could identify diabetic retinopathy (DR) from retinal pictures and evaluate photos of skin cancer more accurately than a dermatologist.

3.3 AI for Medical Applications

After capturing the attention of various industries, artificial intelligence (AI) is now being heralded as a technology that will aid in ensuring that everyone has access to high-quality medical care, including the advancement and improvement of diagnostics, individualized medical care, illness prevention, and the discovery of novel treatments (Parry et al., 2020). According to Uzir et al. (2021), the development of AI has completely changed many facets of human existence, including healthcare. According to (Mehraj, Ul Islam, et al., 2023) research, this development has led to changes in a number of areas, including follow-up scheduling, health applications, big data, data storage, data collection, and analysis. According to Tian et al. (2019), artificial intelligence (AI) technology is rapidly changing the travel and health sectors. (Parry et al., 2020) states that artificial intelligence (AI) for health encompasses the following: machine learning (a fundamental AI component that enables systems to learn and improve on their own without explicit programming), natural language processing (NLP), speech recognition (text-to-speech and speech-to-text), image recognition and machine vision, expert systems (a type of computer system that mimics a human expert's decision-making process), robotics, and planning, scheduling, and optimization systems. Hospital management now offers numerous potentials as a result of the development of AI and data science. According to Yang et al.'s research from 2022, artificial intelligence (AI) is being applied in hospital administration for Medical Imaging and Diagnosis, Drug Discovery, Hospital Workflow, Human Resource, Medical Device, and In-Patient administration (including patient discharge and bed assignment and scheduling). This suggests that implementing AI and data science correctly is enhancing patient care and healthcare administration.

Figure 1. Expenditure per year
Source: https://www.precedenceresearch.com/augmented-reality-and-virtual-reality-market

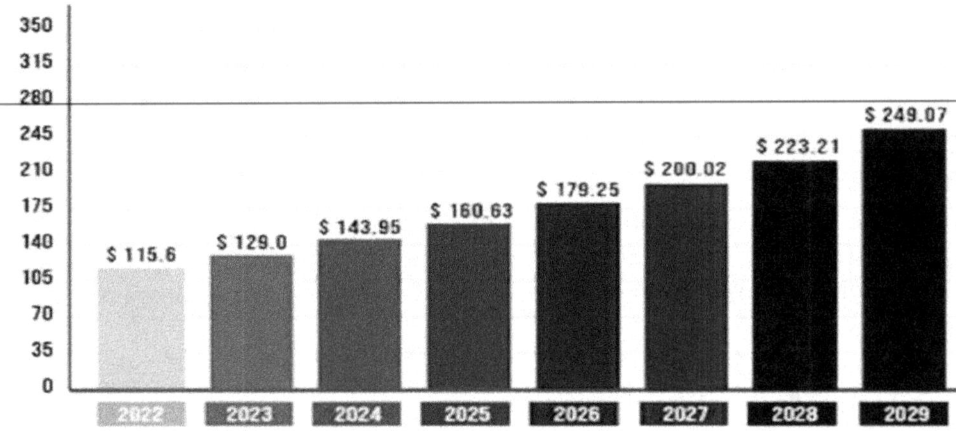

3.4 Medical Tourism's Intelligent Healthcare System

The idea of smart health tourism emerged from the confluence of these technological services in healthcare and smart tourism, which was fueled by the integration of information and communication technology (ICT) and propelled traditional tourism into the era of smart tourism (Nazir et al., 2019). Balandina et al. (2015) have noted that the concepts of smart tourism provide conceptual and practical support for the idea of a Smart Healthcare System (SHS). According to Kelly et al. (2020), a SHS is simply a network of wireless and networked wearable devices containing electronics, software, sensors, and actuators made possible by artificial intelligence (AI) and connectivity to allow for the collection, storage, and sharing of data. This networked system could improve the overall effectiveness of the medical destination, include the local community, and make it easier for medical tourists to receive the correct diagnosis, consultation, and treatment (Mehraj, Qureshi, et al., 2023). The use of technology in the healthcare system is a reflection of a larger movement in the medical tourism sector to develop more intelligent and effective solutions.

3.5 Applying AI to apply Hs in medical travel

An analysis of the technology used in Smart Health (SH) demonstrates its major impact on medical tourism and the network integration of multiple stakeholders to improve the overall medical tourism experience, The report emphasizes a number of significant technologies uses in relation to growing medical travel via Hs. These AI-powered applications include telemedicine, blockchain, cloud, Internet of Medical Things (IoMT), virtual reality, and mobile health. The aforementioned technologies are pivotal in molding and propelling the terrain of medical tourism, presenting novel opportunities for enhanced healthcare provisions, stakeholder communication, and coordination. According to Deepthi and Bansal (2023), the use of AI in medical tourism has greatly enhanced the customer experience, which has resulted in a broad acceptance of chatbots and intelligent travel assistants powered by AI. According to previous studies by Gretzel (2011) and Tussyadiah and Miller (2019), artificial intelligence (AI) helps customers access more relevant information, improves mobility, helps them make better decisions, and ultimately results in a better travel experience (Basheer, Farooq, Hassan, et al., 2023).

According to Tian et al. (2019), artificial intelligence is rapidly advancing in the travel and health sectors, opening up new opportunities for technology-enabled medical tourism services and individualized care. Uzir et al. (2021) go on to say that AI can be extremely helpful in tracking medical tourists, cutting down on medical expenses, and optimizing the information flow for medical tourists.

4. GOAL AND TECHNIQUE OF THE STUDY

This chapter explores the promising future of medical tourism, highlighting the critical role that artificial intelligence (AI) plays in improving the overall experience for those who travel abroad to receive medical care. The story talks about how medical tourists seeking high-quality care abroad might have a more enjoyable and efficient experience thanks to AI.

The study is qualitative in character and uses a desk research methodology. A thorough literature study of recent research from reliable sources, including journals, reports, magazines, websites, and other publications, was used to collect data. Content analysis was used in the literature review to pinpoint important ideas on medical tourism and artificial intelligence. The search was conducted with particular attention to concepts such as medical tourism, artificial intelligence, robotics, smart healthcare, medical tourists, internet of medical things (IoMT), blockchain technology, medical tourism facilitator, technologies in healthcare, applications of smart technologies, smart tourism, mobile health, telehealth, online consultation, wearable health devices, etc. The aim is to utilize these recognized notions to create an integrated framework for the nexus between medical tourism and artificial intelligence.

Figure 2. Impacts on technology medical travel
Source: https://www.precedenceresearch.com/augmented-reality-and-virtual-reality-market

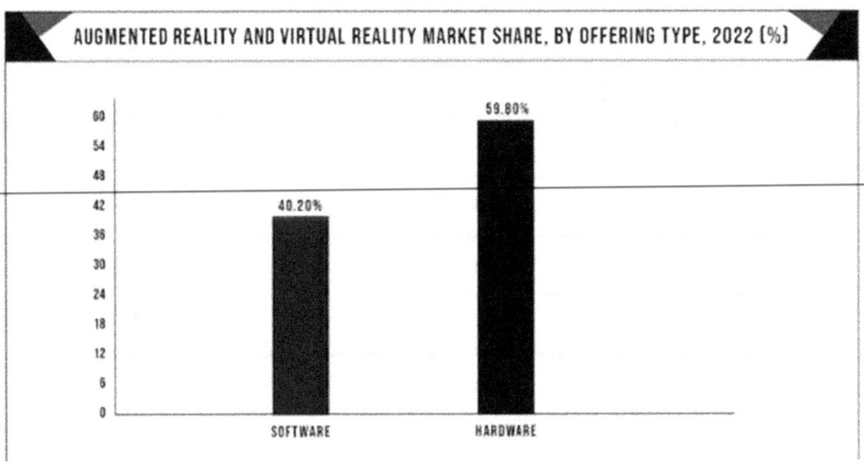

4.1 The State of AI Integration in the Medical Tourism Sector at the Moment

A study by Parekh et al. (2021) found that medical tourists use digital technology to find their way about and learn about Medical Tourist Facilitators (MFTs), medical service providers, and the healthcare options available in the destination nation. Balasubramanian et al. (2022) also discovered that AI, blockchain, and cloud technologies are essential for guaranteeing data security and enabling safer financial transactions. By reducing issues with bank intermediation, currency fluctuations, and transaction delays, these technologies help to improve the general security and efficiency of financial transactions in the context of medical tourism.

Research suggests that AI-enabled smartwatches are used to help consumers self-monitor and self-manage their health, according to Uzir et al. (2021).

These gadgets use artificial intelligence to give consumers features and tools for measuring their own health, empowering them to take an active role in keeping an eye on and maintaining their wellbeing. Numerous research have examined how Artificial Intelligence (AI) affects user experiences in a variety of areas in the past. Digital AI health tools, according to Goel (2019), are essential for improving the patient experience in the medical tourism industry. These aids accomplish this by facilitating clearer information access, enhancing patient-provider communication and interaction, and assisting with the recuperation process when a patient returns home from a medical procedure.

Tussyadiah and Miller (2019) suggest that Artificial Intelligence (AI) is a vital tool for aiding users as it facilitates their access to more pertinent and superior information, improves their mobility, and ultimately enhances their overall tourism experience. In a similar vein, Elsoud and Morsy (2022) contended that artificial intelligence applications streamline travel planning, assist travelers in making well-informed judgments about their upcoming travel plans, and significantly improve the quality of the entire tourist experience. Furthermore, a study by Benckendorff et al. (2019) shows that personalization is essential for helping travelers discover new destinations. Furthermore, it was discovered that the AI's automatic translation capabilities made it easier for visitors to navigate the location, allowing them to explore and partake in a variety of activities. This implies that AI-enabled language translation and personalized experiences add to a more seamless and fulfilling travel experience for medical travelers. These viewpoints demonstrate the many ways in which AI has benefited several facets of the travel and tourism sector.

Research by Luxton (2016) suggests that Artificial Intelligence (AI) makes it easier for caregivers and health professionals to communicate remotely. According to Ali et al. (2022), data from recorded doctor-patient consultations is collected, transferred, and kept on the blockchain. This data includes photos, speech, text, videos, clinical data, and more. Then, using explainable AI models, this data is used for disease diagnosis and prognosis. Virtual reality (VR), according to Wong and Hazley (2020), is used to give medical travelers a sense of familiarity by enabling them to virtually visit different places, including their hometowns. During treatment, medical tourists can also be distracted by colorful virtual reality scenery. VR is also used to educate and clarify treatment alternatives to medical tourists. The Internet of Medical Things (IOMT), an AI strategy that combines healthcare IT through networked hardware, software, and physical infrastructure, is covered by Pradhan et al. (2021). This makes it possible for wireless medical devices to safely connect to the internet, which speeds up and simplifies the processing of medical data applied technologies during the procedural stage of a medical tourist.

Figure 3. Virtual reality benefits in medical field
Source: https://cegh.net/article/S2213-3984%2819%2930429-4/fulltext

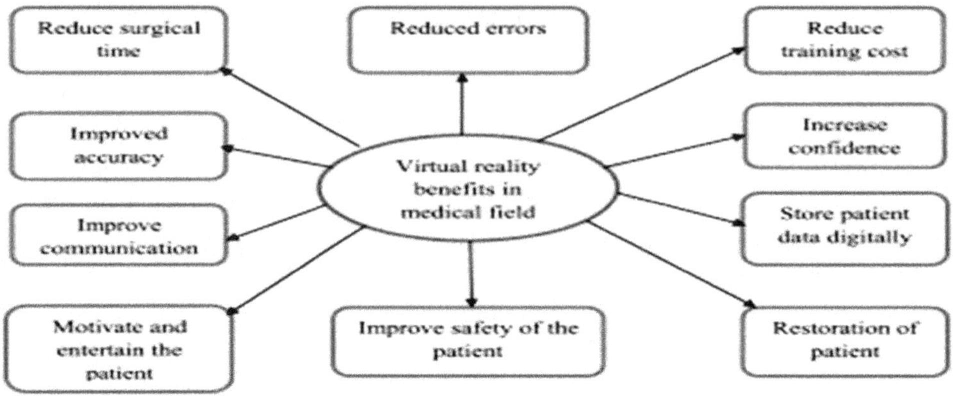

4.2 Potential Impact of AI

The impact of artificial intelligence (AI) and its applications has led to a notable rise in medical tourism, as noted by Wong and Hazley (2020). This can be ascribed to the use of customized adverts on people's web browsers and social media accounts. One example of a ground-breaking effort spearheaded by a healthcare regulator is the AI Lab, which was initiated by the Department of Health in Abu Dhabi. This AI Lab's goal is to hasten the creation and uptake of technological solutions for healthcare. According to Verma (2020), Abu Dhabi has become a top medical tourism destination in recent years thanks to its technological developments in healthcare. AI-enabled technologies used in healthcare and medical services have the potential to outperform human physicians in making accurate assessments for disease identification, according to Gärtner et al. (2020). This increased precision is thought to be essential for enhancing the general client experience. This idea is further reinforced by Ameen et al. (2021), who claim that the accuracy and precision supplied by AI technology enhance the user experience by boosting confidence and promoting a better comprehension of the medical information presented. This implies that the use of AI in healthcare services greatly raises the satisfaction and trust of patients traveling abroad for medical care in the context of medical tourism.

4.3 Medical Tourism and Healthcare Destinations

Sundara Vadivel et al. (2017) assert that the use of artificial intelligence (AI) techniques is essential for anticipating future medical conditions and averting drug shortages. This use of AI not only makes medical tourism destinations more efficient overall, but it also helps healthcare take preventative steps, making better preparation for future health issues and maximizing the use of pharmaceutical resources. According to Yamin (2018), AI-enabled robots have the potential for major breakthroughs in the field of medical tourism. This entails quicker recovery periods, less invasive surgery, greater precision in medical treatments, better diagnostics, and shorter patient wait times. Tian et al. (2019) also draw attention to the function of robotic nurses in hospitals, where they support human nurses by carrying out duties like obtaining supplies and delivering them to medical tourists' rooms, bringing samples to labs, obtaining objects, and even cleaning rooms in preparation for new admissions. The aforementioned applications demonstrate the progressive function of artificial intelligence and robotics in augmenting the effectiveness and caliber of healthcare provisions in the medical tourism domain.

4.4 Barriers and Moral Issues With AI in medical Tourism

It is expected that AI applications, in conjunction with big data and robotics technologies, will cause revolutionary and disruptive shifts in the travel and healthcare industries. Predictive and preventive medicine, insurance, mental health and well-being, medicines, hospitals and hospital administration, and other fields are anticipated to be affected. The integration of these apps poses additional risks and hurdles, notwithstanding their potential benefits. The creation of institutional and policy frameworks to direct the development and application of AI in healthcare will be necessary to allay these worries. This highlights how crucial it is to set up governance frameworks to guarantee the ethical and responsible application of AI technology in the changing healthcare environment. According to Uzir et al. (2021), the impersonal healthcare provided by AI-enabled gadgets has the potential to displace in-person services, upending conventional beliefs about doctor-patient consultations and creating trust concerns. Users may become more concerned about the efficacy and safety of AI technology in delivering precise diagnosis and prescriptions as a result of the trend toward AI in healthcare. Customers' pleasure with the adoption and use of technology is greatly influenced by trust, a dynamic and multifaceted concept. This emphasizes how crucial it is to handle trust-related issues when implementing AI in healthcare in order to guarantee that these technologies are well-received and successfully integrated.

According to a report by, businesses handling sensitive data for AI purposes have a critical duty to protect citizens' health data. Healthcare companies need to take strong action in response to the growing cybersecurity issues they face. It is the duty of policymakers to pass laws that guarantee careful management and security protocols for data that is stored. An example of the negative effects of improperly managing health data is the 2017 backlash against Google DeepMind's collaboration with the Royal Free London NHS (National Health Service) Foundation Trust. The cooperation was severely criticized for improperly disclosing private patient information and using it in the Streams app, which is intended to diagnose, notify users to, and identify acute kidney damage. This instance highlights the critical necessity for strict guidelines and moral behavior while managing health-related data in the context of AI applications in medical tourism and healthcare. Algorithms are biased by nature because of the biases in the training data they use, and AI technologies frequently display biases that are representative of the high-income countries in which they are produced. This is because, in order to produce accurate outputs that are suited to the particular geography and population, these algorithms require millions of historical health datapoints, which may not be available in low-resource settings (USAID, 2019). This acknowledgment emphasizes how crucial it is to design and use AI systems with transparency and thoughtfulness, especially in the healthcare industry.

5. FUTURE OF AI BASED MEDICAL TOURISM

According to Wong and Hazley (2020), artificial intelligence (AI) has been around for a while, but its use in the healthcare sector is still in its infancy. Patients' travel choices and habits are about to undergo a radical change with the introduction of IR 4.0 technologies into the larger healthcare sector, especially during initial examinations and follow-up phases. This change makes healthcare more accessible and becomes especially important for the elderly and people with mobility impairments. Medical Tourism Magazine (2020) states that Artificial Intelligence (AI) technology is anticipated to improve pre- and post-operative care procedures from a medical and customer service standpoint. AI technology is highly relevant and helpful for health tourism. This highlights how artificial intelligence (AI) has the potential to significantly improve a number of healthcare service-related features in the context of medical tourism, leading to better patient care experiences and increased operational efficiency.

According to Wong and Hazley (2020), the use of AI technology may also make in-person appointments for health tourists more convenient, particularly for follow-up visits after they return home. This strategy will make receiving healthcare more affordable and convenient. It is anticipated that the availability of IR 4.0 technologies would also improve experts' virtual mobility, including that of physicians, nurses, educators, and caretakers. This development makes underdeveloped countries more prepared to receive health tourists, which makes healthcare services more easily accessible and effective. Investment in connected and digitalized healthcare has surged recently, and technology-enabled healthcare is becoming more common.

The idea of "Smart Hospitals" is becoming a reality because to this trend (Frost and Sullivan, 2017), which points to a larger movement toward more sophisticated and networked healthcare systems. According to (Lobo, 2020; Tim, 2018), patients usually view the use of precision and timely applications of technology in healthcare as advantageous. Critics, however, voice worries about things like the seeming lack of empathy and their lack of confidence in robots' ability to make decisions. Concerns regarding cybersecurity, privacy, data leaks, and over-commercialization are among other issues that may come up in the context of health tourism (Medical Technology, 2020). These concerns emphasize how crucial it is to address security, privacy, and ethical issues when integrating technology into health tourism in order to guarantee responsible and patient-centered procedures.

6. DISCUSSION

The use of virtual reality (VR) and augmented reality (AR) in medical tourism represents a dramatic advancement in patient care and the general healthcare experience. These immersive technologies have enormous potential to improve medical tourism in many ways, from the first patient interaction to the follow-up after treatment. The use of AR and VR in pre-treatment consultations is one of the paper's main topics. Healthcare practitioners can provide patients a more thorough grasp of the treatment process by using augmented reality (AR) to replicate medical procedures and by providing patients with virtual tours of healthcare facilities. This helps patients feel more at ease in the healthcare setting and also supports them in making well-informed decisions. Another crucial area where AR and VR technologies are having an impact is patient education. Healthcare professionals may now communicate complex medical information in a way that is visually understandable thanks to these immersive tools. Through interaction with three-dimensional models of their anatomy, patients can gain a deeper understanding of their medical issues and the nuances of potential treatments. In order to empower patients and promote a team approach to healthcare decision-making, this educational component is essential.Furthermore, AR and VR play a part in rehabilitation and post-treatment care in addition to the pre-treatment stage. Improved patient outcomes are a result of using VR interventions for therapeutic objectives, including as pain management or physical therapy exercises. This promotes healing and guarantees that patients stay in contact with their medical professionals even after they return to their own nations.

The studies demonstrate how AR and VR technology have been successfully implemented in top medical tourism locations. These actual cases demonstrate the beneficial effects on patient experiences and the ability to draw in a varied international patient base. Healthcare tourism is made more inclusive by AR and VR's capacity to cross geographic boundaries and provide patients with access to state-of-the-art medical care regardless of their location.The conversation concludes by highlighting how AR and VR will fundamentally change medical tourism in the future. Immersion technologies have the potential to revolutionize patient involvement, education, and general pleasure as technology develops, ultimately shaping the future of healthcare services in international locations. Combining AR and VR not only fits in with the technology trends that are reshaping the healthcare sector, but it also establishes medical tourism as a leader in utilizing these innovations to improve patient outcomes globally.

7. CONCLUSION

This book chapter sought to give readers a thorough grasp of how artificial intelligence (AI) could shape medical tourism in the future and how it might enhance patient outcomes, expedite medical procedures, and draw in a worldwide clientele. Artificial intelligence (AI) is being used by the various tourism and travel industries to handle a range of administrative and customer service responsibilities. The advancement of these technologies (robotics, AI, etc.) is crucial for medical tourism since it helps medical organizations link up potential medical tourists with doctors. It also makes it a more viable and trustworthy option for patients, answers inquiries, gives clients useful information, and cuts down on time that would be hard for people to complete. International remote operations would also enable different healthcare facilities to combine their knowledge and provide patients with simple access to a central operating room for remote procedures, all while encouraging patients to support the regional economy.

But as Schmelzer (2019) notes, a major obstacle to fully trusting AI systems is a lack of explain ability and confidence in AI operations. Many technical, resource-related, and ethical issues will need to be resolved when health-related data becomes more widely available and AI is utilized more frequently to analyze it for medical purposes. Concerns around ownership, permission, privacy, governance, quality, and safety present continuous difficulties that have not yet been completely resolved in the application of AI in healthcare. To guarantee the appropriate and moral application of AI in the medical field, comprehensive frameworks and solutions must be established. The current coronavirus outbreak nearly stopped international travel, highlighting the necessity for technology to progress in order for health tourism to grow in the future. This is especially true with IR 4.0 technologies like artificial intelligence (AI). Healthcare is moving away from a localized approach and toward a more globalized one that crosses national boundaries both physically and virtually. This shift emphasizes the critical role that technology—particularly artificial intelligence (AI)—plays in transforming the way that healthcare services are obtained and provided internationally. It also reflects the growing significance of creative solutions in the face of major international issues like the pandemic.

In summary, the implementation of Artificial Intelligence (AI) technology is viewed as a revolutionary move with the potential to improve accessibility, lower costs, raise transparency, improve quality of care, and allow stakeholders more flexibility. These advancements underscore the far-reaching advantages that artificial intelligence (AI) can have for the healthcare industry in the context of medical tourism, since they have a greater positive impact on and absorption into the destination.

REFERENCES

Advani, V. (2021). What Is Artificial Intelligence? How Does AI Work, Types and Future of it? *Medium.* https://medium.com/@mygreatlearning/what-is-artificial-intelligence-how-does-ai-work-and-future-of-it-d6b113fce9be

Ali, S., Abdullah, Armand, T. P. T., Athar, A., Hussain, A., Ali, M., Yaseen, M., Joo, M.-I., & Kim, H.-C. (2023). Metaverse in healthcare integrated with explainable ai and blockchain: Enabling immersiveness, ensuring trust, and providing patient data security. *Sensors (Basel)*, *23*(2), 565. doi:10.3390/s23020565 PMID:36679361

Ameen, N., Tarhini, A., Reppel, A., & Anand, A. (2021). Customer experiences in the age of artificial intelligence. *Computers in Human Behavior*, *114*, 106548. doi:10.1016/j.chb.2020.106548 PMID:32905175

Ayuningtyas, D., & Ariwibowo, D. A. (2020). The strategic role of information communication technology in succeeding medical tourism. *Enfermeria Clinica*, *30*, 170–173. doi:10.1016/j.enfcli.2020.06.039

Balasubramanian, S., Sethi, J. S., Ajayan, S., & Paris, C. M. (2022). An enabling framework for blockchain in tourism. *Information Technology & Tourism*, *24*(2), 165–179. doi:10.1007/s40558-022-00229-6

Basheer, S., Farooq, S., Hassan, V., Malik, Y. M., & Reshi, M. A. (2023). *Augmented Reality and Virtual Reality in Cultural Heritage Tourism Enhancing Visitor Experiences*. doi:10.4018/978-1-6684-9957-3.ch002

Basheer, S., Walia, S., Farooq, S., Shah, M. A., & Mir, F. A. (2023). Exploring the metaverse: The future of tourism through ai and virtual reality. In *Influencer Marketing Applications Within the Metaverse* (pp. 195–204). IGI Global. doi:10.4018/978-1-6684-8898-0.ch012

Basheer, S., Walia, S., Mehraj, D., & Reshi, M. (2023). *Creating Transformative Travel Unveiling the Link Between Meaningful Experiences and Sustainable Tourism*. doi:10.4018/979-8-3693-0650-5.ch002

Benckendorff, P. J., Xiang, Z., & Sheldon, P. J. (2019). *Tourism information technology*. Cabi. doi:10.1079/9781786393432.0000

Collins, A., Medhekar, A., Wong, H. Y., & Cobanoglu, C. (2019). Factors influencing outbound medical travel from the USA. *Tourism Review*, *74*(3), 463–479. doi:10.1108/TR-06-2018-0083

Dar, H., & Kashyap, K. (2023). Smart Healthcare System (SHS): Medical Tourism Delivering, Consumption, and Elevating Tool in the Ages of Smart Technologies. *Tourism Planning & Development*, *20*(3), 397–415. doi:10.1080/21568316.2022.2109206

Deepthi, B., & Bansal, V. (2023). Applications of Artificial Intelligence (AI) in the Tourism Industry: A Futuristic Perspective. In Impact of Industry 4.0 on Sustainable Tourism: Perspectives, Challenges and Future (pp. 31-43). Emerald Publishing Limited. doi:10.1108/978-1-80455-157-820231003

Farooq, S., Farooq, B., Basheer, S., & Walia, S. (2023). Balancing Environmental Sustainability and Privacy Ethical Dilemmas in AI-Enabled Smart Cities. IGI Global. doi:10.4018/979-8-3693-0892-9.ch013

Frost & Sullivan. (2017). Future of smart hospitals – concept definition, application, and growth opportunities, K1FF-01. Frost & Sullivan. https://store.frost.com/future-of-smart-hospitals.html

Gärtner, E., Pirinen, A., & Sminchisescu, C. (2020, April). Deep reinforcement learning for active human pose estimation. *Proceedings of the AAAI Conference on Artificial Intelligence*, *34*(07), 10835–10844. doi:10.1609/aaai.v34i07.6714

Ghasemi, M., Nejad, M. G., & Aghaei, I. (2021). Knowledge management orientation and operational performance relationship in medical tourism (overview of the model performance in the COVID-19 pandemic and post-pandemic era). *Health Services Management Research*, *34*(4), 208–222. doi:10.1177/0951484820971438 PMID:33183088

Goel, P. (2019). *The power of digital health in medical tourism*. LinkedIn. https://www.linkedin.com/pulse/power-digital-health-medical-tourism-pramod-goel/

Gretzel, U. (2011). Intelligent systems in tourism: A social science perspective. *Annals of Tourism Research*, *38*(3), 757–779. doi:10.1016/j.annals.2011.04.014

Hassan, V., & Noaman, S. (2017). Relation between Tourism and Health: Case Study AIDS in Lebanon. In *Atiner Conference Paper Series*.

Health-tourism.com. (2019). Medical tourism statistics and facts. *Health Tourism.* https://www.health-tourism.com/medical-tourism/statistics/

Jangra, P., & Gupta, M. (2021). Internet of Medical Things (IoMT) Based Framework for Smart Healthcare Tourism Sector. *International Journal of Hospitality and Tourism Systems*, 14.

Lancet, T. (2017). Artificial intelligence in health care: Within touching distance. *Lancet*, *390*(10114), 2739. doi:10.1016/S0140-6736(17)31540-4 PMID:29303711

Lobo, F. (2020). Industry 4.0 – manufacturing and the future of medical things. www.asianhhm.com/technology-equipment/industry-manufacturing-future-medical-things

Luxton, D. D. (2016). An introduction to artificial intelligence in behavioral and mental health care. In D. D. Luxton (Ed.), *Artificial Intelligence in Behavioral and Mental Health Care* (pp. 1–26). Academic Press. doi:10.1016/B978-0-12-420248-1.00001-5

Manyika, J., Chui, M., Miremadi, M., Bughin, J., George, K., Willmott, P., & Dewhurst, M. (2017). A future that works: AI, automation, employment, and productivity. McKinsey Global Institute Research. *Tech. Rep*, *60*, 1–135.

Medical Technology. (2020). Medical tourism: how is digital tech reshaping the industry? *Medical Technology.* https://medicaltechnology.nridigital.com/medical_technology_jan20/medical_tourism_how_is_digital_tech_reshaping_the_industry

Medical Tourism Magazine. (2020). Trends in healthcare digital revolution. *Medical Tourism Magazine.* www.magazine.medicaltourism.com/article/trends-healthcare-digital-revolution.

Medical Tourism Magazine. (2021). Will medical tourism survive the coronavirus pandemic? *Medical Tourism Magazine.* https://www.magazine.medicaltourism.com/article/will-medical-tourism survive-the-coronavirus-pandemic

Mehraj, D., Qureshi, I. H., Singh, G., Nazir, N. A., basheer, S., & Nissa, V. (2023, December). Green marketing practices and green consumer behavior: Demographic differences among young consumers. *Business Strategy & Development*, *6*(4), 571–585. doi:10.1002/bsd2.263

Mehraj, D., Ul Islam, M. I., Qureshi, I. H., Basheer, S., Baba, M. M., Nissa, V. u., & Asif Shah, M. (2023). Factors affecting entrepreneurial intention for sustainable tourism among the students of higher education institutions. In *Cogent Business and Management, 10*(3). Cogent OA. doi:10.1080/23311975.2023.2256484

Nazir, S., Ali, Y., Ullah, N., & García-Magariño, I. (2019). Internet of things for healthcare using effects of mobile computing: A systematic literature review. *Wireless Communications and Mobile Computing, 2019*, 1–20. doi:10.1155/2019/5931315

Parekh, J., Jaffer, A., Bhanushali, U., & Shukla, S. (2021). Disintermediation in medical tourism through blockchain technology: An analysis using value-focused thinking approach. *Information Technology & Tourism, 23*(1), 69–96. doi:10.1007/s40558-020-00180-4

Parry, C. M., & Aneja, U. (2020). *Artificial Intelligence for Healthcare: Insights from India.*

Pesapane, F., Volonté, C., Codari, M., & Sardanelli, F. (2018). Artificial intelligence as a medical device in radiology: Ethical and regulatory issues in Europe and the United States. *Insights Into Imaging, 9*(5), 745–753. doi:10.1007/s13244-018-0645-y PMID:30112675

Pradhan, B., Bhattacharyya, S., & Pal, K. (2021). IoT-based applications in healthcare devices. *Journal of Healthcare Engineering, 2021*, 1–18. doi:10.1155/2021/6632599 PMID:33791084

Subasinghe, M., Magalage, D., Amadoru, N., Amarathunga, L., Bhanupriya, N., & Wijekoon, J. L. (2020, November). Effectiveness of artificial intelligence, decentralized and distributed systems for prediction and secure channelling for Medical Tourism. In *2020 11th IEEE Annual Information Technology, Electronics and Mobile Communication Conference (IEMCON)* (pp. 0314-0319). IEEE. 10.1109/IEMCON51383.2020.9284898

Sundaravadivel, P., Kougianos, E., Mohanty, S. P., & Ganapathiraju, M. K. (2017). Everything you wanted to know about smart health care: Evaluating the different technologies and components of the internet of things for better health. *IEEE Consumer Electronics Magazine, 7*(1), 18–28. doi:10.1109/MCE.2017.2755378

Tater, U. R., Jain, A., & Tater, M. J. (2022). Artificial intelligence must be a revolutionary technique of marketing in medical-tourism. *2022 Indian Journal of Business Administration (IJBA).*

Tian, S., Yang, W., Le Grange, J. M., Wang, P., Huang, W., & Ye, Z. (2019). Smart healthcare: Making medical care more intelligent. *Global Health Journal (Amsterdam, Netherlands)*, *3*(3), 62–65. doi:10.1016/j.glohj.2019.07.001

Uzir, M. U. H., Al Halbusi, H., Lim, R., Jerin, I., Hamid, A. B. A., Ramayah, T., & Haque, A. (2021). Applied Artificial Intelligence and user satisfaction: Smartwatch usage for healthcare in Bangladesh during COVID-19. *Technology in Society*, *67*, 101780. doi:10.1016/j.techsoc.2021.101780 PMID:34697510

Vishnoi, S. K., Bagga, T. E. E. N. A., Sharma, A. A. R. U. S. H. I., & Wani, S. N. (2018). Artificial intelligence enabled marketing solutions: A review. *Indian Journal of Economics & Business*, *17*(4), 167–177.

Wong, B. K. M., & Hazley, S. A. S. A. (2020). The future of health tourism in the industrial revolution 4.0 era. *Journal of Tourism Futures*, *7*(2), 267–272. doi:10.1108/JTF-01-2020-0006

Wright, S. A., & Schultz, A. E. (2018). The rising tide of artificial intelligence and business automation: Developing an ethical framework. *Business Horizons*, *61*(6), 823–832. doi:10.1016/j.bushor.2018.07.001

Yamin, M. (2018). IT applications in healthcare management: A survey. *International Journal of Information Technology : an Official Journal of Bharati Vidyapeeth's Institute of Computer Applications and Management*, *10*(4), 503–509. doi:10.1007/s41870-018-0203-3 PMID:32289102

Chapter 13
Growth and Opportunity of Medical Tourism in the Indian Healthcare Sector:
A Critical Review

Amjad Imam Ansari
https://orcid.org/0000-0002-6899-8562
Lovely Professional University, India

Amrik Singh
https://orcid.org/0000-0003-3598-8787
Lovely Professional University, India

ABSTRACT

The aim of the study is to explore the growth and opportunity of medical tourism in context to Indian healthcare. Advancements in science and technology have made it possible for transcontinental healthcare services to exist. In addition to offering standard tourist amenities, medical tourism aims to attract visitors by purposefully promoting its healthcare offerings. Based on extensive literature review, factors impacting inclusive growth and medical tourism in India. Medical tourism—also known as wellness tourism, medical travel, or health tourism—was first used by travel firms and the media to refer to the quickly expanding trend of people traveling abroad for medical treatment. The health care industry is becoming more globalized, and many developing nations like India which have the means and potential to meet this demand. Furthermore, have become interested in the recent surge in demand for affordable, high-quality medical care. Nowadays, because medical care in India is so affordable, people from other nations travel there for procedures including

DOI: 10.4018/979-8-3693-2248-2.ch013

knee transplants, hip replacements, heart surgery, cosmetic surgery, and dental work. While medical tourism is a relatively new phenomenon, it is growing at an exponential rate and is becoming a significant contributor to the global expansion of services exports. Among the leading nations in this sector is India. Medical tourism is a crucial source of income, particularly for underdeveloped countries. Hence, it can be concluded that both the destination countries and the patients benefit from it.

1. INTRODUCTION

The term "medical tourism" describes the practice of visiting other nations to get medical facilities. It also describes the practice of medical service professionals traveling the world to give top-notch medical facilities. The travelers are seeking both straightforward and intricate specialist surgeries, such as dental work, joint replacement (hip/knee), cosmetic procedures, and cardiac surgery. These days, all forms of healthcare are offered, including complementary and alternative therapies, psychotherapy, convalescent care, and even funeral services. The majority of the time, informal channels of communication connection contracts are established between service providers and clients. According to Shanmugam (2013), India began promoting medical tourism after a study revealed the enormous potential of this industry. The government's large-scale, low-budget marketing effort to draw tourists, "Incredible India," was launched by the Ministry of Tourism in the same year. The next year, Jaswant Singh, the finance minister at the time, pushed for the development of airport infrastructure to facilitate the entry and exit of medical tourists and demanded that the nation become a "global health destination."

Figure 1. India's medical tourism industry structure
Source: *(Dawn, 2011)*

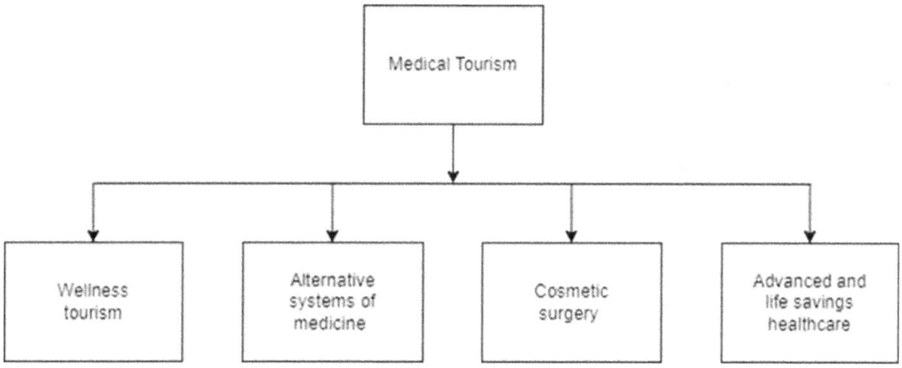

- **Wellness tourism:** Spas, Stress relief, rejuvenation centres
- **Alternative systems of medicine:** Ayurvedics, Siddha treatment for diseases e.g. Arthritis, Rheumatism
- **Cosmetic surgery:** Dental Care, Plastic Surgery, Breast enhancement, Tummy reduction, Skin Treatment
- **Advanced and life savings healthcare:** Open transplants, cardio vascular surgery, Eye treatment, Hip Replacement, In vitrio fertilization

India is renowned for its cultural and religious diversity as well as its wide range of physical monuments. It has a very old civilization dating back more than 5000 years. Her ability to draw tourists is enhanced by her traditional arts and crafts. In India, visitors are regarded as gods, as stated in the well-known phrase "Atithi Devo Bhava." Apart from contemporary medicine, traditional or indigenous medical practitioners are still in practice throughout India (Hazarika, 2010a). Popular indigenous medical practices include yoga, naturopathy, Siddha, Unani, and Ayurveda. Ayurveda offers a comprehensive preventative medicine and healthcare system, and its efficacy has been demonstrated over an extended period in India. The understanding of the human constitution is the foundation of Ayurvedic science. India is a desirable location for travelers looking for medical care because of a number of factors. These include the availability of super specialty centers, the use of cutting-edge diagnostic technology, a large population of competent English-speaking medical staff, well-trained healthcare professionals, a good mix of allopathic and alternative medical systems, and—above all—the availability of these top-notch services at competitive prices (Hazarika, 2010b). A traveler who is primarily there to receive medical attention is known as a medical tourist. When traveling for medical treatments, such as heart surgery, cancer treatment, or other surgeries, the patient may choose to travel alone or with one or more support people, such as a spouse, parent, or friend. As a result, the location will gain from having multiple visitors who would stay and spend their "tourist" money while there. More is being done for this group of travelers as medical tourism gains more attention. Prestigious hotels frequently have a tight relationship with private medical facilities or other establishments that cater to medical tourists. Packages designed to accommodate specific medical tourists (Lee, 2006). India's medical tourism industry is still in its infancy and has a lot of room to expand given the topics we will cover in this essay. The government of India is promoting health as a reason to visit the country through a number of programs, as it is a recently identified industry with great growth potential. There has been little research done on the state of the medical tourism sector in India, despite the fact that there are many surveys and significant literature on the subject in nations like Cuba and Switzerland. The official websites of some states, including Kerala and Gujarat, actively promote the health tourism packages. The Indian Ministry of Tourism has declared several incentives aimed at stimulating the industry (A. Singh & Hassan, 2023). Thirty-one villages nationwide have been designated to serve as tourism hubs. These villages are located in the following states: Madhya Pradesh, Andhra Pradesh, Kerala, Tamil Nadu, Orissa, Assam, Sikkim, Rajasthan, West Bengal, Himachal Pradesh, Gujarat, Maharashtra, Bihar, Karnataka, Madhya Pradesh, and Andhra Pradesh. This portion of the article examines the contributions made by a select few Indian states to the field of medical tourism (S. R. & B. A. Garg, 2012).

According to Saxena (2014), said that in India, medical tourism packages were typically offered in conjunction with the medical services. These medical tourism packages include every aspect, from initial assessment and treatment to the post-treatment phase and travel arrangements, including lodging, entertainment, and translation. In addition, some companies launched a medical conservation package, such as Vedic company developed medical tourism packages for the Nordic seniors for 16 months health care services in India. They were required to pay a three-month pension and were free to select the places. In general, these medical services catered to three categories of medical tourists:

- Patients who traveled to India expressly for medical care, including organ transplants and plastic surgery;
- Those who wanted to experience the mysterious traditional Indian medicine and traditional Indian medical services; and
- Those who wanted to indulge in leisure activities.

2. LITERATURE REVIEW

The Indian Ministry of Tourism is aggressively pushing medical tourism by means of international roadshows wherein medical and wellness tourism service providers receive market development assistance (MDA) to stimulate international promotion. According to (Sagar & Latha (2014), The current infrastructure in the local health care system is insufficient to meet the demands of Indians, and the influx of international patients would increase competition and create new challenges. Public hospitals are congested, while star hospitals' capacity is underutilized. A technology-based strategy will particularly hurt the average Indian, since many of them cannot afford to buy cutting-edge devices from the West. The contemporary SWOT analysis of medical tourism in Indian healthcare sector highlighted in Table 1.

Table 1. The contemporary SWOT analysis of medical tourism in Indian healthcare sector

Strength	Weakness
• High-quality care at a reasonable cost. • An abundance of highly skilled medical professionals. • A strong presence in cutting-edge medical specialties, such as cardiology and organ transplants with a high success rate in operations. • Hospitals' and physicians' international reputations. • A variety of travel experiences and locations.	• Lack of strong government support / initiative to promote medical tourism. • Low coordination amongst the industry's several actors, including hotels, hospitals, and airline carriers. • Consumer perception of the nation as being unclean. • Lack of accreditation and regulating mechanism for hospitals. • Lack of consistent hospital pricing practices.
Opportunity	**Threats**
• Greater need for healthcare services from aging population nations (the United States and the United Kingdom). • The desire for alternative therapies and wellness travel is rising due to fast-paced lifestyles. • Lack of supplies in nations like the United Kingdom and Canada's national health systems • Demand from countries with weak healthcare services. • Demand for retirement residences for older individuals especially Japanese.	• Strong rivalry from countries like Thailand, Malaysia, Singapore. • Absence of worldwide recognition. • Insurance companies do not pay medical treatment received abroad. • Insufficient funding for healthcare facilities.

Source: (A. Kumar, 2023; (Sagar & Latha, 2014)

2.1 Medical Tourism

The term "medical tourism" describes the practice of visiting other nations to get medical facilities. The Indian medical tourism market was valued at US$ 2.89 billion in 2020 and is predicted to reach US$ 13.42 billion by 2026. The India Tourism Statistics at a Glance 2020 study states that in FY19, about 697,300 foreign visitors traveled to India for medical care. The Medical Tourism Association has ranked India ninth out of 46 locations in the Medical Tourism Index (MTI) for 2020–21. India is one of the top global destinations for foreign patients seeking cutting-edge treatment, with US$ 5–6 billion in medical value travel (MVT) and 500,000 international patients yearly. By 2025, the e-health market is projected to be worth US$10.6 billion (India Brand Equity Foundation, 2023). A growing and popular niche sector, according to earlier research, is healthcare and medical travel. To construct an integrated theoretical model, detailed examination of the preceding theoretical framework and review of literature on medical tourism was taken into consideration in Figure 2.

Figure 2. Conceptual model of medical tourism based on extensive literature review
Source: *Author's own based on extensive literature review*

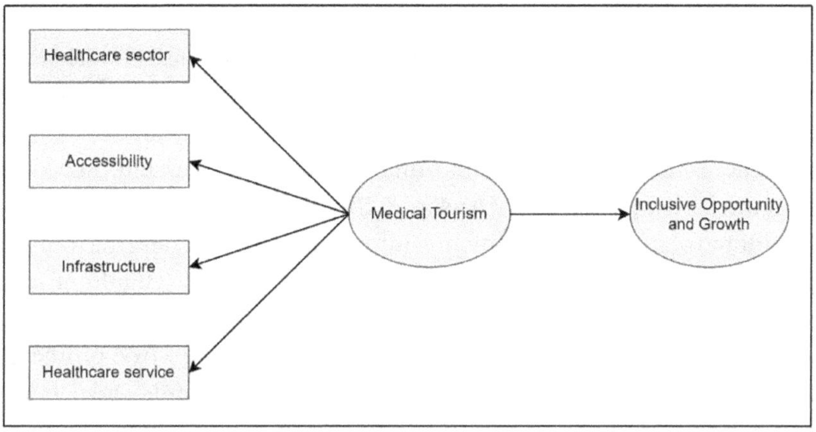

2.2 Healthcare Sector

The medical tourism (MT) industry has experienced significant expansion on a global scale in the last ten years. It has developed into a significant economic industry for numerous nations across the globe. The global MT market was estimated by Transparency Market Research (2016) to have reached a value of US$46 billion in 2016, and over the projected period of 2017–2025, a compound annual growth rate of 14.9% is anticipated (Ebrahim & Ganguli, 2019). Wellness tourism encompasses programs and therapies aimed at enhancing and harmonizing all the principal domains of human existence. A person's entire way of life is affected by their physical, mental, emotional, professional, intellectual, and spiritual well-being. Engaging in preventive and active lifestyle enhancement activities, such as luxurious living, healing treatments, fitness, nutritious food, and relaxation, is the primary driving force behind wellness tourism (Oberoi et al., 2022). It has been perceived as a threat to national health systems as well as a boon to the economy. Potentially detrimental effects include the expansion of disparities brought about by the unequal distribution of health care, the financial deprivation of medical professionals who provide top-notch care at third-world prices, the subsidization of the costs of care for patients from wealthy nations, and the reallocation of resources from the underprivileged and impoverished to foreign patients (Hadi, 2009).

2.3 Accessibility

According to Wapmuk et al. (2015), the medical tourism business has been impacted by the forces of globalization. The concept of globalization as it relates to medical tourism emphasizes the significance of lowering the costs associated with travel time and distance, and it calls for special consideration of the internet's significant influence. There are currently more than one billion people with internet connection, and this figure is rising every day (Ansari & Singh, 2023). The internet provides an incredibly global platform for advertising, information sharing, and the transportation of medical knowledge that knows no geographical bounds (Bhalla et al., 2023). Instantaneous communication and information sharing between physicians and patients is possible because to the internet, which also offers free online chat and video conferencing. Medical histories, X-rays, and other complex tests may be sent over email for practically no cost.

2.4 Infrastructure

According to Mordor Intelligence (2024), in general, improving healthcare quality and accessibility across the country was made easier for patients to receive the required services and treatments because to India's growing healthcare infrastructure. India's healthcare infrastructure is expected to increase further in the upcoming years with sustained investment and government backing, which would contribute to better healthcare results for patients all throughout the nation in figure 3.

Figure 3. Number of healthcare facility in India (2019-2027)
Source: *(Mordor Intelligence, 2024)*

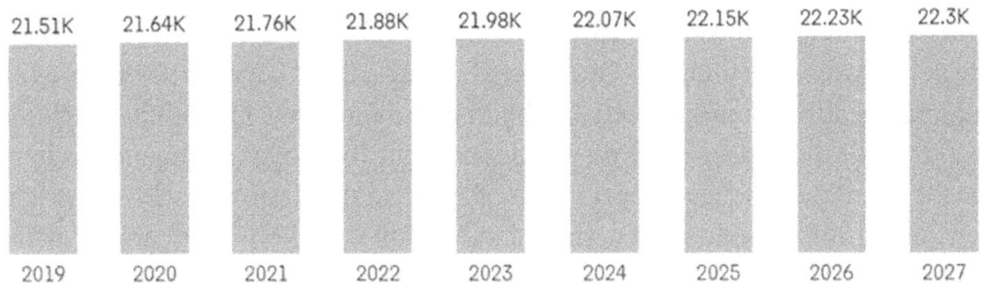

India's medical infrastructure is on par with that of wealthy countries, and hospitals offer special sections just for patients from other countries. The availability of effective infrastructures and technologies in the Indian healthcare sector makes

India a preferred destination for patients. Although India has top-notch hospitals with cutting-edge infrastructure and top-notch medical supplies, there are still some concerning difficulties with public utilities, roads, and cleanliness, among other things (Jain & Ajmera, 2018). Traveling to another country for medical treatment is referred to as "medical tourism". organizational structures required for a society or business to run, or the resources and amenities required for an economy to grow. The phrase usually refers to the infrastructure—roads, water supplies, sewage systems, electricity grids, telecommunications, and so on—that keeps society running. Traveling to and lodging in locations outside of one's normal surroundings for pleasure, business, or other reasons is referred to as tourism. According to (R. Garg, 2021), the establishments that supply the goods and services that the tourism infrastructure requires are regarded as being a part of the tourism sector. Furthermore, although they do not constitute a significant portion of the tourism infrastructure, these businesses also sell their goods primarily to tourists (V. Singh & Singh, 2024). The development of tourism depends on access and connectivity between various locations, and the development of lodging and tourist transportation are the fundamental prerequisites for the development of tourism infrastructure.

2.5 Healthcare Services

According to (Anvekar, 2012), allopathy medicine, homeopathic, naturopathy, Ayurveda, unani, natural care, pranic healing, reiki, and other practices are regulated by the Consumer Protection Act of 1986. Hospitals are the primary service products offered by this industry. Allied services include pathological labs, ambulance services, private mortuaries, and pharmacy centers. Paramedical services include fitness centers and beauty salons that treat skin conditions and obesity. Academic institutions include medical colleges, nursing schools, research centers, and others. Hospitals often offer diagnosis, treatment, and post-operative care as part of their health care service offerings. Additional services include case history management, registration, billing, and support functions like laundry, diet, pharmacy, stores, security, and transportation (Ancy et al., 2020). The service process includes two directions: in-person patient services and out-of-office patient services. It also includes telemedicine consultation and lifestyle coaching. Over the past ten years, India has seen an increase in medical tourism. An increasing number of international patients are choosing India for their medical care due to the establishment of numerous new corporate hospitals offering the best quaternary care in metros and mini-metros like Ahmedabad, Rajkot, Surat, etc. (S. Kumar, 2009).

Additionally, it is crucial to ascertain the factors that influence medical tourists' decisions to come to India for treatment and why they choose India as their preferred location. This needs to be taken into consideration while doing research to develop a formula for gauging customer satisfaction and figuring out how to adjust service standards in an effort to draw in more patients from abroad. Through the implementation of potential solutions to enhance the performance of medical tourism, the research aims to minimize shortcomings and create a suitable framework to benchmark medical tourism in India for efficient decision-making (Priya Rao & Roy Choudhury, 2017). Medical tourism is a rapidly expanding phenomenon that offers both social and economic advantages and disadvantages. It is poised to emerge as the new global business. Importing countries, meaning those from which medical tourists might benefit from lower costs and decreased waiting lists, increased service quality risk and dispute resolution systems. However, exporting nations—primarily those that offer medical services to patients from other countries—can act as a stable source of income and stop medical professionals from migrating to other nations, which reduces the possibility of an unjust healthcare system in which out-of-country patients receive better treatment than in-country ones (Carmen, 2014). Measuring medical tourists' experiences with the range of services they receive at medical locations has gotten less attention to date, despite the focus of this research on exploring different structural components of experiential reactions. Although numerous academics have used the well-known SERVQUAL measure to evaluate service qualities in medical tourism (Ghosh & Mandal, 2019). The global health care industry is expanding and is a dynamic field. People living in more developed countries are taking advantage of the cheaper access to high-quality healthcare that underdeveloped nations offer. While many medical procedures are essential and most people cannot afford them in their home country, other surgeries and treatments are personal choices that fall outside the purview of both public and commercial health care systems. The country's economy is starting to include medical tourism (Mishra et al., 2013).

3. RESEARCH METHODOLOGY

Based on extensive literature review, the study examines how the factors of medical tourism are significant with inclusive opportunity and growth medical tourism in Indian healthcare sector. The critical literature survey was adopted in this study in order to fully comprehend the SWOT analysis of medical tourism in Indian healthcare sector. The data was gathered from secondary sources, which are different published sources about medical tourism and associated topics. For analysis purposes, a variety of books, journals, and reports about medical tourism have been gathered.

4. CONCLUSION

India is well-positioned to benefit from global opportunities in the medical tourism industry. It is essential that the government take part in the expansion of medical tourism. The role of the government in promoting private sector investment in healthcare should be dual. It is necessary to create procedures that allow foreign visitors to receive visa applications more quickly for medical reasons. Patients should be able to contact the Immigration Department at any point of entry to expedite the process. A few of the initiatives that might be used are tax advantages for service providers, duty reductions on medical equipment, and committees to support and encourage medical tourism. India is well-positioned to benefit from global opportunities in the medical tourism industry. It is essential that the government take part in the expansion of medical tourism. The role of the government in promoting private sector investment in healthcare should be dual. It is necessary to create procedures that allow foreign visitors to receive visa applications more quickly for medical reasons. Patients should be able to contact the Immigration Department at any point of entry to expedite the process. A few of the initiatives that might be used are tax advantages for service providers, duty reductions on medical equipment to support and encourage medical tourism including transportation services is also necessary to facilitate tourism in India.

REFERENCES

Ancy, R. J., Shenoy, R. P., Jodalli, P. S., & Pasha, I. M. (2020). Benefits of medical and dental tourism: A review. *IOSR Journal of Dental and Medical Sciences*, *19*(3), 26–31.

Ansari, A. I., & Singh, A. (2023). Application of Augmented Reality (AR) and Virtual Reality (VR) in Promoting Guest Room Sales: A Critical Review. *Impact of Industry 4.0 on Sustainable Tourism*, 95–104.

Anvekar, S. R. (2012). *Medical Tourism in India: A Strategic Approach Towards Effective Branding for Health Care Services Marketing.*

Bhalla, A., Singh, P., & Singh, A. (2023). Technological Advancement and Mechanization of the Hotel Industry. *Application and Adoption of Robotic Process Automation for Smart Cities*, 57–76.

Carmen. (2014). *Medical tourism industry challenges in the context of globalization.*

Dawn, S. K., & P. S. (2011). Medical tourism in India: Issues, opportunities and designing strategies for growth and development. *International Journal of Multidisciplinary Research, 1*(3), 7–10.

Ebrahim, A. H., & Ganguli, S. (2019). A comparative analysis of medical tourism competitiveness of India, Thailand and Singapore. *Tourism: An International Interdisciplinary Journal, 67*(2), 102–115.

Garg, R. (2021). *Infrastructure Development-An Indispensable Component of Tourism Destination Planning*. Research Gate. https://www.researchgate.net/publication/361306587

Garg, S. R., & B. A. (2012). Indian medical tourism industry: Growth opportunities and challenges. *Multi Disciplinary Edu Global Quest, 1*(1), 115–135.

Ghosh, T., & Mandal, S. (2019). Medical Tourism Experience: Conceptualization, Scale Development, and Validation. *Journal of Travel Research, 58*(8), 1288–1301. doi:10.1177/0047287518813469

Hadi, A. (2009). Globalization, medical tourism and health equity. *Implications of Medical Tourism for Canadian Health and Health Policy, 13*, 1–29.

Hazarika, I. (2010a). Medical tourism: Its potential impact on the health workforce and health systems in India. *Health Policy and Planning, 25*(3), 248–251. doi:10.1093/heapol/czp050 PMID:19926658

Hazarika, I. (2010b). Medical tourism: Its potential impact on the health workforce and health systems in India. *Health Policy and Planning, 3*(3), 248–251. doi:10.1093/heapol/czp050 PMID:19926658

India Brand Equity Foundation. (2023). *Healthcare Industry in India*. IBEF.

Jain, V., & Ajmera, P. (2018). Modelling the factors affecting Indian medical tourism sector using interpretive structural modeling. *Benchmarking, 25*(5), 1461–1479. doi:10.1108/BIJ-03-2017-0045

Kumar, A. (2023). *A Critical Evaluation of Medical Tourism in India: Growth.* Drivers and Challenges.

Kumar, S. (2009). Designing promotional strategies for medical tourism in India: A case study of an ophthalmic hospital in NCR. *Health and Population; Perspectives and Issues, 32*(2), 86–95.

Lee, C. (2006). Medical tourism, an innovative opportunity for entrepreneurs. *Journal of Asia Entrepreneurship and Sustainability, 3*(1), 3–5.

Mishra, S., Kant Rai, S., Pandey, A., & Jaiswal, V. (2013). An implicit effect of the health insurance portability and accountability act on medical tourism in India. In *UTMS Journal of Economics, 4*(3). http://hdl.handle.net/10419/105295

Mordor Intelligence. (2024). *India's Medical Tourism Market*. Mordor Intelligence.

Oberoi, S., Kansra, P., & Gupta, S. L. (2022). A Systematic Review on Determinants Inciting Sustainable E-Medical Tourism. *International Journal of Reliable and Quality E-Healthcare*, *11*(2), 1–13. doi:10.4018/IJRQEH.299962

Priya Rao, V., & Roy Choudhury, R. (2017). *A study of the factors influencing customer satisfaction in medical tourism in India*. IASET. www.iaset.us

Sagar, R., & Latha, P. V. (2014). Medical tourism–An emerging industry in India. *Journal of Medical Sciences Research*, *2*(3), 169–177.

Saxena, R. (2014). Indian medical tourism industry-products and destinations: challenges and opportunities. Empyreal Institute of Higher Education, 14.

Shanmugam, K. R. (2013). *Medical tourism in India: Progress, opportunities and challenges*. Madras School of Economics.

Singh, A., & Hassan, S. C. (2023). Identifying Religious Tourism Potential Through Sustainable Approach: A Study on Punjab State of India. *Prospects and Challenges of Global Pilgrimage Tourism and Hospitality*, 180–192.

Singh, V., & Singh, A. (2024). Digital Health Revolution: Enhancing Well-Being Through Technology. *Implementing Sustainable Development Goals in the Service Sector*, 213–219.

Wapmuk, S., Wapmuk, A. E., & Gbajabiamila, T. A. (2015). Medical tourism and Nigeria-India relations. *Journal of Tourism Research*, *12*(1), 44–65.

Compilation of References

Abadi, M., Agarwal, A., Barham, P., Brevdo, E., Chen, Z., Citro, C., & Zheng, X. (2016). Tensorflow: Large-scale machine learning on heterogeneous distributed systems. *arXiv preprint arXiv:1603.04467.*

Abd Mutalib, N. S., Ming, L. C., Yee, S. M., Wong, P. L., & Soh, Y. C. (2016). Medical tourism: Ethics, risks and benefits. *Indian Journal of Pharmaceutical Education and Research, 50*(2), 261–270. https://www.ijper.org/sites/default/files/10.5530ijper.50.2.6.pdf

Abdul-Rahman, M. N., Hassan, T. H., Abdou, A. H., Abdelmoaty, M. A., Saleh, M. I., & Salem, A. E. (2023). Responding to tourists' intentions to revisit medical destinations in the post-COVID-19 era through the promotion of their clinical trust and well-being. *Sustainability (Basel), 15*(3), 2399. doi:10.3390/su15032399

Abo-Elsoud, E., & Morsy, J. (2022). The Role of Artificial Intelligence in Improving Service and Strengthening Tourist Experience in Egypt. *ãÌáÉ ˜áíÉ ÇáÓíÇÍÉ æÇáÝäÇÏÞ. ÌÇãÚÉ ÇáãäÕæÑÉ, 2*(11), pp.821-855.

Adedinsewo, D., Hardway, H. D., Morales-Lara, A. C., Wieczorek, M. A., Johnson, P. W., Douglass, E. J., Dangott, B. J., Nakhleh, R. E., Narula, T., Patel, P. C., Goswami, R. M., Lyle, M. A., Heckman, A. J., Leoni-Moreno, J. C., Steidley, D. E., Arsanjani, R., Hardaway, B., Abbas, M., Behfar, A., & Yamani, M. (2023). Non-invasive detection of cardiac allograft rejection among heart transplant recipients using an electrocardiogram based deep learning model. *European Heart Journal. Digital Health, 4*(2), 71–80. doi:10.1093/ehjdh/ztad001 PMID:36974261

Advani, V. (2021). What is Artificial Intelligence? How does AI work, Types and Future of it. *mygreatlearning.*

Advani, V. (2021). What Is Artificial Intelligence? How Does AI Work, Types and Future of it? *Medium.* https://medium.com/@mygreatlearning/what-is-artificial-intelligence-how-does-ai-work-and-future-of-it-d6b113fce9be

Afzal, N., Sohn, S., Abram, S., Scott, C. G., Chaudhry, R., Liu, H., Kullo, I. J., & Arruda-Olson, A. M. (2017). Mining peripheral arterial disease cases from narrative clinical notes using natural language processing. *Journal of Vascular Surgery, 65*(6), 1753–1761. doi:10.1016/j.jvs.2016.11.031 PMID:28189359

Ahuja, A. S. (2019). The impact of artificial intelligence in medicine on the future role of the physician. *PeerJ*, *7*(10), e7702. doi:10.7717/peerj.7702 PMID:31592346

Akoijam, S. L., & Khan, T. (2020). Prospects and Challenges of Medical Tourism: Evidences From Manipur. *Global Developments in Healthcare and Medical Tourism*, 265-276.

Akpojaro, J., & Bello, R. W. (2019). Smart healthcare system: A primer. *International Journal for Innovative Research in Multidisciplinary Field*, *5*(3), 143–149.

Aksu, A. A., Gürsoy, D., & Aksu, G. M. (2016). *Sağlık turizminde keşfedildikçe büyüyen ülke Türkiye: Pazar analizi ve izlenmesi gereken stratejiler*. Türkiye Turizm Yatırımcıları Derneği.

Aksu, A., & Bayar, K. (2019). Development of Health Tourism in Türkiye: SWOT Analysis of Antalya Province. *Journal of Tourism Management Research*, *6*(2), 134–154. doi:10.18488/journal.31.2019.62.134.154

Al Adwan, A. (2020). The impact of motivation factors and intention to adopt Jordan as a destination for medical tourism in the Middle East. *Innovative Marketing*, *16*(2), 146–158. doi:10.21511/im.16(2).2020.11

Al'Aref, S. J., Maliakal, G., Singh, G., van Rosendael, A. R., Ma, X., Xu, Z., Alawamlh, O. A. H., Lee, B., Pandey, M., Achenbach, S., Al-Mallah, M. H., Andreini, D., Bax, J. J., Berman, D. S., Budoff, M. J., Cademartiri, F., Callister, T. Q., Chang, H.-J., Chinnaiyan, K., & Shaw, L. J. (2020). Machine learning of clinical variables and coronary artery calcium scoring for the prediction of obstructive coronary artery disease on coronary computed tomography angiography: Analysis from the CONFIRM registry. *European Heart Journal*, *41*(3), 359–367. doi:10.1093/eurheartj/ehz565 PMID:31513271

Al-Azzam, A. F. M. (2016). A study of the impact of marketing mix for attracting medical tourism in Jordan. *International Journal of Marketing Studies*, *8*(1), 139–149. doi:10.5539/ijms.v8n1p139

Ali, S., Abdullah, Armand, T. P. T., Athar, A., Hussain, A., Ali, M., Yaseen, M., Joo, M.-I., & Kim, H.-C. (2023). Metaverse in healthcare integrated with explainable ai and blockchain: Enabling immersiveness, ensuring trust, and providing patient data security. *Sensors (Basel)*, *23*(2), 565. doi:10.3390/s23020565 PMID:36679361

Al-Khaled, T., Valikodath, N., Cole, E., Hallak, J., Campbell, J. P., Chiang, M. F., & Chan, R. V. P. (2020). Evaluation of physician perspectives of artificial intelligence in ophthalmology: A pilot study. *Investigative Ophthalmology & Visual Science*, *61*(7), 2023–2023.

Alkhatib, G., & Bayouq, S. T. (2021). A TAM-Based Model of Technological Factors Affecting Use of E-Tourism. *International Journal of Tourism and Hospitality Management in the Digital Age*, *5*(2), 50–67. doi:10.4018/IJTHMDA.20210701.oa1

Alowais, S. A., Alghamdi, S. S., Alsuhebany, N., Alqahtani, T., Alshaya, A. I., Almohareb, S. N., & Albekairy, A. M. (2023). Revolutionizing healthcare: The role of artificial intelligence in clinical practice. *BMC Medical Education*, *23*(1), 689. doi:10.1186/s12909-023-04698-z PMID:37740191

Alrahbi, D., Khan, M., & Hussain, M. (2021). Exploring the motivators of technology adoption in healthcare. *International Journal of Healthcare Management*, *14*(1), 50–63. doi:10.1080/20479700.2019.1607451

Ameen, N., Hosany, S., & Tarhini, A. (2021). Consumer interaction with cutting-edge technologies: Implications for future research. *Computers in Human Behavior*, *120*, 106761. doi:10.1016/j.chb.2021.106761

Ameen, N., Tarhini, A., Reppel, A., & Anand, A. (2021). Customer experiences in the age of artificial intelligence. *Computers in Human Behavior*, *114*, 106548. doi:10.1016/j.chb.2020.106548 PMID:32905175

Ameen, S., Gupta, R., & Singh, P. (2021). Bridging the gap: The role of wearables and AI in proactive mental health monitoring. *Journal of Technological Advances in Mental Health*, *7*(1), 12–28.

Amin, R., Al Ghamdi, M. A., Almotiri, S. H., & Alruily, M. (2021). Healthcare techniques through deep learning: Issues, challenges, and opportunities. *IEEE Access : Practical Innovations, Open Solutions*, *9*, 98523–98541. doi:10.1109/ACCESS.2021.3095312

Amiri, M., & Safariolyaei, N. (2017). Medical tourism in India: Current scenario. *International Journal of Management and Social Sciences*, *5*(3), 106–111.

Amisha, M. P., Pathania, M., & Rathaur, V. K. (2019). Overview of artificial intelligence in medicine. *Journal of Family Medicine and Primary Care*, *8*(7), 2328–2331. doi:10.4103/jfmpc.jfmpc_440_19 PMID:31463251

Amjad, A., Kordel, P., & Fernandes, G. (2023). A Review on Innovation in Healthcare Sector (Telehealth) through Artificial Intelligence. *Sustainability, 15*(8), 6655. doi:10.3390/su15086655

Ancy, R. J., Shenoy, R. P., Jodalli, P. S., & Pasha, I. M. (2020). Benefits of medical and dental tourism: A review. *IOSR Journal of Dental and Medical Sciences*, *19*(3), 26–31.

Andrikopoulou, E. (2023). The rise of AI in telehealth. In *Emerging Practices in Telehealth* (pp. 183–207). Academic Press. doi:10.1016/B978-0-443-15980-0.00011-9

Ansari, A. I., & Singh, A. (2023). Application of Augmented Reality (AR) and Virtual Reality (VR) in Promoting Guest Room Sales: A Critical Review. *Impact of Industry 4.0 on Sustainable Tourism*, 95–104.

Antoniou, G., Papadakis, E., & Baryannis, G. (2022). Mental health diagnosis: A case for explainable artificial intelligence. *International Journal of Artificial Intelligence Tools*, *31*(03), 2241003. doi:10.1142/S0218213022410032

Anvekar, S. R. (2012). *Medical Tourism in India: A Strategic Approach Towards Effective Branding for Health Care Services Marketing*.

Aslan, I., Çınar, O., & Özen, Ü. (2014). Developing strategies for the future of Healthcare in Türkiye by benchmarking and SWOT analysis. *Procedia: Social and Behavioral Sciences*, *150*, 230–240. doi:10.1016/j.sbspro.2014.09.043

Aurelia-Felicia, S., Andreea, B., & Ion, P. (2014). Clarifications Regarding Medical Tourism Destination Marketing. *Balneo Research Journal*, *5*(3), 84–98. doi:10.12680/balneo.2014.1068

Aydin, G., & Karamehmet, B. (2017). Factors affecting health tourism and international healthcare facility choice. *International Journal of Pharmaceutical and Healthcare Marketing*, *11*(1), 16–36. doi:10.1108/IJPHM-05-2015-0018

Aykin, Ö., Uluhan, F., Gümüş, İ., Çabuk, Ş., Bozbayir, U., Duran, V., Özbay, C., Özbay, B., Gündoğmuş, Ş. E., Türkegün Şengül, M., & Övey, İ. S.AYKIN. (2023). Artificial Intelligence And Telemedicine Applications In Health Tourism Marketing. *Eurasian Journal of Health Technology Assessment*, *7*(2), 134–149. doi:10.52148/ehta.1396111

Ayuningtyas, D., Ariwibowo, D. (2020). The strategic role of information communication technology in succeeding medical tourism. *Enfermería Clínica,Volume 30*(6).

Ayuningtyas, D., & Ariwibowo, D. A. (2020). The strategic role of information communication technology in succeeding medical tourism. *Enfermeria Clinica*, *30*, 170–173. doi:10.1016/j.enfcli.2020.06.039

Baghbanian, A., Chen, Y., & Wang, Q. (2021). Global trends in the medical tourism industry: An overview of market projections and growth drivers. *International Journal of Medical Tourism Research*, *17*(4), 301–318.

Baghbanian, A., Safdari, R., Erfannia, L., & Zokaee, M. (2021). The medical tourism industry in Iran: A review of websites designed for cross-border patients. *Health Scope*, *10*(1). Advance online publication. doi:10.5812/jhealthscope.108482

Balasubramanian, S., Sethi, J. S., Ajayan, S., & Paris, C. M. (2022). An enabling framework for blockchain in tourism. *Information Technology & Tourism*, *24*(2), 165–179. doi:10.1007/s40558-022-00229-6

Balcombe, L., & De Leo, D. (2022, February). Human-computer interaction in digital mental health. In Informatics (Vol. 9, No. 1, p. 14). MDPI. doi:10.3390/informatics9010014

Banerjee, N. (2022). *India's medical tourism is gearing up for AI*. SME. https://smefutures.com/indias-medical-tourism-is-gearing-up-for-ai/

Baran, Z., & Karaca, S. (2023). Next-Generation Technologies in Health Tourism. In *Global Perspectives on the Opportunities and Future Directions of Health Tourism* (pp. 138–164). IGI Global., doi:10.4018/978-1-6684-6692-6.ch006

Barbaz-I, R. (2014). *Experimental determining of the elastic modulus and strength of composites reinforced with two nanoparticles* [Doctoral dissertation, Iran University of Science and Technology, Tehran, Iran].

Barca, M., Akdeve, E., Balay, İ. G. (2013). Türkiye sağlık turizm sektörünün analizi ve strateji önerileri. *İşletme araştırmaları dergisi*, *5*(3), 64-92.

Basheer, S., Walia, S., Mehraj, D., & Reshi, M. (2023). Creating Transformative Travel Unveiling the Link Between Meaningful Experiences and Sustainable Tourism. IGI Global. doi:10.4018/979-8-3693-0650-5.ch002

Basheer, S., Farooq, S., Hassan, V., Malik, Y. M., & Reshi, M. A. (2023). *Augmented Reality and Virtual Reality in Cultural Heritage Tourism Enhancing Visitor Experiences.*, doi:10.4018/978-1-6684-9957-3.ch002

Basheer, S., Walia, S., Farooq, S., Shah, M. A., & Mir, F. A. (2023). *Exploring the Metaverse.*, doi:10.4018/978-1-6684-8898-0.ch012

Bassil, C., Saleh, A. S., & Anwar, S. (2019). Terrorism and tourism demand: A case study of Lebanon, Türkiye, and Israel. *Current Issues in Tourism*, *22*(1), 50–70. doi:10.1080/13683500.2017.1397609

Basu, K., Sinha, R., Ong, A., & Basu, T. (2020). Artificial Intelligence: How is It Changing Medical Sciences and Its Future? *Indian Journal of Dermatology*, *65*(5), 365. doi:10.4103/ijd.IJD_421_20 PMID:33165420

Bauer, I. L. (2023). Robots in travel clinics: Building on tourism's use of technology and robots for infection control during a pandemic. *Tropical Diseases, Travel Medicine and Vaccines*, *9*(1), 10. doi:10.1186/s40794-023-00197-7 PMID:37525269

Bellini, V., Valente, M., Turetti, M., Del Rio, P., Saturno, F., Maffezzoni, M., & Bignami, E. (2022). Current applications of artificial intelligence in bariatric surgery. *Obesity Surgery*, *32*(8), 2717–2733. doi:10.1007/s11695-022-06100-1 PMID:35616768

Benckendorff, P. J., Xiang, Z., & Sheldon, P. J. (2019). *Tourism information technology*. Cabi. doi:10.1079/9781786393432.0000

Bentley, P., Ganesalingam, J., Jones, A. L. C., Mahady, K., Epton, S., Rinne, P., & Rueckert, D. (2014). Prediction of stroke thrombolysis outcome using CT brain machine learning. *NeuroImage. Clinical*, *4*, 635–640. doi:10.1016/j.nicl.2014.02.003 PMID:24936414

Berezina, K., Ciftci, O., & Cobanoglu, C. (2019). Robots, artificial intelligence, and service automation in Han, H., 2013. The healthcare hotel: Distinctive attributes for international medical travelers. *Tourism Management*, *36*, 257–268.

Berg, B., & Lune, H. (2015). *Sosyal bilimlerde nitel araştırma yöntemleri*. Eğitim Yayınevi.

Bernasconi, S. (2016). The role of medical technology in the 4th Industrial Revolution. *Med Tech Views*. www.medtechviews.eu/article/role-medicaltechnology- 4th-industrial-revolution

Bhaidkar, A., & Goswami, A. (2017). Medical tourism in India and the Health policy of the NDA government: An overview. *Aweshkar Research Journal, 22*(1), 82-89. Retrieved from https://web.s.ebscohost.com/abstract?direct=true&profile=ehost&scope=site&authtype=crawler&jrnl=09741119&AN=124266520&h=o5zzf875if8NBE19DOCDYQTUt0I6VtEIctiyrSzrHUWnMxQvKTBcmrN5Q2Wg XdJLxbWjvkbqiVHf%2bzSh9shFIQ%3d%3d&crl=c&resultNs=AdminWebAuth&resultLocal=

Bhalla, A., Singh, P., & Singh, A. (2023). Technological Advancement and Mechanization of the Hotel Industry. *Application and Adoption of Robotic Process Automation for Smart Cities*, 57–76.

Bhinder, B., Gilvary, C., Madhukar, N. S., & Elemento, O. (2021). Artificial Intelligence in Cancer Research and Precision Medicine. *Cancer Discovery*, *11*(4), 900–915. doi:10.1158/2159-8290.CD-21-0090 PMID:33811123

Bi, N., Yin, C. Y., & Chen, Y. (2020). A bittersweet experience! The effect of mixed emotions on business tourists' revisit intentions. *Journal of Travel & Tourism Marketing*, *37*(6), 695–710. doi:10.1080/10548408.2020.1795047

Binoy, T. A. (2017). Dynamics of medical tourism in India-A microscopic analysis. *Asia Pacific Journal of Research*, *1*(58), 161–167.

Bi, W. L., Hosny, A., Schabath, M. B., Giger, M. L., Birkbak, N. J., Mehrtash, A., Allison, T., Arnaout, O., Abbosh, C., Dunn, I. F., Mak, R. H., Tamimi, R. M., Tempany, C. M., Swanton, C., Hoffmann, U., Schwartz, L. H., Gillies, R. J., Huang, R. Y., & Aerts, L. (2019). Artificial intelligence in cancer imaging: Clinical challenges and applications. *CA: a Cancer Journal for Clinicians*, *69*(2), 127–157. doi:10.3322/caac.21552 PMID:30720861

Boccanfuso, L., Barney, E., Foster, C., & Ahn, Y. A., Chawarska, K., Scassellati, B., & Shic, F. (2016). Emotional Robot to Examine Differences in Play Patterns and Affective Response of Children with and Without ASD. In *The Eleventh ACM/IEEE International Conference on Human Robot Interaction*, (pp. 19–26). IEEE Press. https://dl.acm.org/citation.cfm?id=2906831.2906837

Boczar, D., Sisti, A., Oliver, J. D., Helmi, H., Restrepo, D. J., Huayllani, M. T., Spaulding, A. C., Carter, R., Rinker, B. D., & Forte, A. J. (2020). Artificial intelligent virtual assistant for plastic surgery patient's frequently asked questions: A pilot study. *Annals of Plastic Surgery*, *84*(4), 16–21. doi:10.1097/SAP.0000000000002252 PMID:32032120

Boguszewicz-Kreft, M., Kuczamer-Kłopotowska, S., Kozłowski, A., Ayci, A., & Abuhashesh, M. (2020). The theory of planned behaviour in medical tourism: International comparison in the young consumer segment. *International Journal of Environmental Research and Public Health*, *17*(5), 1626. doi:10.3390/ijerph17051626 PMID:32138240

Bohr, A., & Memarzadeh, K. (2020). The rise of artificial intelligence in healthcare applications. *Artificial Intelligence in Healthcare*, *25*, 25–60. doi:10.1016/B978-0-12-818438-7.00002-2

Bolger, R. K. (2021). Finding wholes in the metaverse: Posthuman mystics as agents of evolutionary contextualization. *Religions*, *12*(9), 768. doi:10.3390/rel12090768

Bonner, A. (2021). Four ways artificial intelligence will affect your eye care practice. Eyecare Leaders. https://eyecareleaders.com/artificial-intelligence-ophthalmology/.

Bouaziz, J., Mashiach, R., Cohen, S., Kedem, A., Baron, A., Zajicek, M., Feldman, I., Seidman, D., & Soriano, D. (2018). How artificial intelligence can improve our understanding of the genes associated with endometriosis: Natural language processing of the PubMed Database. *BioMed Research International*, *2018*, 20186217812. doi:10.1155/2018/6217812 PMID:29750165

Boucher, E. M., Harake, N. R., Ward, H. E., Stoeckl, S. E., Vargas, J., Minkel, J., & Zilca, R. (2021). Artificially intelligent chatbots in digital mental health interventions: a review. *Expert Review of Medical Devices*, *18*(sup1), 37-49. doi:10.1080/17434440.2021.2013200

Briceño, J. (2020). Artificial Intelligence and Organ Transplantation: Challenges and Expectations. *Current Opinion in Organ Transplantation*, *25*(4), 393–398. doi:10.1097/MOT.0000000000000775 PMID:32487888

Brophy, E., Jones, K., & Smith, L. (2021). Insights into mental health states through wearable technology: A systematic review. *Journal of Health Informatics*, *15*(2), 78–95.

Brophy, K., Davies, S., Olenik, S., Çotur, Y., Ming, D., Van Zalk, N., & Yetisen, A. K. (2021). *The future of wearable technologies*. Imperial College London.

Buckley, R. (2020). Nature tourism and mental health: Parks, happiness, and causation. *Journal of Sustainable Tourism*, *28*(9), 1409–1424. doi:10.1080/09669582.2020.1742725

Buckley, R. (2023). Tourism and mental health: Foundations, frameworks, and futures. *Journal of Travel Research*, *62*(1), 3–20. doi:10.1177/00472875221087669

Buhalis, D., & Karatay, N. (2022). Mixed Reality (MR) for Generation Z in Cultural Heritage Tourism Towards Metaverse. In Information and Communication Technologies in Tourism 2022. doi:10.1007/978-3-030-94751-4_2

Buhalis, D., Leung, D., & Lin, M. (2023). Metaverse as a disruptive technology revolutionising tourism management and marketing. In *Tourism Management* (Vol. 97). Elsevier Ltd. doi:10.1016/j.tourman.2023.104724

Bulatovic, I., & Iankova, K. (2021). Barriers to medical tourism development in the United Arab Emirates (UAE). *International Journal of Environmental Research and Public Health*, *18*(3), 1365. doi:10.3390/ijerph18031365 PMID:33540935

Bulchand-Gidumal. J. (2020). Impact of Artificial Intelligence in Travel, Tourism, and Hospitality. Handbook of E-Tourism, 1–20. doi:10.1007/978-3-030-05324-6_110-1

Bulte, C. (2000). New Product Diffusion Acceleration: Measurement and Analysis. *Marketing Science*, *19*(4), 366–380. doi:10.1287/mksc.19.4.366.11795

Bushel, R., & Sheldon, P. J. (Eds.). (2009). *Wellness and Tourism: Mind, Body, Spirit, Place.* Cognizant Communication Corporation.

Buzcu, Z., & Birdir, K. (2019). Türkiye'de Medikal Turizm İncelemesi: Özel Hastanelerde Bir Çalışma. *Gaziantep Üniversitesi Sosyal Bilimler Dergisi*, *18*(1), 311–327.

Camaréna, S. (2021). Engaging with artificial intelligence (AI) with a bottom-up approach for the purpose of sustainability: Victorian farmers market association, Melbourne Australia. *Sustainability (Basel)*, *13*(16), 9314. doi:10.3390/su13169314

Campra, M., Riva, P., Oricchio, G., & Brescia, V. (2022). Bibliometrix analysis of medical tourism. *Health Services Management Research*, *35*(3), 172–188. doi:10.1177/09514848211011738 PMID:33957813

Cao, Y., Montgomery, S., Ottosson, J., Näslund, E., & Stenberg, E. (2020). Deep learning neural networks to predict serious complications after bariatric surgery: Analysis of scandinavian obesity surgery registry data. *JMIR Medical Informatics*, *8*(5), e15992. doi:10.2196/15992 PMID:32383681

Carmen. (2014). *Medical tourism industry challenges in the context of globalization.*

Carrera, P. M., & Bridges, J. F. P. (2006). Globalization and Healthcare: Understanding health and medical tourism. *Expert Review of Pharmacoeconomics & Outcomes Research*, *6*(4), 447–454. doi:10.1586/14737167.6.4.447 PMID:20528514

Carrera, P., & Bridges, J. (2006). Health and medical tourism: What they mean and imply for health care systems. *Geneva Association Information Newsletter*, *6*(4), 11–15.

Carrera, P., & Lunt, N. (2010). A European perspective on Medical Tourism: The need for a knowledge base. *International Journal of Health Services*, *40*(3), 469–484. doi:10.2190/HS.40.3.e PMID:20799671

Carron, M., Leppariello, G., Safaee, F. B., & Foletto, M. (2020). Perioperative care of the obese patient: A successful approach. *British Journal of Surgery*, *107*(6), 772–773. doi:10.1002/bjs.11541 PMID:32339281

Carr, S. (2020). 'AI gone mental': Engagement and ethics in data-driven technology for mental health. *Journal of Mental Health (Abingdon, England)*, *29*(2), 125–130. doi:10.1080/09638237.2020.1714011 PMID:32000544

Casey, V., Crooks, V. A., Snyder, J., & Turner, L. (2013). Knowledge brokers, companions, and navigators: A qualitative examination of informal caregivers' roles in medical tourism. *International Journal for Equity in Health*, *12*(1), 1–10. https://sci-hub.se/10.1186/1475-9276-12-94. doi:10.1186/1475-9276-12-94 PMID:24289231

Casillo, M., Clarizia, F., Colace, F., Lombardi, M., Pascale, F., & Santaniello, D. (2019). An approach for recommending contextualized services in e-tourism. *Information (Basel)*, *10*(5), 180. doi:10.3390/info10050180

Castro, V. M., Dligach, D., Finan, S., Yu, S., Can, A., Abd-El-Barr, M., Gainer, V., Shadick, N. A., Murphy, S., Cai, T., Savova, G., Weiss, S. T., & Du, R. (2017). Large-scale identification of patients with cerebral aneurysms using natural language processing. *Neurology*, *88*(2), 164–168. doi:10.1212/WNL.0000000000003490 PMID:27927935

Cath, C. (2018). Governing artificial intelligence: Ethical, legal and technical opportunities and challenges. *Philosophical Transactions. Series A, Mathematical, Physical, and Engineering Sciences*, *376*(2133), 20180080. doi:10.1098/rsta.2018.0080 PMID:30322996

CGI. (2023). *Consulate General of India - Dubai : Medical Tourism in India*. https://www.cgidubai.gov.in/page/medical-tourism-in-india/

Chakma, R., Paul, J., & Dhir, S. (2021). Organizational ambidexterity: A review and research agenda. *IEEE Transactions on Engineering Management*, 1–17. doi:10.1109/TEM.2021.3114609

Chakraborty, I., Edirippulige, S., & Ilavarasan, P. V. (2023). The role of telehealth startups in healthcare service delivery: A systematic review. *International Journal of Medical Informatics*, *105048*, 105048. Advance online publication. doi:10.1016/j.ijmedinf.2023.105048 PMID:36963322

Cham, T. H., Cheng, B. L., Low, M. P., & Cheok, J. B. C. (2020). Brand image as the competitive edge for hospitals in medical tourism. *European Business Review*, *33*(1). doi:10.1108/EBR-10-2019-0269

Cham, T. H., Lim, Y. M., Sia, B. C., Cheah, J. H., & Ting, H. (2021). Medical tourism destination image and its relationship with the intention to revisit: A study of Chinese medical tourists in Malaysia. *Journal of China Tourism Research*, *17*(2), 163–191. https://sci-hub.se/10.1080/19388160.2020.1734514. doi:10.1080/19388160.2020.1734514

Cham, T. H., Lim, Y. M., & Sigala, M. (2022). Marketing and social influences, hospital branding, and medical tourists' behavioural intention: Before-and after-service consumption perspective. *International Journal of Tourism Research*, *24*(1), 140–157. doi:10.1002/jtr.2489

Chandra, J., Rangaswamy, M., Banerjee, B., Prajapati, A., Akhtar, Z., Sakauye, K., & Joseph, A. (2022). Applications of artificial intelligence to neurological disorders: current technologies and open problems. In *Augmenting Neurological Disorder Prediction and Rehabilitation Using Artificial Intelligence* (pp. 243–272). Academic Press. doi:10.1016/B978-0-323-90037-9.00005-9

Charlie, M., Degrande, A., Villamor, G., Van Damme, P., Tchoundjeu, Z., & Sygnola, T. (2015). Factors affecting the adoption of agricultural innovation: The case of a Ricinodendron heudelotii kernel extraction machine in southern Cameroon. *Agroforestry Systems*, *89*(5), 799–811. Advance online publication. doi:10.1007/s10457-015-9813-y

Chartier, C., Watt, A., Lin, O., Chandawarkar, A., Lee, J., & Hall-Findlay, E. (2022). BreastGAN: Artificial intelligence-enabled breast augmentation simulation. *Aesthetic Surgery Journal. Open Forum*, *4*, 52. doi:10.1093/asjof/ojab052 PMID:35072073

Chawla, A. (2020). Revolutionizing mental health treatment through AI-powered chatbots and virtual assistants. *AI in Healthcare*, *5*(3), 210–225.

Chawla, N. (2020). AI, IOT and Wearable Technology for Smart Healthcare-A Review. International. *Journal of Recent Research Aspects*, *7*(1).

Chella, A., Iocchi, L., Macaluso, I., & Nardi, D. (2006). Artificial Intelligence and Robotics. *Intelligenza Artificiale*, *3*(1-2), 87–93.

Chen, M., & Decary, M. (2020). Artificial intelligence in healthcare: An essential guide for health leaders. In *Healthcare management forum* (pp. 10–18). SAGE Publications. doi:10.1177/0840470419873123

Chen, M., Wei, S., Hu, J., Yuan, J., & Liu, F. (2017). Does time-lapse imaging have favourable results for embryo incubation and selection compared with conventional methods in clinical in vitro fertilization? A meta-analysis and systematic review of randomized controlled trials. *PLoS One*, *12*(6), e0178720. doi:10.1371/journal.pone.0178720 PMID:28570713

Chinai, R., & Goswami, R. (2007). Medical visas mark growth of Indian medical tourism. *Bulletin of the World Health Organization*, *85*(3), 164–165. doi:10.2471/BLT.07.010307 PMID:17486202

Chinski, H., Lerch, R., Tournour, D., Chinski, L., & Caruso, D. (2022). An artificial intelligence tool for image simulation in rhinoplasty. *Facial Plastic Surgery*, *38*(2), 201–206. doi:10.1055/s-0041-1729911 PMID:34051703

Chinzei, K. (2019). Safety of surgical robots and IEC 80601-2-77: The first international standard for surgical robots. *Acta Polytechnica Hungarica*, *16*(8), 171–184.

Chow, D. J. X., Wijesinghe, P., Dholakia, K., & Dunning, K. R. (2021). Does Artificial intelligence have a role in IVF clinics? *Reproduction & Fertility*, *2*(3), 29–34. doi:10.1530/RAF-21-0043 PMID:35118395

Clarke, A. (2010). Wellness and tourism: Mind, body, spirit. *Annals of Tourism Research*, *37*(1), 276–278. doi:10.1016/j.annals.2009.10.004

Cohen, E. (2012). Medical Travel and the Quality-of-Life (Uysal, M., Perdue, R.R., Sirgy, M.J., Eds.), Handbook of Tourism and Quality-of-Life Research: Enhancing the Lives of Tourists and Residents of Host Communities. Springer Science+Business Media.

Cohen, E. (2014). Recreational Hunting: Ethics, Experiences and Commoditization. *Tourism Recreation Research*, *39*(1), 3–17. doi:10.1080/02508281.2014.11081323

Collado, S., & Boden-Stuart, Z. (2023). The Performative Narrative Interview: A creative strategy for data production drawing on dialogical narrative theory. *Qualitative Research*, *23*(5), 1418–1438. doi:10.1177/14687941221082264

Collins, A., Medhekar, A., Wong, H. Y., & Cobanoglu, C. (2019). Factors influencing outbound medical travel from the USA. *Tourism Review*, *74*(3), 463–479. doi:10.1108/TR-06-2018-0083

Comito, C., Falcone, D., & Forestiero, A. (2022). Convergence between iot and ai for smart health and predictive medicine. In *Integrating Artificial Intelligence and IoT for Advanced Health Informatics: AI in the Healthcare Sector* (pp. 69–84). Springer International Publishing., doi:10.1007/978-3-030-91181-2_5

Connell, J. (2006). Medical tourism: Sea, sun, sand, and surgery. *Tourism Management*, *27*(6), 1093–1100. doi:10.1016/j.tourman.2005.11.005

Connell, J. (2013). Contemporary medical tourism: Conceptualisation, culture and commodification. *Tourism Management*, *34*, 1–13. https://sci-hub.se/10.1016/j.tourman.2012.05.009. doi:10.1016/j.tourman.2012.05.009

Connell, J., & Page, S. (2019). Tourism, ageing and the demographic time bomb-the implications of dementia for the visitor economy: A perspective paper. *Tourism Review*, *75*(1), 81–85. Advance online publication. doi:10.1108/TR-02-2019-0070

Creed, T. A., Salama, L., Slevin, R., Tanana, M., Imel, Z., Narayanan, S., & Atkins, D. C. (2022). Enhancing the quality of cognitive behavioral therapy in community mental health through artificial intelligence generated fidelity feedback (Project AFFECT): A study protocol. *BMC Health Services Research*, *22*(1), 1177. doi:10.1186/s12913-022-08519-9 PMID:36127689

Creswell, J. W. (2014). *Research design. Qualitative, quantitative, and mixed methods approaches* (3rd ed.). Sage Publications.

Crooks, V. A., Turner, L., Snyder, J., Johnston, R., & Kingsbury, P. (2011). Promoting medical tourism to India: Messages, images, and the marketing of international patient travel. *Social Science & Medicine*, *72*(5), 726–732. doi:10.1016/j.socscimed.2010.12.022 PMID:21310519

Cubric, M. (2020). Drivers, barriers and social considerations for AI adoption in business and management: A tertiary study. *Technology in Society*, *62*, 101257. doi:10.1016/j.techsoc.2020.101257

Dahl, T. S., & Kamel Boulos, M. N. (2013). Robots in health and social care: A complementary technology to home care and telehealthcare? *Robotics (Basel, Switzerland)*, *3*(1), 1–21. doi:10.3390/robotics3010001

Dalkiran, G. B. (2023). Health Tourism Components and Intermediary Institutions as Supplier Businesses in the Context of Health 4.0. In *Health 4.0 and Medical Supply Chain* (pp. 147–162). Springer Nature Singapore. doi:10.1007/978-981-99-1818-8_12

Damanpouer, F. (1991). Organizational Innovation: A Meta-Analysis of the Effects of Determinants and Moderators. *Academy of Management Journal*, *34*(3), 555–590. doi:10.2307/256406

Dang, H. S., Nguyen, T. M. T., Wang, C. N., Day, J. D., & Dang, T. M. H. (2020). Grey system theory in the study of medical tourism industry and its economic impact. *International Journal of Environmental Research and Public Health*, *17*(3), 961. doi:10.3390/ijerph17030961 PMID:32033177

Dania. (2023). *The Role of Artificial Intelligence in Medical Tourism: Enhancing Patient Experience.* Dania Medicare Dania Medicare. https://www.daniamedicare.com/blogs/the-role-of-artificial-intelligence-in-medical-tourism-enhancing-patient-experience/

Darcy, A.M., Louie, A.K., & Roberts, L.W. (2016). Machine Learning and the Profession of Medicine. *JAMA, 2*(1), 315-551.

Dash, A. (2020). Exploring visit intention to India for medical tourism using an extended theory of planned behaviour. *Journal of Hospitality and Tourism Insights*, *4*(4), 418–436. doi:10.1108/JHTI-03-2020-0037

Das, R. (2017). Medical marketing in promoting medical tourism in India. *ELK Asia Pacific Journal of Marketing and Retail Management*, *8*(3), 1–10.

Daştan, İ. (2014). Türkiye'de Sağlık Turizmi: Türkiye ve Özelinde İzmir'de Sağlık Turizminin Mevcut Durum Analizi ve Strateji Önerileri. *Mehmet Akif Ersoy Üniversitesi Sosyal Bilimler Enstitüsü Dergisi*, *6*(10), 143–163.

Dawn, S. K., & P. S. (2011). Medical tourism in India: Issues, opportunities and designing strategies for growth and development. *International Journal of Multidisciplinary Research*, *1*(3), 7–10.

De Geyter, C., Wyns, C., Calhaz-Jorge, C., de Mouzon, J., Ferraretti, A. P., Kupka, M., Nyboe Andersen, A., Nygren, K. G., & Goossens, V. (2020). 20 years of the European IVF-monitoring consortium registry: What have we learned? A comparison with registries from two other regions. *Human Reproduction (Oxford, England)*, *35*(12), 352832–352849. doi:10.1093/humrep/deaa250 PMID:33188410

De La Hoz-Correa, A., & Muñoz-Leiva, F. (2023). The role of information sources and image on the intention to visit a medical tourism destination: A cross-cultural analysis. *Journal of Travel & Tourism Marketing*, *36*(2), 204–219. doi:10.1080/10548408.2018.1507865

De la Hoz-Correa, A., Muñoz-Leiva, F., & Bakucz, M. (2018). Past themes and future trends in medical tourism research: A co-word analysis. *Tourism Management*, *65*, 200–211. doi:10.1016/j.tourman.2017.10.001

Deepthi, B., & Bansal, V. (2023). Applications of Artificial Intelligence (AI) in the Tourism Industry: A Futuristic Perspective. In Impact of Industry 4.0 on Sustainable Tourism: Perspectives, Challenges and Future (pp. 31-43). Emerald Publishing Limited. doi:10.1108/978-1-80455-157-820231003

Deepthi, B., & Bansal, V. 2023. Applications of Artificial Intelligence (AI) in the Tourism Industry: A Futuristic Perspective. In Impact of Industry 4.0 on Sustainable Tourism: Perspectives, Challenges and Future (pp. 31-43). Emerald Publishing Limited.

Dehghani, M., Abubakar, A. M., & Pashna, M. (2022). Market-driven management of start-ups: The case of wearable technology. *Applied computing and informatics*, *18*(1/2), 45-60. doi:10.1016/j.aci.2018.11.002

Dehghani, M., Kim, J., & Lee, S. (2022). The projected growth of the wearable technology industry: A comprehensive analysis. *Journal of Wearable Technology*, *8*(1), 32–47.

Deloitte. (2008) *Medical Tourism. Consumers in Search of Value*. Deloitte Center for Health Solutions. https://www.globalwellnesssummit.com/wp-content/uploads/Industry-Research/Americas/2008-deloitte-medical-tourism-consumers.pdf (Access: 2 Jan 2024).

Deloitte. (2018). *A journey towards smart health: The impact of digitalization on medical tourist experience*. Deloitte.

Dilsizian, S. E., & Siegel, E. L. (2014). Artificial intelligence in medicine and cardiac imaging: Harnessing big data and advanced computing to provide personalized medical diagnosis and treatment. *Current Cardiology Reports*, *16*(1), 441. doi:10.1007/s11886-013-0441-8 PMID:24338557

Dimitrov, D. V. (2016). Medical internet of things and big data in healthcare. *Healthcare Informatics Research*, *22*(3), 156–163. doi:10.4258/hir.2016.22.3.156 PMID:27525156

Ding, H., Wu, J., Zhao, W., Matinlinna, J. P., Burrow, M. F., & Tsoi, J. K. H. (2023). Artificial Intelligence in dentistry – A review. *Frontiers in Dental Medicine*, *4*, 1085251. doi:10.3389/fdmed.2023.1085251

Dini, M., & Pencarelli, T. (2021). Wellness tourism and the components of its offer system: A holistic perspective. *Tourism Review*, *77*(2), 394–412. doi:10.1108/TR-08-2020-0373

Diwakar, M., Singh, P., & Shankar, A. (2021). Multi-modal medical image fusion framework using co-occurrence filter and local extrema in NSST domain. *Biomedical Signal Processing and Control*, *68*, 102788. doi:10.1016/j.bspc.2021.102788

Doborjeh, Z., Hemmington, N., Doborjeh, M., & Kasabov, N. (2022). Artificial intelligence: A systematic review of methods and applications in hospitality and tourism. *International Journal of Contemporary Hospitality Management*, *34*(3), 1154–1176. doi:10.1108/IJCHM-06-2021-0767

Dong, L., Yang, Q., Zhang, R. H., & Wei, W. B. (2021). Artificial intelligence for the detection of age-related macular degeneration in colour fundus photographs: A systematic review and meta-analysis. *EClinicalMedicine*, *35*, 100875. doi:10.1016/j.eclinm.2021.100875 PMID:34027334

Donkor, E. S. (2018). Stroke in the century: A snapshot of the burden, epidemiology, and quality of life. *Stroke Research and Treatment*, 2018. PMID:30598741

Downs, G. W., & Mohr, L. B. (1976). Conceptual Issues in the Study of Innovations. *Administrative Science Quarterly*, *21*(4), 700–714. doi:10.2307/2391725

Drigas, A., & Ioannidou, R. (2012). Artificial Intelligence in Special Education: A Decade Review. *International Journal of Engineering Education*, *28*(6), 1366–1372.

Dryglas, D., & Lubowiecki-Vikuk, A. (2019). Image of Poland as perceived by German and British medical tourists. *Tourism Review*, *74*(4), 861–871. doi:10.1108/TR-07-2018-0105

Dryglas, D., & Lubowiecki-Vikuk, A. (2019). The Attractiveness of Poland as a Medical Tourism Destination from the Perspective of German and British Consumers. *Entrepreneurial Business and Economics Review*, *7*(2), 45–62. doi:10.15678/EBER.2019.070203

Dubai Health Experience. (2022). *About DXH*. DXH. https://dxh.ae/en- US/about.

Ebrahim, A. H., & Ganguli, S. (2019). A comparative analysis of medical tourism competitiveness of India, Thailand and Singapore. *Tourism: An International Interdisciplinary Journal*, *67*(2), 102–115.

Ehrbeck, T., Guevara, C., & Mango, P. D. (2008). Mapping the market for medical travel. *The McKinsey Quarterly*, 11.

Eissler, L. A., & Casken, J. (2013). Seeking health care through international medical tourism. *Journal of Nursing Scholarship*, *45*(2), 177–184. doi:10.1111/jnu.12014 PMID:23489601

Elbagoury, B. M., Vladareanu, L., Vlădăreanu, V., Salem, A. B., Travediu, A. M., & Roushdy, M. I. (2023). A Hybrid Stacked CNN and Residual Feedback GMDH-LSTM Deep Learning Model for Stroke Prediction Applied on Mobile AI Smart Hospital Platform. *Sensors (Basel)*, *23*(7), 3500. doi:10.3390/s23073500 PMID:37050561

Elemento, O., Leslie, C., Lundin, J., & Tourassi, G. (2021). Artificial intelligence in cancer research, diagnosis, and therapy. *Nature Reviews. Cancer*, *21*(12), 747–752. doi:10.1038/s41568-021-00399-1 PMID:34535775

Elkhwesky, Z., Castañeda-García, J. A., Abuelhassan, A. E., & Tag-Eldeen, A. (2023). A systematic and critical review of restaurants' business performance: Future directions for theory and practice. *Tourism and Hospitality Research*, *23*(3), 441–458. doi:10.1177/14673584221104983

Ellethy, H., Chandra, S. S., & Nasrallah, F. A. (2022). Deep neural networks predict the need for CT in pediatric mild traumatic brain injury: A corroboration of the PECARN rule. *Journal of the American College of Radiology*, *19*(6), 769–778. doi:10.1016/j.jacr.2022.02.024 PMID:35381190

Enderwick, P., & Nagar, S. (2011). The competitive challenge of emerging markets: The case of medical tourism. *International Journal of Emerging Markets*, *6*(4), 329–350. doi:10.1108/17468801111170347

Esteva, A., Kuprel, B., Novoa, R. A., Ko, J., Swetter, S. M., Blau, H. M., & Thrun, S. (2017). Dermatologist-level classification of skin cancer with deep neural networks. *Nature*, *542*(7639), 115–118. doi:10.1038/nature21056 PMID:28117445

Fada, R., Farhadi Babadi, N., Azimi, R., Karimian, M., & Shahgholi, M. (2021). Mechanical properties improvement and bone regeneration of calcium phosphate bone cement, Polymethyl methacrylate and glass ionomer. *Journal of Nanoanalysis*, *8*(1), 60–79.

Farhadi, S., Afzal, P., Boveiri Konari, M., Daneshvar Saein, L., & Sadeghi, B. (2022). Combination of Machine Learning Algorithms with Concentration-Area Fractal Method for Soil Geochemical Anomaly Detection in Sediment-Hosted Irankuh Pb-Zn Deposit, Central Iran. *Minerals (Basel)*, *12*(6), 689. doi:10.3390/min12060689

Farooq, S., Farooq, B., Basheer, S., & Walia, S. (2023). *Balancing Environmental Sustainability and Privacy Ethical Dilemmas in AI-Enabled Smart Cities*., doi:10.4018/979-8-3693-0892-9.ch013

Fazlollahi, M., Morovvati, M. R., & Mollaei Dariani, B. (2019). Theoretical, numerical and experimental investigation of hydro- mechanical deep drawing of steel/polymer/steel sandwich sheets. *Proceedings of the Institution of Mechanical Engineers. Part B, Journal of Engineering Manufacture*, *233*(5), 1529–1546. doi:10.1177/0954405418780173

Fernandez, E. I., Ferreira, A. S., Cecilio, M. H. M., Cheles, D. S., de Souza, R. C. M., Nogueira, M. F. G., & Rocha, J. C. (2020). Artificial intelligence in the IVF laboratory: Overview through the application of different types of algorithms for the classification of reproductive data. *Journal of Assisted Reproduction and Genetics*, *37*(10), 372359–372376. doi:10.1007/s10815-020-01881-9 PMID:32654105

Filieri, R., D'Amico, E., Destefanis, A., Paolucci, E., & Raguseo, E. (2021). Artificial intelligence (A.I.) for tourism: A European-based study on successful A.I. tourism start-ups. *International Journal of Contemporary Hospitality Management*, *33*(11), 4099–4125. doi:10.1108/IJCHM-02-2021-0220

Fiszman, M., Chapman, W. W., Aronsky, D., Evans, R. S., & Haug, P. J. (2000). Automatic detection of acute bacterial pneumonia from chest X-ray reports. *Journal of the American Medical Informatics Association : JAMIA*, *7*(6), 593–604. doi:10.1136/jamia.2000.0070593 PMID:11062233

Frost & Sullivan. (2017). *Future of smart hospitals – concept definition, application, and growth opportunities, K1FF-01*. Frost & Sullivan.

Frost & Sullivan. (2017). Future of smart hospitals – concept definition, application, and growth opportunities, K1FF-01. Frost & Sullivan. https://store.frost.com/future-of-smart-hospitals.html

Gada, N. (2017). Four technological advancements that could change the medical tourism landscape as we know it. *Medical Travel Today*. https://medicaltraveltoday.com/four-technological-advancements-that-could-change-the- medicaltourism-landscape-as-we-know-it/

Gajić, T., Minasyan, L. A., Petrović, M. D., Bakhtin, V. A., Kaneeva, A. V., & Wiegel, N. L. (2023). Travelers'(in) Resilience to Environmental Risks Emphasized in the Media and Their Redirecting to Medical Destinations: Enhancing Sustainability. *Sustainability (Basel)*, *15*(21), 15297. doi:10.3390/su152115297

Ganguli, S., & Ebrahim, A. H. (2017). A qualitative analysis of Singapore's medical tourism competitiveness. *Tourism Management Perspectives*, *21*, 74–84. doi:10.1016/j.tmp.2016.12.002

García-Madurga, M. Á., & Grilló-Méndez, A. J. (2023). Artificial Intelligence in the Tourism Industry: An Overview of Reviews. *Administrative Sciences*, *13*(8), 172. doi:10.3390/admsci13080172

García-Pola, M., Pons-Fuster, E., Suárez-Fernández, C., Seoane-Romero, J., Romero-Méndez, A., & López-Jornct, P. (2021). Role of Artificial Intelligence in the Early Diagnosis of Oral Cancer. A Scoping Review. *Cancers (Basel)*, *13*(18), 4600. doi:10.3390/cancers13184600 PMID:34572831

Garg, R. (2021). *Infrastructure Development-An Indispensable Component of Tourism Destination Planning*. Research Gate. https://www.researchgate.net/publication/361306587

Garg, M., & Patil, H. S. (2021). *Comparative Study On AI in Mental Health and Wellbeing-Current Applications and Trends*. International Research Journal Of Engineering And Technology.

Garg, S. R., & B. A. (2012). Indian medical tourism industry: Growth opportunities and challenges. *Multi Disciplinary Edu Global Quest, 1*(1), 115–135.

Gärtner, E., Pirinen, A., & Sminchisescu, C. (2020, April). Deep reinforcement learning for active human pose estimation. *Proceedings of the AAAI Conference on Artificial Intelligence, 34*(07), 10835–10844. doi:10.1609/aaai.v34i07.6714

Gaur, L., Afaq, A., Singh, G., & Dwivedi, Y. K. (2021). Role of artificial intelligence and robotics to foster the touchless travel during a pandemic: A review and research agenda. *International Journal of Contemporary Hospitality Management, 33*(11), 4079–4098. doi:10.1108/IJCHM-11-2020-1246

GAUTAM, P., & BHATTA, K. (2020). Medical Tourism in India: Possibilities and problems of alternative medical treatment. *International Journal of Health Management and Tourism, 5*(3), 181–207.

Genç, R. (2023). Risks and benefits of medical tourism in terms of quality of life: Physical, economic, psychological and social aspects. In Handbook of Tourism and Quality-of-Life Research II: Enhancing the Lives of Tourists, Residents of Host Communities and Service Providers (pp. 181-196). Springer International Publishing.

Ghadirinejad, N., Nejad, M. G., & Alsaadi, N. (2021). A fuzzy logic model and a neuro-fuzzy system development on supercritical CO2 regeneration of Ni/Al2O3 catalysts. *Journal of CO2 Utilization, 54,* 101706.

Ghasemi, M., & Hasanpoor, E. (2021). *COVID-19 attack on the world's tourism industry.* Evidence Based Health Policy, Management and Economics.

Ghasemi, M., Nejad, M. G., & Aghaei, I. (2021). Knowledge management orientation and operational performance relationship in medical tourism (overview of the model performance in the COVID- 19 pandemic and post-pandemic era). *Health Services Management Research, 34*(4), 208–222. doi:10.1177/0951484820971438 PMID:33183088

Ghasemi, M., Nejad, M. G., Alsaadi, N., Abdel-Jaber, M. T., Yajid, A., Shukri, M., & Habib, M. (2022). Performance measurment and lead- time reduction in epc project-based organizations: A mathematical modeling approach. *Mathematical Problems in Engineering, 2022*, 2022. doi:10.1155/2022/5767356

Ghasemi, P., Mehdiabadi, A., Spulbar, C., & Birau, R. (2021). Ranking of sustainable medical tourism destinations in Iran: An integrated approach using fuzzy SWARA-PROMETHEE. *Sustainability (Basel), 13*(2), 683. doi:10.3390/su13020683

Ghassemi, M., Nejad, M. G., & Aghaei, I. (2020). Knowledge management orientation and operational performance relationship in medical tourism (overview of the model performance in the COVID-19 pandemic and post-pandemic era). *Health Services Management Research*, 1–15. PMID:33183088

Ghosh, T., & Mandal, S. (2019). Medical Tourism Experience: Conceptualization, Scale Development, and Validation. *Journal of Travel Research*, *58*(8), 1288–1301. doi:10.1177/0047287518813469

Ghubaish, A., Salman, T., Zolanvari, M., Unal, D., Al-Ali, A., & Jain, R. (2020). Recent advances in the internet-of-medical-things (IoMT) systems security. *IEEE Internet of Things Journal*, *8*(11), 8707–8718. doi:10.1109/JIOT.2020.3045653

Gillies, R. J., Kinahan, P. E., & Hricak, H. (2016). Radiomics: Images are more than pictures, they are data. *Radiology*, *278*(1), 563–577. doi:10.1148/radiol.2015151169 PMID:26579733

Glynda, J. D., Bernie, G., & Leanne, M. C. (2014). Integrating mobile devices into nursing curricula: Opportunities for implementation using Rogers' Diffusion of Innovation model. *Nurse Education Today*, *34* (5), 775-782. doi:10.1016/j.nedt.2013.10.021

Goel, P. (2019). *The power of digital health in medical tourism*. LinkedIn. https://www.linkedin.com/pulse/power-digital-health-medical-tourism-pramod-goel

Goel, P. (2019). *The power of digital health in medical tourism*. LinkedIn. https://www.linkedin.com/pulse/power-digital-health-medical-tourism-pramod-goel/

Gohary, A., Pourazizi, L., Madani, F., & Chan, E. Y. (2020). Examining Iranian tourists' memorable experiences on destination satisfaction and behavioral intentions. In Current Issues in Tourism, 23(2). Routledge. doi:10.1080/13683500.2018.1560397

González-Rodríguez, M. R., Díaz-Fernández, M. C., & Pino-Mejías, M. Á. (2020a). The impact of virtual reality technology on tourists' experience: A textual data analysis. *Soft Computing*, *24*(18), 13879–13892. doi:10.1007/s00500-020-04883-y

Goodrich, J. N., & Goodrich, G. E. (1987). Healthcare tourism: An exploratory study. *Tourism Management*, *8*(3), 217–222. doi:10.1016/0261-5177(87)90053-7

Gopalan, N. (2023). Ethical and Regulatory Gaps in Aesthetic Medical Practice in Top Asian Medical Tourism Destinations. *Asian Bioethics Review*, 1–29. doi:10.1007/s41649-023-00267-0 PMID:38213989

Görener, A. A. (2016). SWOT-AHP approach for assessment of the medical tourism sector in Türkiye. *Alphanumeric Journal*, *4*(2), 159–170. doi:10.17093/alphanumeric.277740

Gray, D. E. (2013). *Doing Research in The Real World* (3rd ed.). Sage.

Grespan, L., Fiorini, P., & Colucci, G. (2019). *The route to patient safety in robotic surgery*. Springer International Publishing. doi:10.1007/978-3-030-03020-9

Gretzel, U. (2011). Intelligent systems in tourism: A social science perspective. *Annals of Tourism Research, 38*(3), 757–779. doi:10.1016/j.annals.2011.04.014

Gretzel, U., Fuchs, M., Baggio, R., Hoepken, W., Law, R., Neidhardt, J., Pesonen, J., Zanker, M., & Xiang, Z. (2020). e-Tourism beyond COVID-19: A call for transformative research. *Information Technology & Tourism, 22*(2), 187–203. doi:10.1007/s40558-020-00181-3

Gulshan, V., Peng, L., Coram, M., Stumpe, M. C., Wu, D., Narayanaswamy, A., Venugopalan, S., Widner, K., Madams, T., Cuadros, J., Kim, R., Raman, R., Nelson, P. C., Mega, J. L., & Webster, D. R. (2016). Development and validation of a deep learning algorithm for detection of diabetic retinopathy in retinal fundus photographs. *Journal of the American Medical Association, 316*(22), 2402–2410. doi:10.1001/jama.2016.17216 PMID:27898976

Gursoy, D., Malodia, S., & Dhir, A. (2022). The metaverse in the hospitality and tourism industry: An overview of current trends and future research directions. *Journal of Hospitality Marketing & Management, 31*(5), 527–534. doi:10.1080/19368623.2022.2072504

Guru, S., Sinha, A., & Kautish, P. (2022). Determinants of medical tourism: Application of Fuzzy Analytical Hierarchical Process. *International Journal of Emerging Markets, 18*(11), 4819–4842. doi:10.1108/IJOEM-08-2021-1173

Gusenbauer, M., & Haddaway, N. R. (2020). Which academic search systems are suitable for systematic reviews or meta-analyses? Evaluating retrieval qualities of Google Scholar, PubMed, and 26 other resources. *Research Synthesis Methods, 11*(2), 181–217. doi:10.1002/jrsm.1378 PMID:31614060

Guz, A. N., & Rushchitsky, J. J. (2009). Scopus: A system for the evaluation of scientific journals. *International Applied Mechanics, 45*(4), 351–362. doi:10.1007/s10778-009-0189-4

Guzman, B. MS, Metzger, I., MS, Aphinyanaphongs, Y., D., M., D., P., Grover, H., & D, P. (2020). *Assessment of Amazon Comprehend Medical: Medication Information Extraction*. http://arxiv.org/abs/2002.00481

Gyu Ko, T. (2011). Medical tourism system model. *International Journal of Tourism Sciences, 11*(1), 17–51. https://sci-hub.se/10.1080/15980634.2011.11434634. doi:10.1080/15980634.2011.11434634

Habibi, A., Mousavi, M., Jamali, S. M., & Ale Ebrahim, N. (2022). A bibliometric study of medical tourism. *Anatolia, 33*(3), 415–425. doi:10.1080/13032917.2021.1954042

Habuza, T., Navaz, A. N., Hashim, F., Alnajjar, F., Zaki, N., Serhani, M. A., & Statsenko, Y. (2021). AI applications in robotics, diagnostic image analysis and precision medicine: Current limitations, future trends, guidelines on CAD systems for medicine. *Informatics in Medicine Unlocked, 24*, 100596. doi:10.1016/j.imu.2021.100596

Hadi, A. (2009). Globalization, medical tourism and health equity. *Implications of Medical Tourism for Canadian Health and Health Policy, 13*, 1–29.

Hadian, M., Jabbari, A., Mousavi, S. H., & Sheikhbardsiri, H. (2021). Medical tourism development: A systematic review of economic aspects. *International Journal of Healthcare Management, 14*(2), 576–582. https://sci-hub.se/10.1080/20479700.2019.1677977. doi:10.1080/20479700.2019.1677977

Hafeez, K., Keary, K., & Hanneman, R. (2006). e-Business Capabilities Model Validation and Comparison between Adopter and Non-Adopter to e-Business Companies in the U.K. *Journal of Manufacturing Technology Management, 17*(6), 806–828. doi:10.1108/17410380610678819

Hafizullah, D., & Kirti, K. (2023). Smart Healthcare System (SHS): Medical Tourism Delivering, Consumption, and Elevating Tool in the Ages of Smart Technologies. *Tourism Planning & Development, 20*(3), 397–415. doi:10.1080/21568316.2022.2109206

Haleem, A., Javaid, M., Singh, R. P., & Suman, R. (2021). Telemedicine for healthcare: Capabilities, features, barriers, and applications. *Sensors International, 2*, 100117. doi:10.1016/j.sintl.2021.100117 PMID:34806053

Hamdoun, S., Monteleone, R., Bookman, T., & Michael, K. (2023). AI-based and digital mental health apps: Balancing need and risk. *IEEE Technology and Society Magazine, 42*(1), 25–36. doi:10.1109/MTS.2023.3241309

Han, H. (2013). The healthcare hotel: Distinctive attributes for international medical travelers. *Tourism Management, 36*, 257–268. doi:10.1016/j.tourman.2012.11.016

Han, H., & Hyun, S. S. (2014). Medical hotel in the growth of global medical tourism. *Journal of Travel & Tourism Marketing, 31*(3), 366–380. https://sci-hub.se/10.1080/10548408.2013.876955. doi:10.1080/10548408.2013.876955

Han, H., & Hyun, S. S. (2015). Customer retention in the medical tourism industry: Impact of quality, satisfaction, trust, and price reasonableness. *Tourism Management, 46*, 20–29. doi:10.1016/j.tourman.2014.06.003

Haq, A. U., Li, J. P., Khan, S., Alshara, M. A., Alotaibi, R. M., & Mawuli, C. (2022). DACBT: Deep learning approach for classification of brain tumors using MRI data in IoT healthcare environment. *Scientific Reports, 12*(1), 15331. doi:10.1038/s41598-022-19465-1 PMID:36097024

Hartman, A. E. (2016). Transparency of hope: Ethical issues in marketing cross border reproductive care: A multi-region content analysis of clinic websites. *Marketing Intelligence & Planning, 34*(7), 943–963. doi:10.1108/MIP-06-2015-0121

Hartwell, H. (2014). Medical Tourism: The Ethics, Regulation and Marketing of Health Mobility. *Tourism Management, 45*, 1–2. doi:10.1016/j.tourman.2014.04.001

Hartwell, H., Fyall, A., Willis, C., Page, S., Ladkin, A., & Hemingway, A. (2018). Progress in tourism and destination wellbeing research. *Current Issues in Tourism, 21*(16), 1830–1892. https://sci-hub.se/10.1080/13683500.2016.1223609. doi:10.1080/13683500.2016.1223609

Hassan, V. (n.d.). *The role of pro-poor tourism in the development of the local economy: A case study of Tripoli*. Research Gate. https://www.researchgate.net/publication/375869901

Hassan, V., & Bellos, G. (2022). COVID-19: Reshaping Medical Tourism through Artificial Intelligence (AI) and Robotics. *Athens Journal of Tourism*, 77-98.

Hassan, V., & Bellos, G. (2022). COVID-19: Reshaping Medical Tourism through Artificial Intelligence (AI) and Robotics. *Athens Journal of Tourism*, 77-98. doi:https://www.athensjournals.gr/tourism/2022-9-2-2-Hassan.pdf

Hassan, V., & Noaman, S. (2017). *Relation between Tourism and Health: Case Study AIDS in Lebanon*. Atiner Conference Paper Series, No: TOU2017-2262. https://www.atiner.gr/paper-series.

Hassan, V., & Zrnić, M. (2021). Slow tourism: a niche market towards sustainability of moonlight farm house "barouk area". *Turističko poslovanje*, (28), 57-63. https://www.aseestant.ceon.rs/index.php/turpos/article/view/34456/19278

Hassan, V., & Bellos, G. (2022). COVID-19: Reshaping Medical Tourism through Artificial Intelligence (A.I.) and Robotics. *Athens Journal of Tourism*, *9*(2), 77–98. doi:10.30958/ajt.9-2-2

Hassan, V., & Noaman, S. (2017). Relation between Tourism and Health: Case Study AIDS in Lebanon. In *Atiner Conference Paper Series*.

Hazarika, I. (2010a). Medical tourism: Its potential impact on the health workforce and health systems in India. *Health Policy and Planning*, *25*(3), 248–251. doi:10.1093/heapol/czp050 PMID:19926658

Health-tourism.com. (2019). Medical tourism statistics and facts. *Health Tourism.* https://www.health-tourism.com/medical-tourism/statistics/

Heires, K. (2016). Rise of the Robots. *RM Magazine.* https://www.rmmagazine.com/articles/article/2016/09/01/-rise-of-the-robots-

Heung, V. C. S., Küçükusta, D., & Song, H. (2011). Medical tourism development in Hong Kong: An assessment of the barriers. *Tourism Management*, *32*(5), 995–1005. doi:10.1016/j.tourman.2010.08.012

Heung, V. C., Kucukusta, D., & Song, H. (2010). A conceptual model of medical tourism: Implications for future research. *Journal of Travel & Tourism Marketing*, *27*(3), 236–251. doi:10.1080/10548401003744677

Heydari Fard, M., Sanayei, A., & Ansari, A. (2021a). Determinants of Medical Tourists' Revisit and Recommend Intention. *International Journal of Hospitality & Tourism Administration*, *22*(4), 429–454. doi:10.1080/15256480.2019.1650688

Higgins, O., Short, B. L., Chalup, S. K., & Wilson, R. L. (2023). Artificial intelligence (AI) and machine learning (ML) based decision support systems in mental health: An integrative review. *International Journal of Mental Health Nursing*, *32*(4), 966–978. doi:10.1111/inm.13114 PMID:36744684

Hjalager, A. M. (2009). Innovations in travel medicine and the progress of Tourism-Selected narratives. *Technovation*, *29*(9), 596–601. doi:10.1016/j.technovation.2009.05.012

Hjalager, A. M., & Flagestad, A. (2012). Innovations in well-being tourism in the Nordic countries. *Current Issues in Tourism*, *15*(8), 725–740. doi:10.1080/13683500.2011.629720

Ho, D. (2020). Artificial intelligence in cancer therapy. *Science*, *367*(6481), 982–983. doi:10.1126/science.aaz3023 PMID:32108102

Hofer, S., & Stampfli, K. (2008). *Rehabilitationskliniken – Wohin führt die Zukunft*, working paper Zurich University of Applied Sciences, Zürich.

Hofer, S., Honegger, F., & Hubeli, J. (2012). Health tourism: Definition focused on the Swiss market and conceptualization of health(i)ness. *Journal of Health Organization and Management*, *26*(1), 60–80. doi:10.1108/14777261211211098 PMID:22524099

Hollon, T. C., Pandian, B., Adapa, A. R., Urias, E., Save, A. V., Khalsa, S. S. S., Eichberg, D. G., D'Amico, R. S., Farooq, Z. U., Lewis, S., Petridis, P. D., Marie, T., Shah, A. H., Garton, H. J. L., Maher, C. O., Heth, J. A., McKean, E. L., Sullivan, S. E., Hervey-Jumper, S. L., & Orringer, D. A. (2020). Near real-time intraoperative brain tumor diagnosis using stimulated Raman histology and deep neural networks. *Nature Medicine*, *26*(1), 52–58. doi:10.1038/s41591-019-0715-9 PMID:31907460

Hongal, P., & Kshirsagar, Y. I. (2023). Healthcare management in India: Issues, challenges and prospects. *International Journal of Engineering and Management Research*, *13*(2), 73–82. doi:10.31033/ijemr.13.2.11

Hopkins, L., Labonté, R., Runnels, V., & Packer, C. (2021a). *Medical tourism today: what is the state of The United Arab Emirates Government*. Vision 2021. at: https://u.ae/en/more/uae-future/2021

Hopkins, L., Labonté, R., Runnels, V., & Packer, C. (2010). Medical tourism today: What is the state of existing knowledge? *Journal of Public Health Policy*, *31*(2), 185–198. doi:10.1057/jphp.2010.10 PMID:20535101

Horne, B. D., Nevo, D., Adali, S., Manikonda, L., & Arrington, C. (2020). Tailoring heuristics and timing AI interventions for supporting news veracity assessments. *Computers in Human Behavior Reports*, *2*, 100043. doi:10.1016/j.chbr.2020.100043

Horsfall, D. (2020). Medical tourism from the UK to Poland: How the market masks migration. *Journal of Ethnic and Migration Studies*, *46*(20), 4211–4229. doi:10.1080/1369183X.2019.1597470

Hosain, N., Amin, F., Rehman, S., & Koirala, B. (2017). Know thy neighbors: The status of cardiac surgery in the South Asian countries around India. *Indian Heart Journal*, *69*(6), 790–796. doi:10.1016/j.ihj.2017.09.219 PMID:29174261

Howze, K. S. (2006). Medical tourism: Symptom or cure. *Ga. L. Rev.*, *41*, 1013.

Huang, S., & Hsu, C. H. C. (2009). Effects of travel motivation, past experience, perceived constraint, and attitude on revisit intention. *Journal of Travel Research*, *48*(1), 29–44. doi:10.1177/0047287508328793

Huang, S., Yang, J., Fong, S., & Zhao, Q. (2019). Artificial intelligence in cancer diagnosis and prognosis: Opportunities and challenges. *Cancer Letters*, *471*, 61–71. doi:10.1016/j.canlet.2019.12.007 PMID:31830558

Hussain, A., Malik, A., Halim, M. U., & Ali, A. M. (2014). The use of robotics in surgery: A review. *International Journal of Clinical Practice*, *68*(11), 1376–1382. doi:10.1111/ijcp.12492 PMID:25283250

Hwang, S., Lee, D. & Kang, C.Y. (2018). Medical tourism: focusing on patients' prior, current, and post experience. *International journal of quality innovation, 4*(1), pp.1-22.

Hwang, H. W., Park, J. H., Moon, J. H., Yu, Y., Kim, H., Her, S. B., Srinivasan, G., Aljanabi, M. N. A., Donatelli, R. E., & Lee, S.-J. (2020). Automated identification of cephalometric landmarks: Part 2-might it be better than human? *The Angle Orthodontist*, *90*(1), 69–76. doi:10.2319/022019-129.1 PMID:31335162

India Brand Equity Foundation. (2023). *Healthcare Industry in India*. IBEF.

Iorio, M. L., Verma, K., Ashktorab, S., & Davison, S. P. (2014). Medical tourism in plastic surgery: Ethical guidelines and practice standards for perioperative care. *Aesthetic Plastic Surgery*, *38*(3), 602–607. doi:10.1007/s00266-014-0322-6 PMID:24797678

Ishita, S., & Sachin, S. (2022). Blockchain Enabled Biometric Security in Intemet-of-Medical-Things (IoMT) Devices. *International Conference on Augmented Intelligence and Sustainable Systems (ICAISS)*, (pp. 971-979). IEEE.

Itchhaporia, D. (2022). Artificial intelligence in cardiology. *Trends in Cardiovascular Medicine*, *32*(1), 34–41. doi:10.1016/j.tcm.2020.11.007 PMID:33242635

Ivanov, S., & Webster, C. (n.d.). *Robots, artificial intelligence, and service automation in travel, tourism and hospitality*. Emerald Publishing Limited. doi:10.1108/9781787566873

Jain, V., & Ajmera, P. (2018). Modelling the factors affecting Indian medical tourism sector using interpretive structural modeling. *Benchmarking*, *25*(5), 1461–1479. doi:10.1108/BIJ-03-2017-0045

Jaiswal, P., & Bhirud, S. (2021). Study and analysis of an approach towards the classification of tooth wear in dentistry using machine learning technique. *IEEE International conference on technology, research, and innovation for the betterment of society (TRIBES)*. IEEE. 10.1109/TRIBES52498.2021.9751650

Jamin, A., Rahmafitria, F., & Nurazizah, G. R. (2020, December). *Rebuilding health tourism destination image after Covid-19: The case of Malaysia And Indonesia. In 2020 IEEE 8th R10 Humanitarian Technology Conference (R10-HTC)*. IEEE. doi:10.1109/R10-HTC49770.2020.9357057

Jamin, T., Smith, R., & Patel, N. (2020). Mental health challenges in medical tourism: A review of anxiety and depression prevalence among medical tourists. *Journal of Medical Tourism*, *12*(3), 145–162.

Jangra, P., & Gupta, M. (2021). Internet of Medical Things (IoMT) Based Framework for Smart Healthcare Tourism Sector. *International Journal of Hospitality and Tourism Systems*, 14.

Javed, A. R., Ahmed, W., Pandya, S., Maddikunta, P. K. R., Alazab, M., & Gadekallu, T. R. (2023). A Survey of Explainable Artificial Intelligence for Smart Cities. In Electronics (Switzerland), 12(4). MDPI. doi:10.3390/electronics12041020

Jha, S., & Topol, E. J. (2016). Adapting to Artificial Intelligence: Radiologists and pathologists as information specialists. *Journal of the American Medical Association*, *316*(22), 2353–2354. doi:10.1001/jama.2016.17438 PMID:27898975

Jiang, F., Jiang, Y., Zhi, H., Dong, Y., Li, H., Ma, S., Wang, Y., Dong, Q., Shen, H., & Wang, Y. (2017). Artificial Intelligence in Healthcare: Past, Present and Future. *Stroke and Vascular Neurology*, *2*(4), 230–243. doi:10.1136/svn-2017-000101 PMID:29507784

Jiang, L., Wu, H., & Song, Y. (2022). Diversified demand for health tourism matters: From a perspective of the intra-industry trade. *Social Science & Medicine*, *293*, 114630. doi:10.1016/j.socscimed.2021.114630 PMID:34954675

Johnston, R., Crooks, V. A., Adams, K., Snyder, J., & Kingsbury, P. (2011). An industry perspective on Canadian patients' involvement in Medical Tourism: Implications for public health. *BMC Public Health*, *11*(1), 416. doi:10.1186/1471-2458-11-416 PMID:21627830

Johnston, R., Crooks, V. A., & Snyder, J. (2012). "I didn't even know what I was looking for": A qualitative study of the decision-making processes of Canadian medical tourists. *Globalization and Health*, *8*(1), 1–12. doi:10.1186/1744-8603-8-23 PMID:22769723

Johnston, R., Crooks, V. A., Snyder, J., & Kingsbury, P. (2010). What is known about the effects of medical tourism in destination and departure countries? A scoping review. *International Journal for Equity in Health*, *9*(1), 1–13. doi:10.1186/1475-9276-9-24 PMID:21047433

Joshi, N. (2019, June). How Far Are We From Achieving Artificial General Intelligence? *Forbes*, *10*. https://www.forbes.com/sites/cognitiveworld/2019/06/10/how-far-are-wefrom-achieving-artificial-general-intelligence/#8f678606dc4d

Junata, M., & Tong, R. (2018). Wearable technology in medicine and health care: introduction. In R. Tong (Ed.), *Wearable Technology in Medicine and Health Care* (pp. 1–5). Academic Press. doi:10.1016/B978-0-12-811810-8.00001-4

Kagiyama, N., Shrestha, S., Farjo, P. D., & Sengupta, P. P. (2019). Artificial intelligence: Practical primer for clinical research in cardiovascular disease. *Journal of the American Heart Association*, *8*(17), e012788. doi:10.1161/JAHA.119.012788 PMID:31450991

Kamarian, S., Bodaghi, M., Isfahani, R. B., & Song, J. I. (2022). *A comparison between the effects of shape memory alloys and carbon.*

Kandel, M. E., Rubessa, M., He, Y. R., Schreiber, S., Meyers, S., Matter Naves, L., Sermersheim, M. K., Sell, G. S., Szewczyk, M. J., Sobh, N., Wheeler, M. B., & Popescu, G. (2020). Reproductive outcomes predicted by phase imaging with computational specificity of spermatozoon ultrastructure. *Proceedings of the National Academy of Sciences of the United States of America*, *117*(31), 11718302–11718309. doi:10.1073/pnas.2001754117 PMID:32690677

Kangping, W., Haiyan, K., Naipeng, B., Honggen, X., Xuejie, Q., & Jinghan, L. (2022). A.I. in health tourism: Developing a measurement scale. *Asia Pacific Journal of Tourism Research*, *27*(9), 954–966. doi:10.1080/10941665.2022.2142620

Kansal, V., Ranjan, R., Sinha, S., Tiwari, R., & Wickramasinghe, N. (2021). *Healthcare and Knowledge Management for Society 5.0: Trends, Issues, and Innovations*. CRC Press. doi:10.1201/9781003168638

Karatzia, L., Aung, N., & Aksentijevic, D. (2022). Artificial intelligence in cardiology: Hope for the future and power for the present. *Frontiers in Cardiovascular Medicine*, *9*, 945726. doi:10.3389/fcvm.2022.945726 PMID:36312266

Kashihara, H., Nakayama, T., Hatta, T., Takahashi, N., & Fujita, M. (2016). Evaluating the quality of website information of private-practice clinics offering cell therapies in Japan. *Interactive Journal of Medical Research*, *5*(2), e5479. doi:10.2196/ijmr.5479 PMID:27222494

Kaul, V., Enslin, S., & Gross, S. A. (2020). History of artificial intelligence in medicine. *Gastrointestinal Endoscopy*, *92*(4), 807–812. doi:10.1016/j.gie.2020.06.040 PMID:32565184

Kaushal, V., & Srivastava, V. (2021). Hospitality and tourism industry amid COVID-19 pandemic: Perspectives on challenges and learnings from India. *International Journal of Hospitality Management*, *92*, 102707. doi:10.1016/j.ijhm.2020.102707 PMID:33024348

Kazemha, A. S., & Dehkordi, K. S. (2017). The Principles of Designing Hospital Hotel with the Approach of Health Tourism in Kish Island. *Tarih Kültür ve Sanat Arastirmalari Dergisi*, *6*(1), 515–531. doi:10.7596/taksad.v6i1.758

Kelly, J. T., Campbell, K. L., Gong, E., & Scuffffham, P. (2020). The internet of things: Impact and implications for health care delivery. *Journal of Medical Internet Research*, *22*(11), e20135. doi:10.2196/20135 PMID:33170132

Keng, J. (2020). *Autonomous Diabetic Retinopathy Screening System Gains Approval*. American Academy of Ophthalmology.

Khaksarshahmirzadi, E., Saner, T., Attaeyan, A., Abdollahi, S., Mohammadi, M., & Heydari, S. (2022). Evaluation of the use of robotic surgery with remote simulation for use in the field of Medical Tourism and Pro-Tourism development. *Journal of Simulation and Analysis of Novel Technologies in Mechanical Engineering*, *14*(1), 47–59. 20.1001.1.27834441.2022.14.1.5.5

Khan, A. A., Laghari, A. A., & Awan, S. A. (2021). Machine learning in computer vision: A review. *EAI Endorsed Scal Inf Syst*, *8*(32), e4. doi:10.4108/eai.21-4-2021.169418

Khandan, A., & Esmaeili, S. (2019). Fabrication of polycaprolactone and polylactic acid shapeless scaffolds via fused deposition modelling technology. *Journal of Advanced Materials and Processing*, *7*(4), 16–29.

Khan, M. J., Chelliah, S., & Haron, M. S. (2016). Medical Tourism Destination Image Formation Process: A Conceptual Model. *International Journal of Healthcare Management*, *9*(2), 134–143. doi:10.1080/20479700.2016.1142046

Khedkar, S., Gandhi, P., Shinde, G., & Subramanian, V. (2020). Deep learning and explainable AI in healthcare using EHR. In Deep learning techniques for biomedical and health informatics (pp. 129-148). doi:10.1007/978-3-030-33966-1_7

Kılavuz, E. (2018). Medical tourism competition: The case of Türkiye. *International Journal of Health Management and Tourism*, *3*(1), 42–58. doi:10.31201/ijhmt.372364

Kim, E. H., Kim, S., Kim, H. J., Jeong, H. O., Lee, J., Jang, J., Joo, J.-Y., Shin, Y., Kang, J., Park, A. K., Lee, J.-Y., & Lee, S. (2020). Prediction of chronic periodontitis severity using machine learning models based on salivary bacterial copy number. *Frontiers in Cellular and Infection Microbiology*, *10*, 698. doi:10.3389/fcimb.2020.571515 PMID:33304856

Kim, H. L., & Hyun, S. S. (2022). The future of medical tourism for individuals' health and well-being: A case study of the relationship improvement between the UAE (United Arab Emirates) and South Korea. *International Journal of Environmental Research and Public Health*, *19*(9), 5735. doi:10.3390/ijerph19095735 PMID:35565130

Kim, M., Koo, D. W., Shin, D. J., & Lee, S. M. (2017). From Servicescape to loyalty in the medical tourism industry: A medical Clinic's service perspective. *Inquiry*, *54*, 0046958017746546. doi:10.1177/0046958017746546 PMID:29233057

Kim, W. O. N. J. O. N. G., & Oh, J. (2021). A Study on the Antecedents of Revisit Intention for Korean Medical Tourism: Focusing on Chinese Medical Tourists. *ICIC Express Lett. Part B Appl. Int. J. Res. Surv*, *12*, 741–750. doi:10.24507/icicelb.12.08.741

Kiss, K. (2015). The challenges of developing health tourism in the Balkans. *Tourism: An International Interdisciplinary Journal*, *63*(1), 97–110.

Klebanoff, L. M., & Safdieh, J. E. (2023). Modern Neurology Training Is Failing Outpatients. *JAMA Neurology*, *80*(4), 333–334. doi:10.1001/jamaneurol.2022.5490 PMID:36848066

Kolker, E., Özdemir, V., & Kolker, E. (2016). How Healthcare can refocus on its Super-Customers (Patients, n =1) and Customers (Doctors and Nurses) by Leveraging Lessons from Amazon, Uber, and Watson. *OMICS: A Journal of Integrative Biology*, *20*(1), 329–333. doi:10.1089/omi.2016.0077 PMID:27310474

Kong, X. T., Luo, H., Huang, G. Q., & Yang, X. (2019). Industrial wearable system: The human-centric empowering technology in Industry 4.0. *Journal of Intelligent Manufacturing*, *30*(8), 2853–2869. doi:10.1007/s10845-018-1416-9

Koutsouleris, N., Hauser, T. U., Skvortsova, V., & De Choudhury, M. (2022). From promise to practice: Towards the realisation of AI-informed mental health care. *The Lancet. Digital Health*, *4*(11), e829–e840. doi:10.1016/S2589-7500(22)00153-4 PMID:36229346

Krishnamurthi, R., Gopinathan, D., & Kumar, A. (2021). Wearable devices and COVID-19: state of the art, framework, and challenges. *Emerging Technologies for Battling Covid-19: Applications and Innovations*, (pp. 157-180). Springer. doi:10.1007/978-3-030-60039-6_8

Ku, E. C. S., & Der Chen, C. (2015a). Cultivating travellers' revisit intention to e-tourism service: The moderating effect of website interactivity. *Behaviour & Information Technology*, *34*(5), 465–478. doi:10.1080/0144929X.2014.978376

Kumar, S., Kumar, V., & Attri, K. (2021). Impact of artificial intelligence and service robots in tourism and hospitality sector: current use & future trends. *Administrative Development: A Journal of HIPA, 8*, 59-83.

Kumar, Y., & Mahajan, M. (2020). Recent advancement of machine learning and deep learning in the field of healthcare system. *Computational intelligence for machine learning and healthcare informatics, 1*, 77.

Kumar, A. (2023). *A Critical Evaluation of Medical Tourism in India: Growth*. Drivers and Challenges.

Kumar, P., Sharma, S. K., & Dutot, V. (2023). Artificial intelligence (AI)-enabled CRM capability in healthcare: The impact on service innovation. *International Journal of Information Management*, *69*, 102598. doi:10.1016/j.ijinfomgt.2022.102598

Kumar, S. (2009). Designing promotional strategies for medical tourism in India: A case study of an ophthalmic hospital in NCR. *Health and Population; Perspectives and Issues*, *32*(2), 86–95.

Kumar, S. S., Muthukumaran, V., Devi, A., Geetha, V., & Yadav, P. N. (2023). A Quantitative Approach of Purposive Sampling Techniques for Security and Privacy Issues in IoT Healthcare Applications. In *Handbook of Research on Advancements in AI and IoT Convergence Technologies* (pp. 281–299). IGI Global. doi:10.4018/978-1-6684-6971-2.ch016

Kumar, Y., Koul, A., Singla, R., & Ijaz, M. F. (2023). Artificial intelligence in disease diagnosis: A systematic literature review, synthesizing framework and future research agenda. *Journal of Ambient Intelligence and Humanized Computing*, *14*(7), 8459–8486. doi:10.1007/s12652-021-03612-z PMID:35039756

Kusunose, K. (2021). Steps to use artificial intelligence in echocardiography. *Journal of Echocardiography*, *19*(1), 21–27. doi:10.1007/s12574-020-00496-4 PMID:33044715

Kwon, H., An, S., Lee, H. Y., Cha, W. C., Kim, C., Cho, M., & Kong, H. J. (2022). Review of smart hospital services in real healthcare environments. *Healthcare Informatics Research*, *28*(1), 3–15. doi:10.4258/hir.2022.28.1.3 PMID:35172086

Kye, B., Han, N., Kim, E., Park, Y., & Jo, S. (2021). Educational applications of metaverse: Possibilities and limitations. In Journal of Educational Evaluation for Health Professions, 18. doi:10.3352/jeehp.2021.18.32

Lai, T., Shi, Y., Du, Z., Wu, J., Fu, K., Dou, Y., & Wang, Z. (2023). Psy-LLM: Scaling up Global Mental Health Psychological Services with AI-based Large Language Models. *arXiv preprint arXiv*:2307.11991.

Lai, M. C., Brian, M., & Mamzer, M. F. (2020). Perceptions of artificial intelligence in Healthcare: Findings from a qualitative survey study among actors in France. *Journal of Translational Medicine*, *18*(1), 14. doi:10.1186/s12967-019-02204-y PMID:31918710

Lamba, D. (2023). *Revolutionizing Medical Tourism with Artificial Intelligence*. Invest India. https://www.investindia.gov.in/team-india-blogs/transforming-medical-tourism-harnessing-power-artificial-intelligence

Lanfranco, A. R., Castellanos, A. E., Desai, J. P., & Meyers, W. C. (2004). Robotic surgery: A current perspective. *Annals of Surgery*, *239*(1), 14–21. doi:10.1097/01.sla.0000103020.19595.7d PMID:14685095

Lareyre, F., Behrendt, C. A., Chaudhuri, A., Lee, R., Carrier, M., Adam, C., Lê, C. D., & Raffort, J. (2023). Applications of artificial intelligence for patients with peripheral artery disease. *Journal of Vascular Surgery*, *77*(2), 650–658. doi:10.1016/j.jvs.2022.07.160 PMID:35921995

Laws, E. (1996). Health tourism: A business opportunity approach. In S. Clift & S. J. Page (Eds.), *Health and the international tourist* (pp. 199–214). Routledge.

Learning, M. U. S. M. (2020). Obstructive sleep apnea: a prediction model using supervised machine learning method. *The Importance of Health Informatics in Public Health during a Pandemic, 272*, 387.

Lee, C. (2006). Medical tourism, an innovative opportunity for entrepreneurs. *Journal of Asia Entrepreneurship and Sustainability*, *3*(1), 3–5.

Lee, E. E., Torous, J., De Choudhury, M., Depp, C. A., Graham, S. A., Kim, H. C., Paulus, M. P., Krystal, J. H., & Jeste, D. V. (2021). Artificial intelligence for mental health care: Clinical applications, barriers, facilitators, and artificial wisdom. *Biological Psychiatry: Cognitive Neuroscience and Neuroimaging*, *6*(9), 856–864. doi:10.1016/j.bpsc.2021.02.001 PMID:33571718

Lew, A. A., Cheer, J. M., Haywood, M., Brouder, P., & Salazar, N. B. (2020). Visions of travel and tourism after the global COVID-19 transformation of 2020. *Tourism Geographies*, *22*(3), 455–466. doi:10.1080/14616688.2020.1770326

Liang, F., Wang, S., Zhang, K., Liu, T. J., & Li, J. N. (2022). Development of artificial intelligence technology in diagnosis, treatment, and prognosis of colorectal cancer. *World Journal of Gastrointestinal Oncology*, *14*(1), 124–152. doi:10.4251/wjgo.v14.i1.124 PMID:35116107

Liaqat, S., Akhunzada, A., Shaikh, F. S., Giannetsos, A., & Jan, M. A. (2020). SDN orchestration to combat evolving cyber threats in Internet of Medical Things (IoMT). *Computer Communications*, *160*, 697–705. doi:10.1016/j.comcom.2020.07.006

Li, J., & Carayon, P. (2021). Health Care 4.0: A vision for smart and connected health care. *IISE Transactions on Healthcare Systems Engineering*, *11*(3), 171–180. doi:10.1080/24725579.2021.1884627 PMID:34497970

Li, M., Yin, D., Qiu, H., & Bai, B. (2021). A systematic review of AI technology-based service encounters: Implications for hospitality and tourism operations. *International Journal of Hospitality Management*, *95*, 102930. doi:10.1016/j.ijhm.2021.102930

Lim, J. E. (2020). Understanding the discrimination experienced by customers with disabilities in the tourism and hospitality industry: The case of Seoul in South Korea. *Sustainability (Basel)*, *12*(18), 7328. doi:10.3390/su12187328

Li, R. C., Asch, S. M., & Shah, N. H. (2020). Developing a delivery science for artificial intelligence in healthcare. *NPJ Digital Medicine*, *3*(1), 107. doi:10.1038/s41746-020-00318-y PMID:32885053

Liu, L., Wu, Y., Wei, W., Cao, W., Sahin, S., & Zhang, Q. (2018, July). Benchmarking deep learning frameworks: Design considerations, metrics and beyond. In *2018 IEEE 38th International Conference on Distributed Computing Systems (ICDCS)* (pp. 1258-1269). IEEE.

Liu, Y., Xia, S., Nie, J., Wei, P., Shu, Z., Chang, J. A., & Jiang, X. (2022). aimse: Toward an ai-based online mental status examination. *IEEE Pervasive Computing*, *21*(4), 46–54. doi:10.1109/MPRV.2022.3172419

Lixia, Z., & Zhongli, S. (2019). The Application of Artificial Intelligence Technology in the Tourism Industry of Jinan. *Journal of Physics: Conference Series*, *1302*(3), 032005. doi:10.1088/1742-6596/1302/3/032005

Li, Y., Han, Q., & Wen, T. (2022). Effect of Frugality and Cognition on Forest Health Tourism Intention–A Mediating Effect Analysis Based on Multigroup Comparison. *Frontiers in Psychology*, *13*, 844628. doi:10.3389/fpsyg.2022.844628 PMID:35572268

Li, Y., & Wen, T. (2022). Impact of cognition and social trust on forest-based health tourism intention during COVID-19. *Sustainability (Basel)*, *15*(1), 714. doi:10.3390/su15010714

Lo, C. (2020). Medical tourism: How is digital tech reshaping the industry? Retrieved from: https://medical-technology.nridigital.com/medical_ technology_jan20/medical_tourism_how_is_digital_tech_reshaping_the_industry

Lobo, F. (2020). *Industry 4.0 – manufacturing and the future of medical things.*

Lobo, F. (2020). Industry 4.0 – manufacturing and the future of medical things. www.asianhhm.com/technology-equipment/industry-manufacturing-future-medical-things

Locsin, R. C., & Ito, H. (2018). Can humanoid nurse robots replace human nurses. *Journal of Nursing (Luton, England)*, *5*(1), 1–6. doi:10.7243/2056-9157-5-1

Long, E., Lin, H., Liu, Z., Wu, X., Wang, L., Jiang, J., & Liu, Y. (2017). An artificial intelligence platform for the multihospital collaborative management of congenital cataracts. *Nature biomedical engineering, 1*(2), 0024.

Lucchini, R., Carnelli, D., Gastaldi, D., Shahgholi, M., Contro, R., & Vena, P. (2012). A damage model to simulate nanoindentation tests of lamellar bone at multiple penetration depth. In *6th European Congress on Computational Methods in Applied Sciences and Engineering, ECCOMAS 2012* (pp. 5919-5924). IEEE.

Luchini, C., Pea, A., & Scarpa, A. (2022). Artificial intelligence in oncology: Current applications and future perspectives. *British Journal of Cancer, 126*(1), 4–9. doi:10.1038/s41416-021-01633-1 PMID:34837074

Lu, H., Li, Y., Chen, M., Kim, H., & Serikawa, S. (2018). Brain intelligence: Go beyond artificial intelligence. *Mobile Networks and Applications, 23*(2), 368–375. doi:10.1007/s11036-017-0932-8

Lunt, N., Smith, R. D., Exworthy, M., Green, S. T., Horsfall, D. G., & Mannion, R. (2011). *Medical tourism: treatments, markets, and health system implications: a scoping review.*

Lunt, N. (2022). Medical tourism. In *Key Concepts in Medical Sociology* (3rd ed., pp. 222–227). SAGE. https://books.google.rs/books?hl=en&lr=&id=j2RXEAAAQBAJ&oi=fnd&pg=PA222&dq=Lunt,+N.+(2022).+Medical+tourism.+In+Key+Concepts+in+Medical+Sociology+(Third+Edition)+(pp.+222-227).+SAGE.&ots=-p5n9eElsM&sig=j55l81Eq7M63jjIWHLAgH8CG8bY&redir_esc=y#v=onepage&q&f=false

Lunt, N., & Carrera, P. (2010). Medical tourism: Assessing the evidence on treatment abroad. *Maturitas, 66*(1), 27–32. doi:10.1016/j.maturitas.2010.01.017 PMID:20185254

Lüthje, C., & Herstatt, C. (2004). The Lead User Method: An Outline of Empirical Findings and Issues for Future Research. *R & D Management, 35*(5), 553–568. doi:10.1111/j.1467-9310.2004.00362.x

Luxton, D. D. (2016). An introduction to artificial intelligence in behavioral and mental health care. In D. D. Luxton (Ed.), *Artificial Intelligence in Behavioral and Mental Health Care* (pp. 1–26). Academic Press. doi:10.1016/B978-0-12-420248-1.00001-5

Magaña-Espinoza, P., Aquino-Santos, R., Cárdenas-Benítez, N., Aguilar-Velasco, J., Buenrostro-Segura, C., Edwards-Block, A., & Medina-Cass, A. (2014). WiSPH: A wireless sensor network-based home care m. *Sensors (Basel), 14*(4), 7096–7119. doi:10.3390/s140407096 PMID:24759112

Ma, H., Chiu, Y. H., Tian, X., Zhang, J., & Guo, Q. (2020). Safety or travel: Which is more important? The impact of disaster events on tourism. *Sustainability (Basel), 12*(7), 3038. doi:10.3390/su12073038

Mahjoory, M., Shahgholi, M., & Karimipour, A. (2022). The effects of initial temperature and pressure on the mechanical properties of reinforced calcium phosphate cement with magnesium nanoparticles: A molecular dynamics approach. *International Communications in Heat and Mass Transfer, 135*, 106067. doi:10.1016/j.icheatmasstransfer.2022.106067

Majeed, S., Zhou, Z., Lu, C., & Ramkissoon, H. (2020). Online tourism information and tourist behavior: A structural equation modeling analysis based on a self-administered survey. *Frontiers in Psychology*, *11*, 599. doi:10.3389/fpsyg.2020.00599 PMID:32373008

Majeed, S., Zhou, Z., & Ramkissoon, H. (2020). Beauty and elegance: Value co-creation in cosmetic surgery tourism. *SAGE Open*, *10*(2), 2158244020932538. doi:10.1177/2158244020932538

Majersik, J. J., Ahmed, A., Chen, I. H. A., Shill, H., Hanes, G. P., Pelak, V. S., Hopp, J. L., Omuro, A., Kluger, B., & Leslie-Mazwi, T. (2021). A shortage of neurologists–we must act now: A report from the AAN 2019 Transforming Leaders Program. *Neurology*, *96*(24), 1122–1134. doi:10.1212/WNL.0000000000012111 PMID:33931527

Malekzadeh, R., Abedi, G., Abedini, E., Haghgoshayie, E., Hasanpoor, E., & Ghasemi, M. (2021). Ethical predictability of patient safety in Iranian hospitals. *International Journal of Risk & Safety in Medicine*, *32*(1), 51–60. doi:10.3233/JRS-200022 PMID:32597821

Manhas, P. S. (2015). Marketing analysis of medical tourism in India. *Enlightening Tourism. A Pathmaking Journal, 5*(1), 1-39.

Manjunath, N. M., & Panda, K. (2022). Evolution of Cancer Patient Health Care in Odisha. *Indian Journal of Surgical Oncology*, *13*(1), 81–82. doi:10.1007/s13193-022-01661-0 PMID:36691494

Mannini, A., Trojaniello, D., Cereatti, A., & Sabatini, A. M. (2016). A machine learning framework for gait classification using inertial sensors: Application to elderly, post-stroke and huntington's disease patients. *Sensors (Basel)*, *16*(1), 134. doi:10.3390/s16010134 PMID:26805847

Manyika, J., Chui, M., Miremadi, M., Bughin, J., George, K., Willmott, P., & Dewhurst, M. (2017). A future that works: AI, automation, employment, and productivity. McKinsey Global Institute Research. *Tech. Rep*, *60*, 1–135.

Marescaux, J., Leroy, J., Rubino, F., Smith, M., Vix, M., Simone, M., & Mutter, D. (2002). Mutter Transcontinental robot-assisted remote telesurgery: Feasibility and potential applications. *Annals of Surgery*, *235*(4), 300–301. doi:10.1097/00000658-200204000-00005 PMID:11923603

Masri, N. W., You, J. J., Ruangkanjanases, A., Chen, S. C., & Pan, C. I. (2020). Assessing the effects of information system quality and relationship quality on continuance intention in e-tourism. *International Journal of Environmental Research and Public Health*, *17*(1), 174. Advance online publication. doi:10.3390/ijerph17010174 PMID:31881762

MayoClinic.org. (2023). *Five ways artificial intelligence promises to transform organ transplant.* Medical Professionals – Transplant Medicine. https://www.mayoclinic.org/medical-professionals/transplant-medicine/news/five-ways-artificial-intelligence-promises-to-transform-organ-transplant/mac-20548221.

Mbunge, E., & Batani, J. (2023). Application of deep learning and machine learning models to improve healthcare in sub-Saharan Africa: Emerging opportunities, trends and implications. *Telematics and Informatics Reports*, 100097.

Mbunge, E., & Muchemwa, B. (2022). Towards emotive sensory Web in virtual health care: Trends, technologies, challenges and ethical issues. *Sensors International*, *3*, 100134. doi:10.1016/j.sintl.2021.100134

McNeil, R. (2019). Coming to terms with AI. *Eye News*, *26*(2), 6–14.

Medhekar, A., & Wong, H. Y. (2020). Medical travellers' perspective on factors affecting medical tourism to India. *Asia Pacific Journal of Tourism Research*, *25*(12), 1295–1310. doi:10.1080/10941665.2020.1837893

Medical Technology. (2020). *Medical tourism: how is digital tech reshaping the industry?* Medical Technology.

Medical Technology. (2020). Medical tourism: how is digital tech reshaping the industry? *Medical Technology*. https://medicaltechnology.nridigital.com/medical_technology_jan20/medical_tourism_how_is_digital_ tech_reshaping_the_industry

Medical Tourism Magazine (2020). Trends in healthcare digital revolution. *Medical Tourism Magazine*.

Medical Tourism Magazine. (2020). Trends in healthcare digital revolution. *Medical Tourism Magazine*. www.magazine.medicaltourism.com/article/trends-healthcare-digital-revolution.

Medical Tourism Magazine. (2021). Will medical tourism survive the coronavirus pandemic? *Medical Tourism Magazine*.

Medical Tourism Magazine. (2021). Will medical tourism survive the coronavirus pandemic? *Medical Tourism Magazine*. https://www.magazine.medicaltourism.com/article/will-medical-tourism survive-the-coronavirus-pandemic

Medical Tourism. (2022). *MTI Overall Ranking*. Medical Tourism. https://www.medicaltourism.com/mti/home

Meghani, Z. (2011). A Robust, Particularist Ethical Assessment of Medical Tourism. *Developing World Bioethics*, *11*(1), 16–29. doi:10.1111/j.1471-8847.2010.00282.x PMID:20491867

Mehraj, D., Qureshi, I. H., Singh, G., Nazir, N. A., basheer, S., & Nissa, V. (2023, December). basheer, S., & Nissa, V. u. (2023). Green marketing practices and green consumer behavior: Demographic differences among young consumers. *Business Strategy & Development*, *6*(4), 571–585. doi:10.1002/bsd2.263

Mehraj, D., Ul Islam, M. I., Qureshi, I. H., Basheer, S., Baba, M. M., Nissa, V., & Asif Shah, M. (2023). Factors affecting entrepreneurial intention for sustainable tourism among the students of higher education institutions. *Cogent Business & Management*, *10*(3), 2256484. doi:10.1080/23311975.2023.2256484

Miller, T. P., Li, Y., Getz, K. D., Dudley, J., Burrows, E., Pennington, J., Ibrahimova, A., Fisher, B. T., Bagatell, R., Seif, A. E., Grundmeier, R., & Aplenc, R. (2017). Using electronic medical record data to report laboratory adverse events. *British Journal of Haematology*, *177*(2), 283–286. doi:10.1111/bjh.14538 PMID:28146330

Mintz, Y., & Brodie, R. (2019). Introduction to artificial intelligence in medicine. *Minimally Invasive Therapy & Allied Technologies*, *28*(2), 73–81. doi:10.1080/13645706.2019.1575882 PMID:30810430

Mishra, S., Kant Rai, S., Pandey, A., & Jaiswal, V. (2013). An implicit effect of the health insurance portability and accountability act on medical tourism in India. In *UTMS Journal of Economics, 4*(3). http://hdl.handle.net/10419/105295

Mishra, P., & Singh, G. (2023). Internet of medical things healthcare for sustainable smart cities: Current status and future prospects. *Applied Sciences (Basel, Switzerland)*, *13*(15), 8869. doi:10.3390/app13158869

Mital, S., & Nguyen, H. V. (2022). Cost-effectiveness of using artificial intelligence versus polygenic risk score to guide breast cancer screening. *BMC Cancer*, *22*(1), 501. doi:10.1186/s12885-022-09613-1 PMID:35524200

Mittal, S., & Hasija, Y. (2020). Applications of deep learning in healthcare and biomedicine. *Deep learning techniques for biomedical and health informatics*, 57-77.

Moghavvemi, S., Ormond, M., Musa, G., Isa, C. R., Thirumoorthi, T., Mustapha, M., & Chandy, J. J. (2017). Connecting with prospective medical tourists online: A cross-sectional analysis of private hospital websites promoting medical tourism in India, Malaysia and Thailand. *Tourism Management*, *58*, 154–163. doi:10.1016/j.tourman.2016.10.010

Monaco, S., & Sacchi, G. (2023). Travelling the Metaverse: Potential Benefits and Main Challenges for Tourism Sectors and Research Applications. *Sustainability (Basel)*, *15*(4), 3348. doi:10.3390/su15043348

Monfared, R. M., Ayatollahi, M. R., & Isfahani, R. B. (2018). Synergistic effects of hybrid MWCNT/nanosilica on the tensile and tribological properties of woven carbon fabric epoxy composites. *Theoretical and Applied Fracture Mechanics*, *96*, 272–284. doi:10.1016/j.tafmec.2018.05.007

Monteiro, M., Newcombe, V. F., Mathieu, F., Adatia, K., Kamnitsas, K., Ferrante, E., Das, T., Whitehouse, D., Rueckert, D., Menon, D. K., & Glocker, B. (2020). Multiclass semantic segmentation and quantification of traumatic brain injury lesions on head CT using deep learning: An algorithm development and multicentre validation study. *The Lancet. Digital Health*, *2*(6), e314–e322. doi:10.1016/S2589-7500(20)30085-6 PMID:33328125

Morales, J. R., Nativi-Nicolau, J., Jang, J., Patel, P., Yip, D., Leoni-Moreno, J., & Goswami, R. (2022). Artificial intelligence 12 lead ECG based heart age estimation and 1-year outcomes after heart transplantation. *The Journal of Heart and Lung Transplantation*, *41*(4), S213. doi:10.1016/j.healun.2022.01.1671

Mordor Intelligence. (2024). *India's Medical Tourism Market*. Mordor Intelligence.

Morovvati, M. R., Mollaei-Dariani, B., & Haddadzadeh, M. (2010). Initial blank optimization in multilayer deep drawing process using GONNS. *Journal of Manufacturing Science and Engineering*, *132*(6), 061014. doi:10.1115/1.4003121

Mortezaei, G., Otaghvar, H. A., Vazifehdoost, H., Saeedi, P., & Pegheh, A. (2021). Prioritization of Brand Equity Components Using an Integrated Structural Equation Modelling and Fuzzy AHP. *Tehnički glasnik, 15*(3), 357-365. doi:10.31803/tg-20210410214620

Mouridsen, K., Thurner, P., & Zaharchuk, G. (2020). Artificial intelligence applications in stroke. *Stroke*, *51*(8), 2573–2579. doi:10.1161/STROKEAHA.119.027479 PMID:32693750

Mueller, H., & Kaufmann, E. L. (2001). Wellness tourism: Market analysis of a special health tourism segment and implications for the hotel industry. *Journal of Vacation Marketing*, *7*(1), 5–17. doi:10.1177/135676670100700101

Murff, H. J., FitzHenry, F., Matheny, M. E., Gentry, N., Kotter, K. L., Crimin, K., Dittus, R. S., Rosen, A. K., Elkin, P. L., Brown, S. H., & Speroff, T. (2011). Automated identification of postoperative complications within an electronic medical record using natural language processing. *Journal of the American Medical Association*, *306*(8), 848–855. doi:10.1001/jama.2011.1204 PMID:21862746

Mursch-Edlmayr, A. S., Ng, W. S., Diniz-Filho, A., Sousa, D. C., Arnould, L., Schlenker, M. B., Duenas-Angeles, K., Keane, P. A., Crowston, J. G., & Jayaram, H. (2020). Artificial intelligence algorithms to diagnose glaucoma and detect glaucoma progression: Translation to clinical practice. *Translational Vision Science & Technology*, *9*(2), 55–55. doi:10.1167/tvst.9.2.55 PMID:33117612

Musa, G., Thirumoorthi, T., & Doshi, D. (2012). Travel behavior among inbound medical tourists in Kuala Lumpur. *Current Issues in Tourism*, *15*(6), 525–543. doi:10.1080/13683500.2011.626847

Nagar, R., Singh, Y., & Jaglan, V., & Meenakshu (2021). A review on machine learning applications in medical tourism. In *2021 Fourth International Conference on Computational Intelligence and Communication Technologies (CCICT)* (pp. 208-215). IEEE. 10.1109/CCICT53244.2021.00048

Natasha Spencer-Jolliffe. (2023). *AI prompts wave of medical tourism*. Medical Device Network. https://www.medicaldevice-network.com/features/ai-prompts-wave-of-medical-tourism/

Navaz, A. N., Serhani, M. A., El Kassabi, H. T., Al-Qirim, N., & Ismail, H. (2021). Trends, technologies, and key challenges in smart and connected healthcare. *IEEE Access : Practical Innovations, Open Solutions*, *9*, 74044–74067. doi:10.1109/ACCESS.2021.3079217 PMID:34812394

Nazir, S., Ali, Y., Ullah, N., & García-Magariño, I. (2019). Internet of things for healthcare using effects of mobile computing: A systematic literature review. *Wireless Communications and Mobile Computing*, *2019*, 1–20. doi:10.1155/2019/5931315

NetHope. (2022). *How IoT Helps Transplant Surgeons Track Organ Shipments*. NetHope. https://www.networkworld.com/article/3098879/how-iot-helps-transplant-surgeons-track-organ-shipments.html.

Ngobeni, C. C. L. (2020). *Identifying the push and pull factors of a medical tourism destination* [Doctoral dissertation, University of Pretoria]. http://hdl.handle.net/2263/77916

Nilashi, M., Samad, S., Manaf, A. A., Ahmadi, H., Rashid, T. A., Munshi, A., & Ahmed, O. H. (2019). Factors influencing medical tourism adoption in Malaysia: A DEMATEL-Fuzzy TOPSIS approach. *Computers & Industrial Engineering*, *137*, 106005. doi:10.1016/j.cie.2019.106005

Nilsson, N. (2009). *The Quest for Artificial Intelligence: A History of Ideas and Achievement*. Cambridge University Press. doi:10.1017/CBO9780511819346

Ning, Z., Dong, P., Wang, X., Hu, X., Guo, L., Hu, B., & Kwok, R. Y. (2020). Mobile edge computing enabled 5G health monitoring for Internet of medical things: A decentralized game theoretic approach. *IEEE Journal on Selected Areas in Communications*, *1*(2), 463–478. doi:10.1109/JSAC.2020.3020645

Nitski, O., Azhie, A., Qazi-Arisar, F. A., Wang, X., Ma, S., Lilly, L., Watt, K. D., Levitsky, J., Asrani, S. K., Lee, D. S., Rubin, B. B., Bhat, M., & Wang, B. (2021). Long-term Mortality Risk Stratification of Liver Transplant Recipients: Real-Time Application of Deep Learning Algorithms on Longitudinal Data. *The Lancet. Digital Health*, *3*(5), e295–e305. doi:10.1016/S2589-7500(21)00040-6 PMID:33858815

Nkurunziza, J. M., Udahemuka, J. C., Dusenge, J. B., & Umutesi, F. (2022). Overview of Trending Medical Technologies. *Global Clinical Engineering Journal*, *16*(46), 6–46. doi:10.31354/globalce.v4i3.142

Noaman, S., & Montargot, N. (2023). Understanding Interactions Between Providers of Cosmetic Tourism and Tourists: The Lebanon Experience. *Tourism, Culture & Communication*, *23*(1), 45–60. doi:10.3727/109830422X16510695152055

Noh, H.-K. (2022). Metaverse-related Issues in Tourism Using News Big Data. *Journal of Tourism and Leisure Research*, *34*(2), 151–166. doi:10.31336/JTLR.2022.2.34.2.151

Norgeot, B., Quer, G., Beaulieu-Jones, B. K., Torkamani, A., Dias, R., Gianfrancesco, M., Arnaout, R., Kohane, I. S., Saria, S., Topol, E., Obermeyer, Z., Yu, B., & Butte, A. J. (2020). Minimum information about clinical artificial intelligence modeling: The MI-CLAIM checklist. *Nature Medicine*, *26*(9), 1320–1324. doi:10.1038/s41591-020-1041-y PMID:32908275

Oberoi, S., & Kansra, P. (2019). Factors influencing medical tourism in India: A critical review. *SAMVAD*, *17*, 9–16. doi:10.53739/samvad%2F2019%2Fv17%2F140676

Oberoi, S., Kansra, P., & Gupta, S. L. (2022). A Systematic Review on Determinants Inciting Sustainable E-Medical Tourism. *International Journal of Reliable and Quality E-Healthcare*, *11*(2), 1–13. doi:10.4018/IJRQEH.299962

Ogink, P. T., Groot, O. Q., Karhade, A. V., Bongers, M. E. R., Oner, F. C., Verlaan, J. J., & Schwab, J. H. (2021). Wide range of applications for machine-learning prediction models in orthopedic surgical outcome: A systematic review. *Acta Orthopaedica*, *92*(5), 526–531. doi:10.1080/17453674.2021.1932928 PMID:34109892

Oleribe, O. O., Momoh, J., Uzochukwu, B. S., Mbofana, F., Adebiyi, A., Barbera, T., Williams, R., & Taylor-Robinson, S. D. (2019). Identifying key challenges facing healthcare systems in Africa and potential solutions. *International Journal of General Medicine*, *12*, 395–403. doi:10.2147/IJGM.S223882 PMID:31819592

Oleribe, O., Momoh, J., & Uzochukwu, B. (2019). Mental health care deficiencies in medical tourism: Challenges and opportunities for improvement. *International Journal of Mental Health Systems*, *13*(1), 45–60.

Oltean, F. D., Gabor, M. R., Stăncioiu, A. F., Kardos, M., Kiss, M., & Marinescu, R. C. (2020). Aspects of marketing in dental tourism—Factor of sustainable development in Romania. *Sustainability (Basel)*, *12*(10), 4320. doi:10.3390/su12104320

Omankwu, O. C., Osodeke, E. C., & Ubah, V. I. (2023). Artificial Intelligence a Paradigm Shift in Healthcare Sector Past, Present and Future Prospects. *Journal of Science and Technology Research*, *5*(3).

Ouyang, D., He, B., Ghorbani, A., Yuan, N., Ebinger, J., Langlotz, C. P., Heidenreich, P. A., Harrington, R. A., Liang, D. H., Ashley, E. A., & Zou, J. Y. (2020). Video-based AI for beat-to-beat assessment of cardiac function. *Nature*, *580*(7802), 252–256. doi:10.1038/s41586-020-2145-8 PMID:32269341

Padilla-Meléndez, A., & Del-Águila-Obra, A. R. (2016). Health Tourism. *Conceptual Framework and Insights from the Case of a Spanish Mature Destination Tourism & Management Studies*, *12*(1), 86–96.

Panda, P. K. P., Chavda, V., Neve, K., Mishra, S., Verma, S. K., & Ahuja, R. (2022). COVID-19: Lesson Learnt from Diagnostics to Therapeutics. In Microbial Engineering for Therapeutics (pp. 345-374). Singapore: Springer Nature Singapore. doi:10.1007/978-981-19-3979-2_16

Pandey, C., Sharma, S., & Matta, P. (2021). Body Sensor Network Architectures in Healthcare Internet-of-Things (HIoT): A Survey. *6th International Conference on Communication and Electronics Systems* (pp. 494-499). ICCES. 10.1109/ICCES51350.2021.9489205

Pang, Z., Tian, J., & Chen, Q. (2014). Intelligent packaging and intelligent medicine box for medication management towards the Internet-of-Things. *16th international conference on advanced communication technology* (pp. 352-360). IEEE.

Pan, T.-Z., & Chen, W. C. (2014). Chinese medical tourists- their perceptions of Taiwan. *Tourism Management*, *44*, 108–112. doi:10.1016/j.tourman.2014.02.008

Pantelis, A. G., Stravodimos, G. K., & Lapatsanis, D. P. (2021). A scoping review of artificial intelligence and machine learning in bariatric and metabolic surgery: Current status and future perspectives. *Obesity Surgery*, *31*(10), 4555–4563. doi:10.1007/s11695-021-05548-x PMID:34264433

Parekh, J., Jaffer, A., Bhanushali, U., & Shukla, S. (2021). Disintermediation in medical tourism through blockchain technology: An analysis using value-focused thinking approach. *Information Technology & Tourism*, *23*(1), 69–96. doi:10.1007/s40558-020-00180-4

Park, S. M., & Kim, Y. G. (2022). A Metaverse: Taxonomy, Components, Applications, and Open Challenges. *IEEE Access : Practical Innovations, Open Solutions*, *10*, 4209–4251. doi:10.1109/ACCESS.2021.3140175

Park, S., & Kim, S. (2022). Identifying World Types to Deliver Gameful Experiences for Sustainable Learning in the Metaverse. *Sustainability (Basel)*, *14*(3), 1361. doi:10.3390/su14031361

Parry, C. M., & Aneja, U. (2020). *Artificial Intelligence for Healthcare: Insights from India.*

Patel, R., Tseng, C. C., Choudhry, H. S., Lemdani, M. S., Talmor, G., & Paskhover, B. (2022). Applying machine learning to determine popular patient questions about mentoplasty on social media. *Aesthetic Plastic Surgery*, *46*(5), 2273–2279. doi:10.1007/s00266-022-02808-8 PMID:35201377

Paul, J., Alhassan, I., Binsaif, N., & Singh, P. (2023). Digital entrepreneurship research: A systematic review. *Journal of Business Research*, *156*, 113507. doi:10.1016/j.jbusres.2022.113507

Paul, J., & Rosado-Serrano, A. (2019). Gradual internationalization vs born-global/international new venture models: A review and research agenda. *International Marketing Review*, *36*(6), 830–858. doi:10.1108/IMR-10-2018-0280

Paul, S., & Kulshreshtha, S. K. (2019). *Global Developments in Healthcare and Medical Tourism.* IGI Global.

Penney, K., Snyder, J., Crooks, V. A., & Johnston, R. (2011). Risk communication and informed consent in the medical tourism industry: A thematic content analysis of Canadian broker websites. *BMC Medical Ethics*, *12*(1), 1–9. doi:10.1186/1472-6939-12-17 PMID:21943392

Peric, M., & Vanja, V. (2021). Tourism Getting Back to Life after COVID-19: Can Artificial Intelligence Help? *Societies (Basel, Switzerland)*, *11*(4), 115. Advance online publication. doi:10.3390/soc11040115

Perry, P. (2016). *How Artificial Intelligence will Revolutionize Healthcare.* Big Think. https://bigthink.com/philip-perry/how-artificial-intelligence-will-revolutionize-healthcare

Pesapane, F., Volonté, C., Codari, M., & Sardanelli, F. (2018). Artificial intelligence as a medical device in radiology: Ethical and regulatory issues in Europe and the United States. *Insights Into Imaging*, *9*(5), 745–753. doi:10.1007/s13244-018-0645-y PMID:30112675

Pessot, E., Spoladore, D., Zangiacomi, A., & Sacco, M. (2021). Natural resources in health tourism: A systematic literature review. *Sustainability (Basel), 13*(5), 2661. doi:10.3390/su13052661

Phuthong, T., Anuntavoranich, P., Chandrachai, A., & Piromsopa, K. (2022). Developing and validating an assessment scale to measure the competitiveness of wellness destinations. *Sustainability (Basel), 14*(7), 4152. doi:10.3390/su14074152

Pighin, M., Alvarez-Risco, A., Del-Aguila-Arcentales, S., Rojas-Osorio, M., & Yáñez, J. A. (2022). Factors of the Revisit Intention of Patients in the Primary Health Care System in Argentina. *Sustainability (Basel), 14*(20), 13021. doi:10.3390/su142013021

Podina, I. R., & Caculidis-Tudor, D. (2023). Increasing Well-Being and Mental Health Through Cutting-Edge Technology and Artificial Intelligence. In *Brain, Decision Making and Mental Health* (pp. 347–364). Springer International Publishing. doi:10.1007/978-3-031-15959-6_17

Podina, I., & Caculidis, C. (2023). Overcoming obstacles: Integrating wearables and AI in mental health therapy. *Journal of Digital Therapy, 6*(4), 189–204.

Popa, A. L., Ţarcă, N. N., Sasu, D. V., Bodog, S. A., Roşca, R. D., & Tarcza, T. M. (2022). Exploring Marketing Insights for Healthcare: Trends and Perspectives Based on Literature Investigation. *Sustainability (Basel), 14*(17), 10499. doi:10.3390/su141710499

Pradhan, B., Bhattacharyya, S., & Pal, K. (2021). IoT-based applications in healthcare devices. *Journal of Healthcare Engineering, 2021*, 1–18. PMID:33791084

Prakash, A. V., & Das, S. (2020). Intelligent conversational agents in mental healthcare services: A thematic analysis of user perceptions. *Pacific Asia Journal of the Association for Information Systems, 12*(2), 1–34. doi:10.17705/1thci.12201

Prawiyogi, A. G., Purnama, S., Meria, L., & Giri, A. (2022). Smart Cities Using Machine Learning and Intelligent Applications. [ITALIC]. *International Transactions on Artificial Intelligence, 1*(1), 102–116. doi:10.33050/italic.v1i1.204

Premarathne, U., Abuadbba, A., Alabdulatif, A., Khalil, I., Tari, Z., Zomaya, A., & Buyya, R. (2016). Hybrid cryptographic access control for cloud-based EHR systems. *IEEE Cloud Computing, 3*(4), 58–64. doi:10.1109/MCC.2016.76

Priporas, C. V., & Mylona, I. (2008). Mobile services: Potentiality of SMS as a new business communication tool in attracting consumers. *International Journal of Mobile Communications, 6*(4), 456–466. doi:10.1504/IJMC.2008.018053

Priya Rao, V., & Roy Choudhury, R. (2017). *A study of the factors influencing customer satisfaction in medical tourism in India*. IASET. www.iaset.us

Psiha, M. M., & Vlamos, P. (2017). IoT applications with 5G connectivity in medical tourism sector management: Third-party service scenarios. *Advances in Experimental Medicine and Biology, 989*, 141–154. doi:10.1007/978-3-319-57348-9_12 PMID:28971423

Punch, K. F. (2005). *Sosyal araştırmalara giriş*. Siyasal Kitabevi.

Pussakham, J. (2023). Social Sciences: Anthropology, Socio-Psychological and Cultural-Educational Motivations Impact of Promoting Sport Tourism in Thailand. *Boletin de Literatura Oral, 10*(1), 2265–2273.

PwC. (2021). *Can AI help us understand AI-powered healthcare?* PwC. https://www.pwc.com/m1/en/publications/strategic-foresight/sector-strategies/healthcare-intellectual-capital/can-ai-help-us-understand-ai-powered-healthcare.html. Accessed: 01/11/2023.

Qin, F., & Gu, J. (2023). Artificial intelligence in plastic surgery: Current developments and future perspectives. *Plastic and Aesthetic Research, 10*(3), 3. doi:10.20517/2347-9264.2022.72

Quazi, S., Saha, P. R., & Singh, M. K. (2022). Applications of Artificial Intelligence in Healthcare. *Journal of Experimental Biology and Agricultural Sciences, 10*(1), 211–226. doi:10.18006/2022.10(1).211.226

Rahman, I., Martin, D. S., & Liu, S. (2022). Outbound medical tourists: The interplay of perceived quality, length of stay, group-size, post-visit destination image and revisit intention. *PLoS One, 17*(5), e0267755. doi:10.1371/journal.pone.0267755 PMID:35536776

Rahman, M. M., Khatun, F., Sami, S. I., & Uzzaman, A. (2022). The evolving roles and impacts of 5G enabled technologies in healthcare: The world epidemic COVID-19 issues. *Array (New York, N.Y.), 14*, 100178. doi:10.1016/j.array.2022.100178 PMID:35571870

Rahman, R. (2019). The privatisation of healthcare system in Bangladesh. *International Journal of Health Care Quality Assurance, 32*(1), 97–107. doi:10.1108/IJHCQA-11-2017-0217 PMID:30859884

Rai, A., & Rai, A. (2019). Medical tourism: An introduction. *Medical Tourism in Kolkata, Eastern India*, 1-41.

Rai, A. (2019). *Medical Tourism in Kolkata, Eastern India*. Springer International Publishing., doi:10.1007/978-3-319-73272-5

Raja, A. (2023). *Reshaping medical tourism with AI*. India AI. https://indiaai.gov.in/article/reshaping-medical-tourism-with-ai

Rajpurkar, P., Irvin, J., Zhu, K., Yang, B., Mehta, H., Duan, T., & Ng, A. Y. (2017). Chexnet: Radiologist-level pneumonia detection on chest x-rays with deep learning. Doi: *arXiv preprint arXiv:1711.05225*.

Ramírez de Arellano, A. B. (2011). Medical tourism in the Caribbean. *Signs (Chicago, Ill.), 36*(2), 289–297. doi:10.1086/655908 PMID:21114073

Rathore, D. K., & Mannepalli, P. K. (2021). Recent Trends in Machine Learning for Health Care Sector. *International Journal of Innovative Research in Technology and Management, 5*(2).

Reddy, S., Fox, J., & Purohit, M. P. (2019). Artificial intelligence-enabled healthcare delivery. *Journal of the Royal Society of Medicine, 112*(1), 22–28. doi:10.1177/0141076818815510 PMID:30507284

Reddy, S., York, V. K., & Brannon, L. A. (2010). Travel for Treatment: Students' Perspective on Medical Tourism. *International Journal of Tourism Research*, *12*(5), 510–522. doi:10.1002/jtr.769

RejebA.KeoghJ. G.TreiblmaierH. (2019). The impact of blockchain on medical tourism. Web2019 workshop on e-business. doi:10.2139/ssrn.3521841

Rennick-Egglestone, S., Morgan, K., Llewellyn-Beardsley, J., Ramsay, A., McGranahan, R., Gillard, S., Hui, A., Ng, F., Schneider, J., Booth, S., Pinfold, V., Davidson, L., Franklin, D., Bradstreet, S., Arbour, S., & Slade, M. (2019). Mental health recovery narratives and their impact on recipients: Systematic review and narrative synthesis. *Canadian Journal of Psychiatry*, *64*(10), 669–679. doi:10.1177/0706743719846108 PMID:31046432

Revathi, J., & Rani, S. J. (2019). Analysis of Medical Tourism and its Economic Impact. *Indian Journal of Public Health Research & Development*, *10*(1), 21. doi:10.5958/0976-5506.2019.00006.8

Robertson, E. M., Moorman, S. W., & Korus, L. J. (2022). Why do canadians travel abroad for cosmetic surgery? A qualitative analysis on motivations for cosmetic surgery tourism. *Plastic Surgery (Oakville, Ont.)*, *30*(4), 353–359. doi:10.1177/22925503211019607 PMID:36212104

Rogers, K. (2023). Medical tourism. *Encyclopedia Britannica*. https://www.britannica.com/science/medical-tourism.

Rogers, E. (1995). *Diffusion of innovations*. The Free Press.

Rohokale, V. M., Prasad, N. R., & Prasad, R. (2011). A cooperative Internet of Things (IoT) for rural healthcare monitoring and control. *2nd international conference on wireless communication, vehicular technology, information theory and aerospace & electronic systems technology (Wireless VITAE)* (pp. 1-6). IEEE.

Rohokale, V. M., Prasad, N. R., & Prasad, R. (2020). A cooperative Internet of Things (IoT) for rural healthcare monitoring and control. *2nd international conference on wireless communication, vehicular technology, information theory and aerospace & electronic systems technology (Wireless VITAE)* (pp. 1-6). IEEE.

Rokach, L. (2010). Ensemble-based classifiers. *Artificial Intelligence Review*, *33*(1–2), 1–39. doi:10.1007/s10462-009-9124-7

Ruiz, J., Desai, S., Paghdar, S., Malkani, S., Nativi, J., Yip, D., Patel, P., Leoni, J., Lyle, M., & Goswami, R. (2023). The Impact of Axillary Mechanical Circulatory Support in Patients Awaiting Heart Transplantation with Pulmonary Hypertension. *The Journal of Heart and Lung Transplantation*, *42*(4), S354. doi:10.1016/j.healun.2023.02.820

Saadatnia, F., & Mehregan, M. R. (2014). Determining and Prioritizing Factors Affecting Customers Attraction of Medical Tourism from the Perspective of Arabic Countries (Case Study: Iran-Mashhad Razavi Hospital). *International Journal of Marketing Studies*, *6*(3), 155–162. doi:10.5539/ijms.v6n3p155

Saeedi, M. R., Morovvati, M. R., & Mollaei-Dariani, B. (2020). Experimental and numerical investigation of impact resistance of aluminum–copper cladded sheets using an energy-based damage model. *Journal of the Brazilian Society of Mechanical Sciences and Engineering*, *42*(6), 1–24. doi:10.1007/s40430-020-02397-0

Saenger, A. K., & Christenson, R. H. (2010). Stroke biomarkers: Progress and challenges for diagnosis, prognosis, differentiation, and treatment. *Clinical Chemistry*, *56*(1), 21–33. doi:10.1373/clinchem.2009.133801 PMID:19926776

Sagar, R., & Latha, P. V. (2014). Medical tourism–An emerging industry in India. *Journal of Medical Sciences Research*, *2*(3), 169–177.

Sahli, A. (2015). Revisiting Perceived Risk and Trust in E-tourism Context: Toward an Extended Technology Acceptance Model. *Journal of Marketing Research & Case Studies*, 1–14. doi:10.5171/2015.516086

Saiprasert, W., Xu, B., & Tavitiyaman, P. (2022). The Relationships Among Perceived Medical Quality, Well-Being Perception, And Behavioral Intention: A Comparison Between Domestic And Overseas Medical Destinations. *Journal of Quality Assurance in Hospitality & Tourism*, *23*(4), 1011–1036. doi:10.1080/1528008X.2021.1943598

Sajjad, M., Zahir, S., Ullah, A., Akhtar, Z., & Muhammad, K. (2020). Human behavior understanding in big multimedia data using CNN based facial expression recognition. *Mobile Networks and Applications*, *25*(4), 1611–1621. doi:10.1007/s11036-019-01366-9

Sakipov, N., & Vikhnin, O. (2021). The Modern Symbolic Artificial Intelligence and Its Use in Medical Tourism. *2021 IEEE 15th International Conference on Application of Information and Communication Technologies (AICT)*, (pp. 1-6). IEEE. 10.1109/AICT52784.2021.9620321

Sandberg, D. S. (2017). Medical tourism: An emerging global healthcare ind. *International Journal of Healthcare Management*, *10*(4), 281–288. doi:10.1080/20479700.2017.1296213

Santosh, K. C., & Gaur, L. (2022). *Artificial intelligence and machine learning in public healthcare: Opportunities and societal impact*. Springer Nature.

Saxena, P. D., MayI, K., Arun, R., Kumar, S. S., Mishra, B. R., & Praveen, K. B. (n.d.). Impact of Artificial Intelligence on Healthcare Informatics: Opportunities and Challenges. *Journal of Informatics Education and Research, 3*(2).

Saxena, R. (2014). Indian medical tourism industry-products and destinations: challenges and opportunities. Empyreal Institute of Higher Education, 14.

Saxena, P. D., Mayi, K., Arun, R., Kumar, S. S., Mishra, B. R., & Praveen, K. B. (2023). Impact of Artificial Intelligence on Healthcare Informatics: Opportunities and Challenges. *Journal of Informatics Education and Research*, *3*(2).

Scanzera, A. C., Shorter, E., Kinnaird, C., Valikodath, N., Al-Khaled, T., Cole, E., Kravets, S., Hallak, J. A., McMahon, T., & Chan, R. P. (2022). Optometrist's perspectives of Artificial Intelligence in eye care. *Journal of Optometry*, *15*, S91–S97. doi:10.1016/j.optom.2022.06.006 PMID:36137899

Schäfer, M. B., Stewart, K. W., & Pott, P. P. (2019). Industrial robots for teleoperated surgery–a systematic review of existing approaches. *Current Directions in Biomedical Engineering*, *153-156*(1), 153–156. Advance online publication. doi:10.1515/cdbme-2019-0039

Schmude, J., Pillmayer, M., Witting, M., & Corradini, P. (2021). Geography Matters, But horizontal ellipsis Evolving Success Factors for Nature-Oriented Health Tourism within Selected Alpine Destinations. *International Journal of Environmental Research and Public Health*, (10). Advance online publication. doi:10.3390/ijerph18105389 PMID:34070110

Schwendicke, F., Rossi, J., Göstemeyer, G., Elhennawy, K., Cantu, A., Gaudin, R., Chaurasia, A., Gehrung, S., & Krois, J. (2021). Cost-effectiveness of artificial intelligence for proximal caries detection. *Journal of Dental Research*, *100*(4), 369–376. doi:10.1177/0022034520972335 PMID:33198554

Sedik, A., Hammad, M., Abd El-Samie, F. E., Gupta, B. B., & Abd El-Latif, A. A. (2021). Efficient deep learning approach for augmented detection of Coronavirus disease. *Neural Computing & Applications*, 1–18. doi:10.1007/s00521-020-05410-8 PMID:33487885

Seide, F., & Agarwal, A. (2016, August). CNTK: Microsoft's open-source deep-learning toolkit. In *Proceedings of the 22nd ACM SIGKDD international conference on knowledge discovery and data mining* (pp. 2135-2135). ACM. 10.1145/2939672.2945397

Seligman, L. (2006). Sensemaking throughout the adoption and the innovation-decision process. *European Journal of Innovation Management*, *9*(1), 108–120. doi:10.1108/14601060610640050

Senbekov, M., Saliev, T., Bukeyeva, Z., Almabayeva, A., Zhanaliyeva, M., Aitenova, N., Toishibekov, Y., & Fakhradiyev, I. (2020). The recent progress and applications of digital technologies in healthcare: A review. *International Journal of Telemedicine and Applications*, *2020*, 1–18. doi:10.1155/2020/8830200 PMID:33343657

Seow, A. N., Choong, Y. O., Moorthy, K., & Chan, L. M. (2017). Intention to visit Malaysia for medical tourism using the antecedents of Theory of Planned Behaviour: A predictive model. *International Journal of Tourism Research*, *19*(3), 383–393. doi:10.1002/jtr.2120

Seyitoğlu, F., & Ivanov, S. (2020). A conceptual study of the strategic role of gastronomy in tourism destinations. *International Journal of Gastronomy and Food Science*, *21*, 100230. https://sci-hub.se/10.1016/j.ijgfs.2020.100230. doi:10.1016/j.ijgfs.2020.100230

Shabarish, P. S., & Shivashankar, K. C. (2019). Impact of technology on medical tourism: A conceptual study. *International Journal of Commerce and Management Research*, *5*(3), 145 148.

Shafii, M., Madadizadeh, F., Askari, R., Zarezadeh, M., Abdoli, A. M., & Bahariniya, S. (2022). Investigating medical tourism development indicators in Yazd Reproductive Sciences Institute: A cross-sectional study. *International Journal of Reproductive BioMedicine*, *20*(12), 1039. 10.18502%2Fijrm.v20i12.12565 PMID:36819203

Shahgholi, M., Oliviero, S., Baino, F., Vitale-Brovarone, C., Gastaldi, D., & Vena, P. (2016). Mechanical characterization of glass-ceramic scaffolds at multiple characteristic lengths through nanoindentation. *Journal of the European Ceramic Society*, *36*(9), 2403–2409. doi:10.1016/j.jeurceramsoc.2016.01.042

Shahmirzadi, E. K. (2012). *Community Based Tourism (CBT) Planning and Possibilities: The Case of Shahmirzad, Iran* [Doctoral dissertation, Eastern Mediterranean University (EMU)].

Shahmirzadi, E. K. (2017). *Pro- Tourism and Anti-Tourism Community Groups at a World Heritage Site in Turkey*. Pamukkale.

Shaikh, T. A., Dar, T. R., & Sofi, S. (2022). A data-centric artificial intelligent and extended reality technology in smart healthcare systems. *Social Network Analysis and Mining*, *12*(1), 122. doi:10.1007/s13278-022-00888-7 PMID:36065420

Shanmugam, K. R. (2013). *Medical tourism in India: Progress, opportunities and challenges*. Madras School of Economics.

Sharma, I., & Sharma, S. (2022). Blockchain Enabled Biometric Security in Intemet-of-Medical-Things (IoMT) Devices. *2022 International Conference on Augmented Intelligence and Sustainable Systems (ICAISS)* (pp. 971-979). IEEE. 10.1109/ICAISS55157.2022.10010716

Sharma, S., Chauhan, Y., & Tyagi, R. (2023). *Artificial Intelligence based Applications in Medical Tourism. 14th International Conference on Computing Communication and Networking Technologies (ICCCNT)*, Delhi, India. 10.1109/ICCCNT56998.2023.10308020

Shaygani, F., Jahangiri, S., Roozegar, M. H. Z., Kavosi, Z., & Marzaleh, M. A. (2023). Investigating the quality of Iranian hospitals' websites and their association to the Province's share of medical tourism. *BMC Research Notes*, *16*(1), 332. doi:10.1186/s13104-023-06619-1 PMID:37964374

Shreve, J. T., Khanani, S. A., & Haddad, T. C. (2022). Artificial Intelligence in Oncology: Current Capabilities, Future Opportunities, and Ethical Considerations. *American Society of Clinical Oncology Educational Book*, *42*(42), 1–10. doi:10.1200/EDBK_350652 PMID:35687826

Shriedeh, F. B. (2019). Strategic factors for building brand equity: Jordan medical tourism. *Jurnal The Messenger*, *11*(1A), 121–137. doi:10.26623/themessenger.v11i1A.821

Shukla, R., Singh, M., & Saxena, S. (2019). Consumer perception of hospitality services in JCI accredited hospitals at Delhi–NCR: An exploratory research on growth of medical tourism. *Humanities & Social Sciences Reviews*, 2395-6518.

Shukla, R., Singh, M., & Saxena, S. (2019). Consumer perception of hospitality services in JCI accredited hospitals at Delhi–NCR: An exploratory research on growth of medical tourism. *Humanities & Social Sciences Reviews*. file:///C:/Users/milos/Downloads/SSRN-id3438296.pdf

Siddharth, G., & Sachin, S. (2022). Augmented Reality in Medical Education: An Industry 4.0 Enabled Innovative Approach. *3rd International Conference on Smart Electronics and Communication (ICOSEC)*, (pp. 29-33). IEEE.

Sigrist, S. (2006a). Zukunftsperspektiven des Gesundheitsmarktes. Gottlieb Duttweiler Institut /GDI im Auftrag in oder of Eidgenössisches Departement des Inneren, Bern.

Sigrist, S. (2006b). *Health Horizons – Guide zu den neuen Gesundheitsmärkten/Guide to the New Healthmarkets, No. 20*. Gottlieb Duttweiler Institut/GDI.

Sikdar, S., & Guha, S. (2020). Advancements of healthcare technologies: Paradigm towards smart healthcare systems. *Recent trends in image and signal processing in computer vision*, 113-132. doi:10.1007/978-981-15-2740-1_9

Singh, A., & Hassan, S. C. (2023). Identifying Religious Tourism Potential Through Sustainable Approach: A Study on Punjab State of India. *Prospects and Challenges of Global Pilgrimage Tourism and Hospitality*, 180–192.

Singh, S., Sharma, S., & Bhadula, S. (2022). Automated Deep Learning based Disease Prediction Using Skin Health Records: Issues Challenges and Future Directions. *International Conference on Electronics and Renewable Systems (ICEARS)* (pp. 638-643). ICEARS. 10.1109/ICEARS53579.2022.9752422

Singh, S., Sharma, S., & Mohan, S. (2023). Industry 4.0 Internet of Medical Things Enabled Cost Effective Secure Smart Patient Care Medicine Pouch. In *New Horizons for Industry 4.0 in Modern Business* (pp. 149-170). International Publishing. Retrieved from https://link.springer.com/chapter/10.1007/978-3-031-20443-2_7

Singh, V., & Singh, A. (2024). Digital Health Revolution: Enhancing Well-Being Through Technology. *Implementing Sustainable Development Goals in the Service Sector*, 213–219.

Singh, N. (2013). Exploring the factors influencing the travel motivations of U.S. medical tourists. *Current Issues in Tourism*, *16*(5), 436–454. doi:10.1080/13683500.2012.695341

Singh, O. P. (2023). Artificial intelligence in the era of ChatGPT-Opportunities and challenges in mental health care. *Indian Journal of Psychiatry*, *65*(3), 297. doi:10.4103/indianjpsychiatry.indianjpsychiatry_112_23 PMID:37204980

Singh, P., Arora, L., & Choudhry, A. (2023). Consumer Behavior in the Service Industry: An Integrative Literature Review and Research Agenda. *Sustainability (Basel)*, *15*(1), 250. doi:10.3390/su15010250

Singh, V. K., Singh, P., Karmakar, M., Leta, J., & Mayr, P. (2021). The journal coverage of Web of Science, Scopus and Dimensions: A comparative analysis. *Scientometrics*, *126*(6), 5113–5142. doi:10.1007/s11192-021-03948-5

Siregar, A. Y. M., Habibie, I., Sihotang, R. B., Allo, R. P., Kusumawardana, R., Sunjaya, D. K., & Remi, S. S. (2023). Costs of providing hospital-based psychotic disorder treatment in Indonesia. *Journal of Mental Health (Abingdon, England)*, *32*(2), 443–451. doi:10.1080/09638237.2021.2022609 PMID:34983298

Slater, G., & Sambo, T. (2023). *Applications of AI in orthopaedic treatments.* Orthopedic Surgeon. https://orthopaedic-surgeon.com.au/applications-of-ai-in-orthopaedic-treatments/#:~:text=AI%20can%20identify%20any%20anomalies,diagnose%20and%20plan%20patient%20treatment.&text=It%20is%20always%20important%20to,procedure%20prior%20to%20its%20implementation. Accessed - 19/11/2023.

Smith, M., & Puczkó, L. (2014). *Health, tourism and hospitality: Spas, wellness and medical travel.* Routledge. https://books.google.rs/books?hl=en&lr=&id=p0bIAgAAQBAJ&oi=fnd&pg=PP1&dq=Smith,+M.,+%26+Puczk%C3%B3,+L.+(2014).+Health,+tourism+and+hospitality:+Spas,+wellness+and+medical+travel.+Routledge.&ots=1L74wfuVW7&sig=m573Qi-bLlxz7quDx7yi1kJ41_A&redir_esc=y#v=onepage&q=Smith%2C%20M.%2C%20%26%20Puczk%C3%B3%2C%20L.%20(2014).%20Health%2C%20tourism%20and%20hospitality%3A%20Spas%2C%20wellness%20and%20medical%20travel.%20Routledge.&f=false

Smith, E., Behrmann, J., Martin, C., & Williams-Jones, B. (2010). Reproductive tourism in Argentina: Clinic accreditation and its implications for consumers, health professionals and policy makers. *Developing World Bioethics*, *10*(2), 59–69. doi:10.1111/j.1471-8847.2009.00256.x PMID:19508291

Smith, M., & Kelly, C. (2006). Wellness tourism. *Tourism Recreation Research*, *31*(1), 1–4. doi:10.1080/02508281.2006.11081241

Smith, M., & Puczko, L. (2008). *Health and wellness tourism.* Butterworth-Heinemann. doi:10.4324/9780080942032

Snyder, H. (2019). Literature review as a research methodology: An overview and guidelines. *Journal of Business Research*, *104*, 333–339. doi:10.1016/j.jbusres.2019.07.039

Soltis-Jarrett, V. (2023). The Future of Psychiatric-Mental Health Nursing: Observe, Reflect, and Take Action to Empower Knowledge for the Greater Good. *Issues in Mental Health Nursing*, *44*(10), 1–9. doi:10.1080/01612840.2023.2270066 PMID:37939371

Somashekhar, S. P., Kumarc, R., Rauthan, A., Arun, K. R., Patil, P., & Ramya, Y. E. (2017). Abstract S6-07: Double-blinded validation study to assess the performance of IBM artificial intelligence platform, Watson for oncology in comparison with manipal multidisciplinary tumour board? The first study of 638 breast cancer cases. *Cancer Research*, *77*(4), S6–S07. doi:10.1158/1538-7445.SABCS16-S6-07

Sony, M., Antony, J., & McDermott, O. (2023). The impact of healthcare 4.0 on the healthcare service quality: A systematic literature review. *Hospital Topics*, *101*(4), 288–304. doi:10.1080/00185868.2022.2048220 PMID:35324390

Sourav, S., Sachin, S., Shuchi, B., & Seshadri, M. (2023). Industry 4.0 Internet of Medical Things Enabled Cost Effective Secure Smart Patient Care Medicine Pouch in New Horizons for Industry 4.0 in Modern Business. Springer International Publishing.

Sousa, M. J., Nunes, F. G., do Nascimento, G., & Chakraborty, C. (2023). *Future Health Scenarios: AI and Digital Technologies in Global Healthcare Systems*. Springer Nature. doi:10.1201/9781003227892

Spencer-Jolliffe, N. (2023). *AI prompts wave of medical tourism.* Medical Advice Network.https://www.medicaldevice-network.com/features/ai-prompts-wave-of-medical-tourism/.

Subasinghe, M., Magalage, D., Amadoru, N., Amarathunga, L., Bhanupriya, N., & Wijekoon, J. L. (2020). Effectiveness of artificial intelligence, decentralized and distributed systems for prediction and secure channeling for Medical Tourism. *11th IEEE Annual Information Technology, Electronics and Mobile Communication Conference (IEMCON)*, (pp. 314-319). IEEE. 10.1109/IEMCON51383.2020.9284898

Su, C. Y., Samani, H., Yang, C. Y., & Fernando, O. N. (2018). Doctor robot with physical examination for skin disease diagnosis and telemedicine application. *International Conference on System Science and Engineering (ICSSE)* (pp. 1-6). 10.1109/ICSSE.2018.8520218

Suess, C., & Mody, M. (2017). Hospitality healthscapes: A conjoint analysis approach to understanding patient responses to hotel-like hospital rooms. *International Journal of Hospitality Management*, *61*, 59–72. https://sci-hub.se/10.1016/j.ijhm.2016.11.004. doi:10.1016/j.ijhm.2016.11.004

Sundaravadivel, P., Kougianos, E., Mohanty, S. P., & Ganapathiraju, M. K. (2017). Everything you wanted to know about smart health care: Evaluating the different technologies and components of the internet of things for better health. *IEEE Consumer Electronics Magazine*, *7*(1), 18–28. doi:10.1109/MCE.2017.2755378

Sung, H., Ferlay, J., Siegel, R. L., Laversanne, M., Soerjomataram, I., Jemal, A., & Bray, F. (2021). Global Cancer Statistics 2020: GLOBOCAN Estimates of Incidence and Mortality Worldwide for 36 Cancers in 185 Countries. *CA: a Cancer Journal for Clinicians*, *71*(3), 209–249. doi:10.3322/caac.21660 PMID:33538338

Swenson, M. S., & Bansal, A. (2024). *Medical, Dental, and Wellness Tourism: A Post-Pandemic Perspective*. CRC Press. doi:10.1201/9781003433569

Swiecicki, A., Li, N., O'Donnell, J., Said, N., Yang, J., Mather, R. C., Jiranek, W. A., & Mazurowski, M. A. (2021). Deep learning-based algorithm for assessment of knee osteoarthritis severity in radiographs matches performance of radiologists. *Computers in Biology and Medicine*, *133*, 104334. doi:10.1016/j.compbiomed.2021.104334 PMID:33823398

Tagde, P., Tagde, S., Bhattacharya, T., Tagde, P., Chopra, H., Akter, R., Kaushik, D., & Rahman, M. H. (2021). Blockchain and Artificial Intelligence Technology in E-Health. *Environmental Science and Pollution Research International*, *28*(38), 52810–52831. doi:10.1007/s11356-021-16223-0 PMID:34476701

Taheri, B., Chalmers, D., Wilson, J., & Arshed, N. (2021). Would you really recommend it? Antecedents of word-of-mouth in medical tourism. *Tourism Management, 83*, 104209. doi:10.1016/j.tourman.2020.104209

TalebiM.Abbasi-RadS.MalekzadehM.ShahgholiM.ArdakaniA. A.FoudehK., & Rad,

Talebi, M., Abbasi-Rad, S., Malekzadeh, M., Shahgholi, M., Ardakani, A. A., Foudeh, K., & Rad, H. S.H. S. (2021). Cortical bone mechanical assessment via free water relaxometry at 3 T. *Journal of Magnetic Resonance Imaging, 54*(6), 1744–1751. doi:10.1002/jmri.27765 PMID:34142413

Tanikawa, C., & Yamashiro, T. (2021). Development of novel artificial intelligence systems to predict facial morphology after orthognathic surgery and orthodontic treatment in Japanese patients. *Scientific Reports, 11*(1), 1–11. doi:10.1038/s41598-021-95002-w PMID:34349151

Tapar, A. V., Dhaigude, A. S., & Jawed, M. S. (2017). Customer experience-based satisfaction and behavioural intention in adventure tourism: Exploring the mediating role of commitment. *Tourism Recreation Research, 42*(3), 344–355. doi:10.1080/02508281.2017.1303564

Tariq, A., Gill, A. Y., & Hussain, H. K. (2023). Evaluating the Potential of Artificial Intelligence in Orthopedic Surgery for Value-based Healthcare. *International Journal of Multidisciplinary Sciences and Arts, 2*(1), 27-35. doi:https://jurnal.itscience.org/index.php/ijmdsa/article/view/2394

Tariq, A., Santos, T., & Banerjee, I. (2022). Natural Language Processing for Cardiovascular Applications. In *Artificial Intelligence in Cardiothoracic Imaging* (pp. 231–243). Springer International Publishing. doi:10.1007/978-3-030-92087-6_23

Tater, U. R., Jain, A., & Tater, M. J. (2022). Artificial intelligence must be a revolutionary technique of marketing in medical-tourism. *2022 Indian Journal of Business Administration (IJBA).*

Tater, Umaid; Jain, A. T. (2022). Artificial Intelligence must be a Revolutionary Technique of Marketing in. [IJBA]. *Indian Journal of Business Administration, 15*(1), 39–45.

Tehseen, K. Y., Imran, N., Qazi, A. S., Ahmad, W., Arslan, F., Ijaz, A., & Ali, M. (2021). Transformative Effects of COVID-19 on Global Econo. *International J. on Emerging Technologies, 12*(2), 66–76.

Temizkan, V., Yetgin, M. A., & Yılmaz, K. (2022). Motivations of Retailers Accepting Cryptocurrency Payments and Their Implications on Retail Customer Experience. *Cumhuriyet University Journal of Economics and Administrative Sciences, 23*(1), 25–48. doi:10.37880/cumuiibf.987656

Tewari, A. (n.d.). The future of prostate cancer care: AI and Machine Learning. *Medical Tourism Magazine.* https://www.magazine.medicaltourism.com/article/the-future-of-prostate-cancer-care-ai-and-machine-learning.

The Lancet. (2017). Artificial intelligence in health care: within touching distance. *The Lancet, 390*(10114), (accessed 30 Mar. 2020). doi:10.1016/S0140-6736(17)31540-4

The United Arab Emirates Government. (2021b). *Robotics and AI Applications*. UAE. https://u.ae/en/about-the-uae/digital-uae/robotics-and-ai-

Thieme, A., Belgrave, D., & Doherty, G. (2020). Machine learning in mental health: A systematic review of the HCI literature to support the development of effective and implementable ML systems. [TOCHI]. *ACM Transactions on Computer-Human Interaction*, *27*(5), 1–53. doi:10.1145/3398069

Tian, S., Yang, W., Le Grange, J. M., Wang, P., Huang, W., & Ye, Z. (2019). Smart healthcare: Making medical care more intelligent. *Global Health Journal (Amsterdam, Netherlands)*, *3*(3), 62–65. doi:10.1016/j.glohj.2019.07.001

Tileva, I. G. (2021). How technologies will affect the future of international medical tourism. *Вестник Пермского национального исследовательского политехнического университета. Социально-экономические науки, 3*(1), 308-318.

Tim, W. (2018). *Health tourism today and into the future.*

Timothy, D. J., & Nyaupane, G. P. (2022). Conclusion: The Experience of Himalayan Tourism. In Tourism and Development in the Himalaya (pp. 308-323). Routledge.

Torkan, E., & Pirmoradian, M. (2019). Efficient higher-order shear deformation theories for instability analysis of plates carrying a mass moving on an elliptical path. *Journal of Solid Mechanics*, *11*(4), 790–808.

Tranfield, D., Denyer, D., & Smart, P. (2003). Towards a methodology for developing evidence-informed management knowledge by means of systematic review. *British Journal of Management*, *14*(3), 207–222. doi:10.1111/1467-8551.00375

Tripathi, R. P. (2023). Harnessing the Potential of AI and ML in Medical Tourism: Challenges, Opportunities, and Ethical Implications. *3rd International Conference on Technological Advancements in Computational Sciences (ICTACS)* (pp. 1068-1072). IEEE. 10.1109/ICTACS59847.2023.10390358

Turner, L. (2012). Beyond" medical tourism": Canadian companies marketing medical travel. *Globalization and Health, 8,* 1-11. file:///C:/Users/milos/Downloads/1744-8603-8-16.pdf

Turner, L. (2010). "Medical tourism" and the global marketplace in health services: US patients, international hospitals, and the search for affordable health care. *International Journal of Health Services*, *40*(3), 443–467. https://sci-hub.se/10.2190/HS.40.3.d. doi:10.2190/HS.40.3.d PMID:20799670

Turner, L. (2012). Beyond "medical tourism": Canadian companies marketing medical travel. *Globalization and Health*, *8*(1), 16. doi:10.1186/1744-8603-8-16 PMID:22703873

Tussyadiah, I., & Miller, G. 2019. Perceived impacts of artificial intelligence and responses to positive behaviour change intervention. In *Information and Communication Technologies in Tourism 2019: Proceedings of the International Conference in Nicosia,* (pp. 359-370). Springer International Publishing. 10.1007/978-3-030-05940-8_28

Tutun, S., Johnson, M. E., Ahmed, A., Albizri, A., Irgil, S., Yesilkaya, I., Ucar, E. N., Sengun, T., & Harfouche, A. (2023). An AI-based decision support system for predicting mental health disorders. *Information Systems Frontiers*, *25*(3), 1261–1276. doi:10.1007/s10796-022-10282-5 PMID:35669335

Twinanda, A. P., Yengera, G., Mutter, D., Marescaux, J., & Padoy, N. (2018). RSDNet: Learning to predict remaining surgery duration from laparoscopic videos without manual annotations. *IEEE Transactions on Medical Imaging*, *38*(4), 1069–1078. doi:10.1109/TMI.2018.2878055 PMID:30371356

Tyan, I., Plaza, A., & Yague, M. (2021). The benefifits of blockchain technology for medical tourism. *Sustainability (Basel)*, *13*(22), 12448–12412. doi:10.3390/su132212448

UCD, University College Dublin. (30 August 2023). *New AI tool makes it possible to predict weight loss after bariatric surgery.* UCD. https://www.ucd.ie/newsandopinion/news/2023/august/30/newaitoolmakesitpossibletopredictweightlossafterbariatricsurgery/#:~:text=New%20AI%20tool%20makes%20it%20possible%20to%20predict%20weight%20loss%20after%20bariatric%20surgery,-Posted%2030%20August&text=First%2Dtime%20patients%20in%20need,bypass%20or%20sleeve%20gastrectomy%20operation

Ulnicane, I. (2022). Artificial Intelligence in the European Union: Policy, ethics and regulation. In The Routledge handbook of European integrations. Taylor & Francis. doi:10.4324/9780429262081-19

Um, T., Kim, H., Kim, H., Lee, J., Koo, C., & Chung, N. (2022). Travel Incheon as a Metaverse: Smart Tourism Cities Development Case in Korea. In Information and Communication Technologies in Tourism 2022. doi:10.1007/978-3-030-94751-4_20

UNWTO. (2020a). *Impact assessment of the COVID-19 outbreak on international tourism.* UN. https://www.unwto.org/impact-assessment-of-the-covid-19-outbreak -on-international-tourism.

UNWTO. (2020b). *International tourism growth continues to outpace the global economy.* UN. https://www.unwto.org/international-tourism-growth-continues-to-outpace-the-economy.

USAID. (2019). *Artificial Intelligence in Global Health: Defining a Collective Path Forward.* USAID. https://www.usaid.gov/cii/ai-in-global-health

Üstün, U. & ve Demir Uslu, Y. X. (2022). Türkiye'nin Sağlık Turizminde Tercih Edilme Nedenleri Üzerine Bir Çalışma: Medikal Turizm Endeksi. *Avrupa Bilim ve Teknoloji Dergisi*, (33), 344–353.

Uusitalo, S., Tuominen, J., & Arstila, V. (2021). Mapping out the philosophical questions of AI and clinical practice in diagnosing and treating mental disorders. *Journal of Evaluation in Clinical Practice*, *27*(3), 478–484. doi:10.1111/jep.13485 PMID:32996664

Uzir, M. U. H., Al Halbusi, H., Lim, R., Jerin, I., Hamid, A. B. A., Ramayah, T., & Haque, A. (2021). Applied Artificial Intelligence and user satisfaction: Smartwatch usage for healthcare in Bangladesh during COVID-19. *Technology in Society*, *67*, 101780. doi:10.1016/j.techsoc.2021.101780 PMID:34697510

Vargo, D., Zhu, L., Benwell, B., & Yan, Z. (2021). Digital technology use during COVID-19 pandemic: A rapid review. *Human Behavior and Emerging Technologies*, *3*(1), 13–24. doi:10.1002/hbe2.242

VerMilyea, M., Hall, J. M. M., Diakiw, S. M., Johnston, A., Nguyen, T., Perugini, D., Miller, A., Picou, A., Murphy, A. P., & Perugini, M. (2020). Development of an artificial intelligence-based assessment model for prediction of embryo viability using static images captured by optical light microscopy during IVF. *Human Reproduction (Oxford, England)*, *35*(4), 35770–35784. doi:10.1093/humrep/deaa013 PMID:32240301

Vespestad, M. K., Lindberg, F., & Mossberg, L. (2019). Value in tourist experiences: How nature-based experiential styles influence value in climbing. *Tourist Studies*, *19*(4), 453–474. doi:10.1177/1468797619837966

Villar, J. R., González, S., Sedano, J., Chira, C., & Trejo-Gabriel-Galan, J. M. (2015). Improving human activity recognition and its application in early stroke diagnosis. *International Journal of Neural Systems*, *25*(04), 1450036. doi:10.1142/S0129065714500361 PMID:25684369

Vinuesa, R., Azizpour, H., Leite, I., Balaam, M., Dignum, V., Domisch, S., Felländer, A., Langhans, S. D., Tegmark, M., & Fuso Nerini, F. (2020). The role of artificial intelligence in achieving the Sustainable Development Goals. In Nature Communications, 11(1). doi:10.1038/s41467-019-14108-y

Vishnoi, S. K., Bagga, T. E. E. N. A., Sharma, A. A. R. U. S. H. I., & Wani, S. N. (2018). Artificial intelligence enabled marketing solutions: A review. Indian Journal of Economics & Business, 17(4), 167-177.

Vishnoi, S. K., Bagga, T. E. E. N. A., Sharma, A. A. R. U. S. H. I., & Wani, S. N. (2018). Artificial intelligence enabled marketing solutions: A review. *Indian Journal of Economics & Business*, *17*(4), 167–177.

Vitthal, P. C., Subhash, A. R., Sharma, B. R., & Ramachandran, M. (2015). Emerging trends and future prospects of medical tourism in India. *Journal of pharmaceutical sciences and research,* 7(5), 248. Retrieved from https://www.proquest.com/openview/60f1635cc6f0bf652e264ee5040645a9/1?cbl=54977&pq-origsite=gscholar&parentSessionId=CSnA2me%2FMJ7z5eQxqemQnO4b7ALEsYCy0qTaXqhL0Kw%3D

Voigt, C. (2014). Wellness tourism: A destination perspective. London: Routledge.

Wang, H., Ghasemi, M., Ghadiri Nejad, M., & Khandan, A. S. (2023). Assessing the Potential Growth of Iran's Hospitals with Regard to the Sustainable Management of Medical Tourism. *Health & Social Care in the Community*, *2023*, 1–12. Advance online publication. doi:10.1155/2023/8734482

Wang, Q., Su, M., Zhang, M., & Li, R. (2021). Integrating digital technologies and public health to fight Covid-19 pandemic: Key technologies, applications, challenges and outlook of digital healthcare. *International Journal of Environmental Research and Public Health*, *18*(11), 6053. doi:10.3390/ijerph18116053 PMID:34199831

Wapmuk, S., Wapmuk, A. E., & Gbajabiamila, T. A. (2015). Medical tourism and Nigeria-India relations. *Journal of Tourism Research, 12*(1), 44–65.

Wen, Z., & Huang, H. (2022). The potential for artificial intelligence in healthcare. *Journal of Commercial Biotechnology, 27*(4). Advance online publication. doi:10.5912/jcb1327

Wiggins, G. A. (2019). A framework for description, analysis and comparison of creative systems. Computational creativity: *The philosophy and engineering of autonomously creative systems*, 21-47. doi:10.1007/978-3-319-43610-4_2

Williams, R. I. Jr, Clark, L. A., Clark, W. R., & Raffo, D. M. (2021). Re-examining systematic literature review in management research: Additional benefits and execution protocols. *European Management Journal, 39*(4), 521–533. doi:10.1016/j.emj.2020.09.007

Wong, B. K. M., & Hazley, S. A. S. A. (2020). The future of health tourism in the industrial revolution 4.0 era. *Journal of Tourism Futures, 7*(2), 267–272. doi:10.1108/JTF-01-2020-0006

Wong, H., & Hazley, L. (2020). Addressing the neglect of mental health care in medical tourism: A call for individualized monitoring and intervention. *Journal of Medical Tourism Practice and Research, 9*(2), 87–104.

Wong, J. K., Mayberg, H. S., Wang, D. D., Richardson, R. M., Halpern, C. H., Krinke, L., Arlotti, M., Rossi, L., Priori, A., Marceglia, S., Gilron, R., Cavanagh, J. F., Judy, J. W., Miocinovic, S., Devergnas, A. D., Sillitoe, R. V., Cernera, S., Oehrn, C. R., Gunduz, A., & Okun, M. S. (2023). Proceedings of the 10th annual deep brain stimulation think tank: Advances in cutting edge technologies, artificial intelligence, neuromodulation, neuroethics, interventional psychiatry, and women in neuromodulation. *Frontiers in Human Neuroscience, 16*, 1084782. doi:10.3389/fnhum.2022.1084782 PMID:36819295

Woodman, J. (2019). Eight drivers for medical tourism. Insights. -https://insights.omnia-health.com/hospital-management/eight-drivers-medical-tourism-2019.

Woodman, J. (2015). *Patients Beyond Borders: Everybody's Guide to Affordable, World-Class Medical Travel*. Healthy Travel Media.

World Bank. (2023). *Data: GDP Growth (annual %)*. World Bank. https://data.worldbank.org/indicator/NY.GDP.MKTP.KD.ZG

World Health Organization (WHO). (2022). *Noncommunicable Diseases*. WHO. https://www.who.int/news-room/fact-sheets/detail/noncommunicable-diseases..

World Health Organization. (2010). *Telemedicine: Opportunities and Developments in Member States. Report on the Second Global Survey on EHealth*. World Health Organization.

Wright, D. W., & Zascerinska, S. (2023). Becoming immortal: Future wellness and medical tourism markets. *Journal of Tourism Futures, 9*(2), 168–195. doi:10.1108/JTF-05-2021-0119

Wright, S. A., & Schultz, A. E. (2018). The rising tide of artificial intelligence and business automation: Developing an ethical framework. *Business Horizons*, *61*(6), 823–832. doi:10.1016/j.bushor.2018.07.001

Wu, H. C., Li, T., & Li, M. Y. (2016). A Study of Behavioral Intentions, Patient Satisfaction, Perceived Value, Patient Trust and Experiential Quality for Medical Tourists. *Journal of Quality Assurance in Hospitality & Tourism*, *17*(2), 114–150. doi:10.1080/1528008X.2015.1042621

Yamin, M. (2018). IT applications in healthcare management: A survey. *International Journal of Information Technology : an Official Journal of Bharati Vidyapeeth's Institute of Computer Applications and Management*, *10*(4), 503–509. doi:10.1007/s41870-018-0203-3 PMID:32289102

Yang, F., Darsey, J. A., Ghosh, A., Li, H. Y., Yang, M. Q., & Wang, S. (2022). Artificial Intelligence and Cancer Drug Development. *Recent Patents on Anti-Cancer Drug Discovery*, *17*(1), 2–8. doi:10.2174/1574892816666210728123758 PMID:34323201

Yang, Y., Siau, K., Xie, W., & Sun, Y. (2022). Smart health: Intelligent healthcare systems in the metaverse, artificial intelligence, and data science era. [JOEUC]. *Journal of Organizational and End User Computing*, *34*(1), 1–14. doi:10.4018/JOEUC.308814

Yasodhara, A., Dong, V., Azhie, A., Goldenberg, A., & Bhat, M. (2021). Identifying Modifiable Predictors of Long-Term Survival in Liver Transplant Recipients with Diabetes Mellitus Using Machine Learning. *Liver Transplantation*, *27*(4), 536–547. doi:10.1002/lt.25930 PMID:33113221

Ye, B. H., Qiu, H. Z., & Yuen, P. P. (2011). Motivations and experiences of Mainland Chinese medical tourists in Hong Kong. *Tourism Management*, *32*(5), 1125–1127. doi:10.1016/j.tourman.2010.09.018

Yener, S., & Köse, İ. (2023). Treatment Prediction Using Artificial Intelligence-Integrated Chatbots in Call Centers in Medical Tourism. *Turkish Studies-Economics, Finance. Politics*, *18*(2).

Yigitcanlar, T., Corchado, J. M., Mehmood, R., Li, R. Y. M., Mossberger, K., & Desouza, K. (2021). Responsible urban innovation with local government artificial intelligence (Ai): A conceptual framework and research agenda. *Journal of Open Innovation*, *7*(1), 1–16. doi:10.3390/joitmc7010071

Yin, J., Chen, Y., & Ji, Y. (2021). Effect of the event strength of the coronavirus disease (COVID-19) on potential online organic agricultural product consumption and rural health tourism opportunities. *Managerial and Decision Economics*, *42*(5), 1156–1171. doi:10.1002/mde.3298 PMID:34149119

Yirik, Ş., Ekici, R., & Baltacı, F. (2015). *Türkiye ve Dünyada sağlık turizmi (Medikal Turizm)*. Detay Yayıncılık.

Yoo, J. H., Jeong, H., & Chung, T. M. (2023). Cutting-Edge Technologies for Digital Therapeutics: A Review and Architecture Proposals for Future Directions. *Applied Sciences (Basel, Switzerland)*, *13*(12), 6929. doi:10.3390/app13126929

Yoon, S. N., & Lee, D. (2018). Artificial intelligence and robots in healthcare: What are the success factors for technology-based service encounters? *International Journal of Healthcare Management*.

Yu, J. Y., & Ko, T. G. (2012). A cross-cultural study of perceptions of medical tourism among Chinese, Japanese, and Korean tourists in Korea. *Tourism Management*, *33*(1), 80–88. doi:10.1016/j.tourman.2011.02.002

Zain, N. A. M., Connell, J., Zahari, M. S. M., & Hanafiah, M. H. (2022). Intra-Regional Medical Tourism Demand in Malaysia: A Qualitative Study of Indonesian Medical Tourists' Rationale and Preferences. *The Malaysian Journal of Medical Sciences : MJMS*, *29*(2), 138–156. doi:10.21315/mjms2022.29.2.13 PMID:35528809

Zandi, D., Reis, A., Vayena, E., & Goodman, K. (2019). New ethical challenges of digital technologies, machine learning and artificial intelligence in public health: A call for papers. *Bulletin of the World Health Organization*, *97*(1), 1–72. doi:10.2471/BLT.18.227686

Zarei, A., Feiz, D., Maleki Minbashrazgah, M., & Maleki, F. (2020). Factors influencing selection of medical tourism destinations: A special niche market. *International Journal of Healthcare Management*, *13*(sup 1), 192-198. doi:10.1080/20479700.2018.1492764

Zeleňák, K., Krajina, A., Meyer, L., & Fiehler, J., & Vukasinovic, I. (2021). How to improve the management of acute ischemic stroke by modern technologies, artificial intelligence, and new treatment methods. *Life (Chicago, Ill.)*, *11*(6), 488. PMID:34072071

Zemmar, A., Lozano, A. M., & Nelson, B. J. (2020). The rise of robots in surgical environments during COVID-19. *Nature Machine Intelligence*, *2*(10), 566–572. doi:10.1038/s42256-020-00238-2

Zeng, Z., Chen, P. J., & Lew, A. A. (2020). From high-touch to high-tech: COVID-19 drives robotics adoption. *Tourism Geographies*, *22*(3), 724–734. doi:10.1080/14616688.2020.1762118

Zhang, L., Li, N., & Liu, M. (2012). Smart tourism basic concepts and theoretical system. *Journal of Tourism*, (5), 66–73.

Zhang, M., Ding, H., Naumceska, M., & Zhang, Y. (2022). Virtual reality technology as an educational and intervention tool for children with autism spectrum disorder: Current perspectives and future directions. *Behavioral Sciences (Basel, Switzerland)*, *12*(5), 138. doi:10.3390/bs12050138 PMID:35621435

Zhou, B., Yang, G., Shi, Z., & Ma, S. (2022). Natural language processing for smart healthcare. *IEEE Reviews in Biomedical Engineering*. PMID:36170385

Zolfagharian, M., Rajamma, R. K., Naderi, I., & Torkzadeh, S. (2018). Determinants of medical tourism destination selection process. *Journal of Hospitality Marketing & Management*, *27*(7), 775–794. https://sci-hub.se/10.1080/19368623.2018.1444527. doi:10.1080/19368623.2018.1444527

Zrnić., M, Đoković, F., Košutić, J., Mašić, S. Hassan, V. (2021). *Planning and development of spa & wellness tourism.* Sitcon. Belgrade. file:///C:/Users/milos/Downloads/Welnesstourism2021.pdf

Zrnić, M., Brdar, I., & Kilibarda, N. (2021). The importance of traditional food quality—The viewpoint of the tourism. Scientific journal. *Metals Technology*, *62*(1), 69–76. doi:10.18485/2021.62.1.7

About the Contributors

Viana Hassan is actively involved in academic and research work in Tourism & Cultural Management in Lebanon and Malta. She is the Founder and CEO of European Tourism and Events training Academy, She got her Ph.D. in Tourism Management and Cultural from Saint Joseph University, Beirut. Her Thesis was entitled "Medical Tourism in Lebanon". She has more than 19 years of tourism experience in many related Travel and Tourism Fields, varying from airline sales experience (MEA), to Co-management several Lebanese Travel agencies. She is an Experienced Lecturer with more than 16 years' experience at several universities in Lebanon (Lebanese University, A.U. L, Lebanese International University, Islamic University), the Institute of Tourism Studies, STC higher education partner of Wolverhmapton University UK and University of Malta (Malta), GBS Malta, Scholars School system (UK) with a demonstrated history of working in the Higher education industry, Skilled in Tourism Management, Customer service, Branding, Event Management, Airline Business Management, Niche Tourism, Business operations management, Customer acquisition and Medical Tourism. Dr. Viana has participated in international conferences (Expo Dubai 2020, 7th Medical Tourism World Summit..) including Athens, Paris, Vienna, Dubai, Valencia, India, Turkey etc....

Ahmad Albattat is an Associate Professor In Graduate School of Management, Post Graduate Centre, Management and Science University, Shah Alam, Selangor, Malaysia. He is a visiting Professor and external examiner in Medan Academy of Tourism (Akpar Medan). He holds a doctoral degree in Hospitality Management "Disaster and Emergency Planning and Preparedness" from University Sains Malaysia (USM). He worked as an Assistant Professor, Ammon Applied University College, Amman, Jordan. Senior Lecturer and research coordinator in School of Hospitality & Creative Arts, Management and Science University, Shah Alam, Selangor, Malaysia, and Researcher at Sustainable Tourism Research Cluster (STRC), Pulau Pinang, Malaysia. He was working for the Jordanian hospitality industry for 17 years. He has participated and presented research papers in a number of academic conferences held in Malaysia, Taiwan, Thailand, Indonesia, Sri Lanka, and Jordan.

He is an active member of Scientific and Editorial Review Board on Hospitality management, hotel, tourism, events, emergency planning, disaster management, human resource for Journal of Tourism Management, Journal of Hospitality Marketing & Management (JHMM), Current Issues in Tourism (CIT), Asia-Pacific Journal of Innovation in Hospitality and Tourism (APJIHT), International Journal of Economics and Management (IJEAM), AlmaTourism, Journal of Tourism, Culture and Territorial Development, International Journal of Tourism and Sustainable Community Development. His latest works have been published in the refereed international journals, conference proceedings, books and book chapters.

* * *

Amjad Imam Ansari is a Research Scholar, currently working as an Assistant Professor at the School of Hotel Management and Tourism at Lovely Professional University, Jalandhar, India. His research interests focus on hotel revenue management, hotel services and hospitality marketing. He has over 8 years of experience in academics and industry.

Zahra Behboodi graduated in Strategic Management from the School of Management, Kharazmi University, Tehran, Iran. Her research interests include crisis management, organizational learning, organizational resiliency, and artificial intelligence in tourism.

Santosh R. Dastane is a highly regarded academic professional and worked as a former professor of Economics and Management at ASM's IBMR, Pune, India. He is also a CCD (Certified Corporate Director). He has over 40 years experience of teaching, research, and consultancy. He holds a BA, MA, and PhD. Dr. Dastane has authored and contributed numerous articles/book reviews/research papers to various journals and newspapers, authored 24 books, and guided over 29 Ph.D. scholars and 10 M.Phill candidates, showcasing his commitment to academia and research.

Olabode Gbobaniyi (PhD) is a senior lecturer at the management department of the Global Banking School. His core academic and research interest is business strategy; however, he has wider research and publication interests in AI, Teaching and Education, and Tourism.

Rita Komalasari is a lecturer at YARSI University. Her current work is focused on Harmonizing Minds and Machines: A Transformative AI-Based Mental Health Care Framework in Medical Tourism

Ashish Raina is working as Dean in the School of Hotel management, Airlines and Tourism, CT University and a chef by profession with specializations in Indian, Italian, and French cuisines. I have worked in hotels and restaurants as a chef. Working as a consultant for designing training kitchens for universities/colleges, bakeries, restaurants, and other similar food establishments is also my forte. Apart from the industrial experience, I have excelled in academics and research and published research articles in reputed, indexed journals. I have also completed my doctorate degree in hospitality management with the title "Promote local food from the destinations as a tourist product," which in turn can tremendously involve and uplift the local communities for the overall development of any nation.

Birendra Kishore Roy is a passionate hotelier and academician having worked with leading International Hotels, Restaurants, and other Hospitality leaders. He has also work with International QSR chains. During his more than 10 years of work experience he has also worked with leading academic institutions & universities of Eastern India and mentored over 500 students who are currently placed in leading Brands. Mr. Roy is a passionate researcher in the field of Tourism, Hospitality, Hotel Management, and Human Resource Management. He has published 3 research article in International and National peer reviewed journals including scopus. He has published two book chapters in the relevant field. Mr. Roy has obtained his Master Degree from IGNOU in Tourism and Travel Management. He also obtained his MBA from SGVU with a Specialization in Human Resources Management. He has obtained his graduation from NCHMCT & IGNOU in B.Sc in Hospitality and Hotel Administration.

Viveka Nand Sharma is a competent visionary and goal-oriented professional with demonstrated exposure & multi skilled experience of over 20 Years in Academics & Administration. An alumnus of the prestigious college "Saptagiri College of Hotel Management-affiliated to Mangalore University, Completed BHM, MBA-HR from Algappa University, Post Graduate Diploma in Hotel Management & Tourism form Periyar University and Masters in Hotel Management from KSOU. Earned PhD from Usha Martin University, Jharkhand. His passion to give back to the society in terms of trained manpower brought him into Academics in 2003. He has been instrumental in opening and establishing some of the finest Hospitality Institutions, starting from Lords Institute of Management-Surat, ESMS-IMS-Ranchi, Department of Hotel Management at IES University, Department of Hospitality & Hotel Administration at Assam Don Bosco University and FTMS College- Kuala Lumpur, Malaysia.

Amrik Singh is working as Professor in the School of Hotel Management and Tourism at Lovely Professional University, Punjab, India. He obtained his Ph.D. degree in Hotel Management from Kurukshetra University, Kurukshetra. He started his academic career at Lovely Professional University, Punjab, India in the year 2007. He has published more than 40 research papers in UGC and peer-reviewed and Scopus/Web of Science) journals. He has published 12 patents and 01 patent has been granted in the inter-disciplinary domain. Dr. Amrik Singh participated and acted as a resource person in various national and international conferences, seminars, research workshops, and industry talks. His area of research interest is accommodation management, ergonomics, green practices, human resource management in hospitality, waste management, AR VR in hospitality, etc. He is currently guiding 8Ph.D. scholars and 2 Ph.D. scholars have been awarded Ph.D.

Prakash Singh is an Assistant Professor in the E-Commerce department at the College of Administrative and Financial Sciences, Saudi Electronic University, Saudi Arabia. He holds a Ph.D. in Marketing Management from Savitribai Phule Pune University. He has over nine years of experience and has authored more than 14 research papers in prestigious journals indexed in Scopus, ABDC, and Web of Science. Dr. Singh's research interests include Digital Entrepreneurship, Consumer Behavior, E-Learning, Social Commerce, Mobile Commerce, Customer Engagement, and the Metaverse. He is also a blogger, E-content developer, case study developer, and researcher. Dr. Singh's teaching approach is highly effective in helping his students learn and succeed. He uses Gamification, Learning By Doing (LBD), and a Case-based teaching approach to ensure his students acquire practical skills relevant to the industry.

Muhammed Ali Yetgin is an Assistant Professor at the University of Karabük which is a State University in Türkiye. He earned his Bachelor of Science in computer science from the State University of New York in the USA, and master's in business administration at Istanbul Bilgi University, and Ph.D. in Business Administration at Turkish Aeronautical Association University (Ankara). He worked in the Ministry of Interior of Türkiye for about 15 years and has been working as an academician since 2019. He teaches undergraduate and graduate (master's and Ph.D. classes) related to Business Administration. His research interests include Digitalization in businesses, the organizational structure of institutions, and technological innovations in companies.

Index

I

L

M

N

P

R

S

T

V

W

www.ingramcontent.com/pod-product-compliance
Lightning Source LLC
LaVergne TN
LVHW080633011026
849160LV00014B/1200